Plant Growth Regulating Chemicals

Volume I

Editor

Louis G. Nickell, Ph.D.
Vice President
Research and Development
Velsicol Chemical Corporation
Chicago, Illinois

CRC Press
Taylor & Francis Group
Boca Raton London New York

CRC Press is an imprint of the
Taylor & Francis Group, an **informa** business

First published 1983 by CRC Press
Taylor & Francis Group
6000 Broken Sound Parkway NW, Suite 300
Boca Raton, FL 33487-2742

Reissued 2018 by CRC Press

Library of Congress Cataloging-in-Publication Data
Main entry under title:

Plant growth regulating chemicals.

Includes bibliographies and index.
1. Plant regulators. I. Nickell, Louis G.,
1921- .
QK745.P56 1983 631.8 82-22832
ISBN 0-8493-5002-6 (v. 1)
ISBN 0-8493-5003-4 (v. 2)

A Library of Congress record exists under LC control number: 82022832

ISBN 13: 978-1-315-89664-9 (hbk)
ISBN 13: 978-1-351-07574-9 (ebk)

Visit the Taylor & Francis Web site at http://www.taylorandfrancis.com and the
CRC Press Web site at http://www.crcpress.com

PREFACE

The need to increase the world food supply substantially by the end of this century poses one of the greatest challenges yet faced by man. Many agricultural scientists believe that this challenge can be met, and it is expected that plant growth regulators will play an increasingly important role in meeting this challenge.

Plant growth regulating chemicals are used to modify crops by changing the rate or pattern or both of their response(s) to the internal and external factors which govern all stages at crop development from germination through vegetative growth, reproductive development, maturity, and senescence or aging, as well as post-harvest preservation.

The purpose of this two-volume work is to make available both to the investigator and to the user, on a crop by crop basis, the latest information on the use of chemicals to regulate plant growth and development. Emphasis is given to the major crops and to those with which the most success has been achieved. Since the degree of practical success with each crop varies, primary attention is given to chemicals registered for specific use(s) with the particular crop discussed. Also included is information concerning chemicals not yet registered, but for which practical results are available. In some cases information concerning active compounds in the exploratory stages is included. Where known and pertinent, information concerning mode of action is included.

The obvious classifications to use in presenting data on effectiveness of plant growth regulating chemicals are (1) by crop, (2) by chemical class, and (3) by plant function or process. Essentially all major summary or survey publications to date have been based on the plant function or process approach. This is primarily an academic approach and is not nearly as useful for practical purposes as a presentation by crops, as is done in this publication.

THE EDITOR

Louis G. Nickell, Ph.D., Vice President of Research and Development, Velsicol Chemical Corporation, Chicago, Illinois, was born July 10, 1921, in Little Rock, Arkansas. He received his B.S. degree in botany from Yale University in 1942. After serving 4 years in the U.S. Marine Corps as a regular commissioned officer, he returned to Yale University, receiving his M.S. in microbiology in 1947 and his Ph.D. in plant physiology in 1949. He is married to Natalie Wills Nickell and has three children and four grandchildren. His first professional experience was as Research Associate at the Brooklyn Botanic Garden from 1949 to 1951 where he was engaged in research on plant tissue culture and plant growth substances. He joined industry in 1951, going to Pfizer, Inc. in Brooklyn as its Plant Physiologist and Assistant Mycologist. There he specialized in antibiotics and their effects in agriculture as well as plant tissue and cell culture. In 1953 he became Head of Pfizer's Phytochemistry Laboratory and received the first patent issued for the use of plant cell cultures for the production of secondary products. In 1961 he moved to Hawaii to become Head of the Plant Physiology and Biochemistry Department of the Hawaiian Sugar Planters' Association, becoming its director of research in 1965. His first commercial success with plant growth regulating chemicals was the registration of diquat for the prevention of flowering in sugarcane in the early 1960's. This was followed successively by the registration of gibberellic acid for increasing the sugar yields in cane and later by the development of the first commercial product for the ripening of sugarcane, glyphosine. In 1975, he joined the Research Division of W. R. Grace & Company as Vice President of its Research Division in charge of agricultural, biological, and medical research, development, and commercialization. In 1978, he joined Velsicol Chemical Corporation as Vice President of Research and Development — his present position. His publications (over 300) and patents (over 30) have been primarily in the area of plant cell and tissue culture and the regulation of plant growth through the use of chemicals. He is the author of a book published in early 1982 entitled *Plant Growth Regulators — Agricultural Uses*.

He has served as President of the Hawaii Academy of Science, as Chairman of the Hawaiian Section of the American Chemical Society, as Vice Chairman and Chairman of the Plant Growth Regulator Society of America, as Council Member of the Society for Economic Botany, and has been the Treasurer of the American Society of Plant Physiologists since 1976. He has served as Chairman of the Governor's Advisory Committee on Science and Technology in Hawaii, as a member of the National Academy of Sciences — National Research Council Committee on Agricultural Production Efficiency, and is Chairman of the "The Forward Edge" Session of CHEMRAWN II, the International Conference on Chemistry and World Food Supplies: The New Frontiers, and a member of the Editorial Board of the Journal of Plant Growth Regulation.

CONTRIBUTORS

James E. Baker, Ph.D.
Research Plant Physiologist
Plant Hormone Laboratory
Plant Physiology Institute
Agricultural Research Service
U.S. Department of Agriculture
Beltsville, Maryland

Duane P. Bartholomew
Associate Agronomist
Department of Agronomy and Soil Science
Unversity of Hawaii
Honolulu, Hawaii

Wolfgang D. Binder, Ph.D.
Tree Physiologist
Research Branch
British Columbia Ministry of Forests
Victoria, British Columbia
Canada

Kenneth Bridge
Product Development Manager
Plant Growth Regulators
Union Carbide Agricultural Products Company, Inc.
Raleigh, North Carolina

George W. Cathey
Plant Physiologist
Cotton Physiology and Genetics Research Unit
Agricultural Research Service
U.S. Department of Agriculture
Stoneville, Mississippi

John A. Considine, Ph.D.
Senior Viticultural Research Officer
Department of Agriculture
Ferntree Gully, Victoria
Australia

Richard A. Criley, Ph.D.
Professor of Horticulture
Department of Horticulture
University of Hawaii
Honolulu, Hawaii

Martha Davis, Ph.D.
Postdoctoral Research Associate
Agronomy Department
University of Arkansas
Fayetteville, Arkansas

E. F. Eastin, Ph.D.
Professor of Weed Science
Texas Agricultural Experiment Station
Beaumont, Texas

Donald M. Elkins, Ph.D.
Professor of Plant and Soil Science
Southern Illinois University
Carbondale, Illinois

Gail Ezra, Ph.D.
Research Associate
Department of Environmental Biology
University of Guelph
Guelph, Ontario
Canada

E. Hayman, Ph.D.
Research Chemist
Agricultural Research Service
Fruit and Vegetable Chemistry Laboratory
U.S. Department of Agriculture
Pasadena, California

W. J. Hsu
Research Chemist
Agricultural Research Service
Fruit and Vegetable Chemistry Laboratory
U.S. Department of Agriculture
Pasadena, California

Johannes Jung, D.Sc.
BASF Agricultural Research Center
Limburgerhof
Federal Republic of Germany

Darold L. Ketring
Plant Physiologist
Agricultural Research Center
U.S. Department of Agriculture
Department of Agronomy
Oklahoma State University
Stillwater, Oklahoma

N. E. Looney, Ph.D.
Pomologist and Plant Physiologist
Agriculture Canada Research Station
Summerland, British Columbia
Canada

Louis G. Nickell, Ph.D.
Vice President
Research and Development
Velsicol Chemical Corporation
Chicago, Illinois

Edward S. Oplinger, Ph.D.
Professor of Agronomy
University of Wisconsin
Madison, Wisconsin

Richard P. Pharis, Ph.D.
Professor of Biology
University of Calgary
Calgary, Alberta
Canada

S. M. Poling
Chemist
Agricultural Research Service
Fruit and Vegetable Chemistry Laboratory
U.S. Department of Agriculture
Pasadena, California

Wilhelm Rademacher, D.Sc.
BASF Agricultural Research Center
Limburgerhof
Federal Republic of Germany

Stephen A. Ross, Ph.D.
Senior Tree Physiologist
British Columbia Ministry of Forests
Research Branch
Victoria, British Columbia
Canada

Otto John Schwarz, Ph.D.
Associate Professor of Botany
University of Tennessee
Knoxville, Tennessee

Gilbert F. Stallknecht, Ph.D.
Superintendent/Agronomist
Montana State University
Huntley, Montana

George L. Steffens
Plant Physiologist
Plant Hormone and Regulators Laboratory
U.S. Department of Agriculture
Beltsville, Maryland

Gerald R. Stephenson, Ph.D.
Professor of Weed Science
Department of Environmental Biology
University of Guelph
Guelph, Ontario
Canada

Charles A. Stutte, Ph.D.
Distinguished Professor
Altheimer Chair for Soybean Research
Agronomy Department
University of Arkansas
Fayetteville, Arkansas

W. C. Wilson
Research Scientist III
Acting Harvest Coordinator
Florida Department of Citrus
Lake Alfred, Florida

S. H. Wittwer, Ph.D.
Director
Agricultural Experiment Station
Michigan State University
East Lansing, Michigan

Henry Yokoyama, Ph.D.
Research Chemist
Fruit and Vegetable Chemistry Laboratory
Agricultural Research Service
U.S. Department of Agriculture
Pasadena, California

TABLE OF CONTENTS

Volume I

Volume II

Chapter 1

GROWTH REGULATOR USAGE IN APPLE AND PEAR PRODUCTION

Norman E. Looney

TABLE OF CONTENTS

I. INTRODUCTION

The range of plant growth regulating chemicals used by apple producers around the world is unmatched by any other crop. Plant growth regulators are used in the production of nursery trees, to regulate the shape and size of orchard trees, to promote flowering, and to increase and reduce fruit set. Other treatments control the growth of unwanted shoots or suckers, and still others prevent preharvest fruit drop, regulate fruit ripening, and influence various aspects of fruit quality. The number of products and practices available to pear growers is smaller but still impressive.

This high degree of interest among apple and pear producers in chemical growth regulation is quite understandable. These are high value crops produced on expensive land with increasingly intensive management systems and costly equipment. A relatively small increase in orchard productivity or improvement in fruit quality will cover the cost of the treatment. Secondly, apples and pears are produced on long-lived trees. The use of a growth regulator to overcome a tree productivity or fruit quality deficiency in an existing planting is almost always more economical than replanting the orchard. Thirdly, there are strong market preferences for a few established apple and pear cultivars and introducing new cultivars is a very slow process. Some growth regulator techniques have been developed specifically to improve the performance of these well-established cultivars which, of course, perpetuates both the cultivar and the growth regulator procedure. And finally, fresh market sales of all fruits depend heavily on their appearance. Treatments that either directly or indirectly enhance fruit appearance and condition are valued highly by producers and their marketing agents.

This discussion of growth regulator usage in apple and pear production is organized by practice rather than by chemical. There are several reasons for choosing this format, an important one being that some materials are used to achieve very diverse aims. For example, naphthaleneacetic acid (NAA) in various formulations promotes the rooting of cuttings, is an ingredient of fruit setting mixtures for some cultivars yet is used to reduce set of both apples and pears, and it prevents abscission of mature fruits of both species. Applied as a paint, NAA prevents the growth of unwanted sprouts and suckers.

The reader is cautioned that some of the practices and several of the materials mentioned are either unimportant or unavailable to some large sections of the fruit growing community. An example is the thinning of pear fruits. Producers in many areas use NAA or naphthaleneacetamide (NAAm) yet this practice is considered unnecessary in California, the state that produces the most pears in North America. Obviously, it is difficult to generalize about growth regulator practices. The materials, and especially the procedures for their use, differ considerably from region to region. On the other hand, it is impossible to be as thorough as the subject deserves. It is hoped that by surveying these practices and regional differences the reader will see some opportunities for expanding the useful role of plant growth regulators in worldwide apple and pear production.

II. GROWTH REGULATOR USAGE IN THE NURSERY

Commercial orchardists usually purchase finished trees from a local nurseryman. These trees are produced by grafting or budding a scion piece of the cultivar of interest onto a seedling or, preferably, a clonal rootstock with known characteristics. This rootstock is 1 or 2 years old before it is budded or grafted and the two-part tree is grown for at least another year before it is sold to the orchardist. The production of self-rooted trees by tissue culture methods may shorten this operation somewhat, but it introduces still other opportunities for error into what is already a highly technical business. Obviously, the successful operation of a fruit tree nursery requires highly skilled technicians and constant attention to detail. It is not surprising that nurserymen are keenly interested and innovative participants in the plant growth regulator field.

FIGURE 1. Commercial production of fruit tree rootstocks by in vitro meristem culture. (Bottom) Mother cultures of various apple and cherry rootstock clones held in a controlled environment room. (Top left) A proliferating culture of EMLA 7 apple ready to "harvest". Rooting of these harvested shoots is accomplished by placing them first on an auxin-containing and then on an auxin-free medium. (Top right) Rooted plants ready for potting. (Courtesy of Dr. D. I. Dunstan, Kelowna Nurseries Ltd., Kelowna, British Columbia.)

A. Production of Clonal Rootstocks and Self-Rooted Trees

The clonal rootstocks so important in modern apple and pear culture are mainly produced by stooling or layering procedures. However, the growth regulator-assisted rooting of hardwood cuttings is growing in importance, a development stemming from problems with the stool-bed procedures ranging from low productivity to a lack of flexibility in responding to changing market demands.

Successful techniques for rooting large winter cuttings of various rootstock clones have been developed by English researchers in the last decade.[1] The specific recipe varies somewhat with cultivar and location, but one essential step is the treatment of the basal end of the cuttings with an auxin — most commonly 0.25 to 0.50% indoyl-3-butyric acid (IBA) — to encourage root initiation. Rooting occurs within 2 weeks in a carefully constituted rooting medium held at 21°C. Ideally, these rooted cuttings can then be handled like rooted

FIGURE 2. Defoliation of Cox's Orange Pippin nursery trees with a 2% CuEDTA spray applied September 25, 1981. The trees in the left row were untreated. The photograph was taken October 14, 1981. (Courtesy of Dr. J. N. Knight, East Malling Research Station, Maidstone, Kent, England.)

layers, but more commonly they require additional attention to encourage the development of a strong root system before being used as rootstocks by the nurseryman. Several important apple rootstock clones, including M26, M27, MM106, and MM111, respond well to this procedure. The quince rootstocks used to control pear tree size and the Old Home × Farmingdale series of *Pyrus communis* rootstocks are also readily propagated by hardwood cuttings.

It is also possible to produce self-rooted trees of a number of commercially important apple and pear cultivars by the rooting of hardwood cuttings.[2] It has been suggested that the cost of finished trees can be reduced by this technique, but commercial application to date has been very limited.

Perhaps the most exciting developments in fruit tree propagation technology are occurring in the in vitro propagation area. Production of virus-free rootstock materials of apple and pear and of own-rooted trees of important cultivars of each species is now a commercial reality in both North America and Europe (Figure 1). This procedure offers great flexibility in adjusting to market demands since plants of desirable clones can be rapidly multiplied from mother cultures maintained by the laboratory.

The procedure[3,4] is to culture surface-sterilized shoot meristems on a nutrient medium

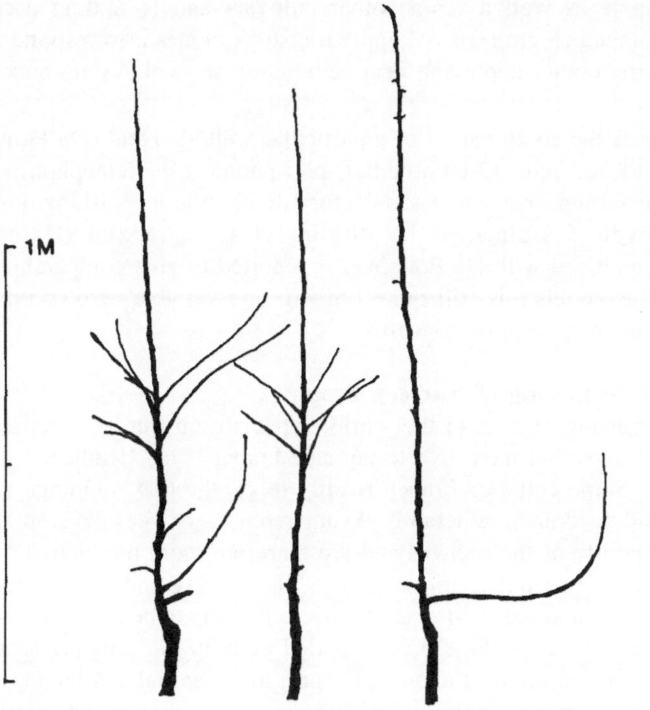

FIGURE 3. Feathering of Gloster apple trees following a single spray of 1000 ppm propyl 1-3-t-butyl phenoxyacetate (M & B 25-105) applied at approximately 60-cm tree height. The tree on the right was not treated. (Courtesy of Dr. Henk van Oosten, Research Station for Fruit Growing, Wilhelminadorp, The Netherlands.)

containing enough 6-benzyladenine (BA) (about 5×10^{-6} *M*) to stimulate shoot proliferation from axillary buds. These shoots are periodically harvested and ''planted'' on a second medium lacking BA but containing NAA or IBA to stimulate root initiation. A third transfer onto a medium free of cytokinins or auxins encourages root development. These small complete plants are eventually potted and grown into salable trees using standard nursery practices.

B. Defoliation of Nursery Stock

Finished nursery trees intended for autumn or spring orchard planting, or rootstock layers intended for sale or further use by the nursery, are dug in the autumn after the plants have stopped growing and are largely defoliated. However, since nursery stock is encouraged to grow very vigorously, these physiological processes are delayed. Natural defoliation may occur too late to make trees available for autumn planting, or in some locations, to permit digging before the soil freezes.

Therefore, there is great interest among nurserymen in chemical treatments to advance defoliation although no universally suitable procedure has been developed yet. Most effective defoliants also cause bark and bud injury from time to time. Furthermore, the results obtained with a given procedure can differ greatly from year to year, and a number of important apple and pear cultivars are remarkably resistant to chemical defoliation. Nonetheless, there are chemical defoliation procedures in commercial use in Europe and North America that are worthy of mention here.

A 1% CuEDTA spray applied in mid-October effectively defoliates nursery trees of Cox's

Orange Pippin apple as well as some other cultivars and is finding acceptance among nurserymen in England (Figure 2). All apple rootstock clones also respond satisfactorily to CuEDTA, but certain other apple and pear scion cultivars will sustain bark and bud injury yet resist defoliation.[5]

In North America the combination of a particular wetting agent (Du Pont® WK at about 1.5%) with 100 to 200 ppm (2-chloroethyl)-phosphonic acid (ethephon), applied two or three times in late autumn, has led to satisfactory defoliation in Washington State.[6] Another material, 2,3-dihydro-5,6-dimethyl-1,4-dithiin-1,1,4,4-tetraoxide (Harvade®, Uniroyal Chemical Co.), combined with Du Pont WK is reported to give comparable results.[7] However, commercial experience is still quite limited, and variable defoliation and occasional tree injury continue to be serious concerns.

C. Promotion of Feathering of Nursery Trees

In some fruit growing regions of the world, especially in Europe, orchardists are accustomed to planting trees that have five to ten side branches or "feathers" when they come from the nursery. Some cultivars branch readily (e.g., the Cox's Orange Pippin shown in Figure 2), while others branch reluctantly. Feathered nursery trees develop a greater fruiting volume early in the life of the orchard and are therefore more productive.[8]

A growth regulator introduced in the mid-1970s by May and Baker Ltd., n-propyl-3-*t*-butylphenoxyacetate marketed as M & B 25-105, is proving to be a highly effective promoter of feathering in nursery trees (Figure 3). Applied early in the growing season of the final nursery year at a concentration of about 750 ppm, this material can increase the number of feathers by threefold or more (Table 1). This treatment is even more beneficial if the feathers developing too low on the trunk are removed by hand.[9]

M & B 25-105 is effective on several important apple and pear cultivars, but is considerably less promising with others.[10] It acts by temporarily interfering with basipetal auxin movement, thus reducing apical dominance.[11] The feathering effect is achieved most satisfactorily on vigorously growing nursery trees.

In North America there appears to be greater interest in a treatment to promote branching of young trees after they are established in the orchard. The growth regulator techniques being developed for this purpose will be discussed in a later section, but this difference in attitude toward the production of well-branched nursery trees warrants a brief discussion here.

Tree planting in Europe occurs in the autumn or very early spring. In the latter case the trees remain in the nursery over winter or are heeled-in in the orchard in the autumn. Thus, cold storage of finished trees is less common than is the case in much of North America. Feathered trees are more difficult and expensive to store than tight bundles of nonbranched trees. Furthermore, the European climate is generally more conducive to the establishment of young trees, and less severe pruning is practiced at the time of planting. Thus, feathered trees are less of a problem for nurserymen and relatively more beneficial to the orchardist.

III. CONTROLLING GROWTH AND MODIFYING THE STRUCTURE OF ORCHARD TREES

The growth regulator treatments discussed in this section are those applied to orchard trees, but generally not to fruits or flowers. They are applied to young nonbearing trees to assist in tree training and to promote spur development; to nonbearing parts of the tree (such as treatments to control the growth of root suckers); or they are applied to the upper tree parts during the dormant season.

A. Regulation of Tree Branching Habit

Unfortunately, materials like M & B 25-105 used by nurserymen in Europe to induce

Table 1
FEATHERING OF NURSERY TREES AS INFLUENCED BY A SINGLE EARLY SUMMER SPRAY OF M & B 25-105

	Number of branches per tree	
	Untreated	M & B 25-105
Apples on MM106		(850 ppm)
Spartan	3.2	9.3
Crispin (Mutsu)	3.2	9.6
Bramley's Seedling	2.0	10.2
Pears on Quince A		(750 ppm)
Conference	0.0	3.5
Williams (Bartlett)	1.3	6.5
Doyenne du Comice	1.1	3.7

From Quinlan, J. D., *Acta Hortic.*, 120, 55, 1981. With permission.

Table 2
STIMULATION OF GROWTH AND BRANCHING OF YOUNG MACSPUR APPLE TREES TREATED WITH MIXTURES OF 6-BENZYLADENINE (BA), GA_{4+7} AND GA_3

No.	Treatment	Concentration (ppm)	No. of new shoots per tree	Average shoot length (cm)	Total growth per tree (cm)
1	No spray	—	5.5	7.61	41.9
2	BA	500	12.0	8.28	99.3
	GA_{4+7}	250			
3	BA	500	20.8	4.30	89.3
	GA_{4+7}	250			
	GA_3	250			

From Costante, J. F., Proc. New Engl. Fruit Meet., 86, 65, 1980. With permission.

feathering of maiden trees are much less effective when applied to trees in the orchard. Therefore, considering that poorly branched nursery trees are the norm in many countries and that the value of early branching is being increasingly recognized, treatments to encourage branching of young orchard trees are beginning to emerge.

With many strong-growing cultivars and strains of apple and pear, not only the number but also the location and angle of attachment of lateral branches can cause concern. When a young unbranched tree is planted, the height at which it is headed (cut back) depends to a large extent upon where the first whorl of lateral branches is desired. Unfortunately, these branches often emerge from a very narrow zone below the heading cut, with two or more upright shoots competing for leadership. These branches are often spaced too closely and are otherwise unsuitable as scaffold branches unless they are mechanically spread.

This unsatisfactory growth habit can be altered by treating dormant buds well below the heading cut with a mixture of benzyladenine (BA) and GA_{4+7} in a lanolin paste.[12] Treatments

FIGURE 4. Modification of growth and fruiting habit of Cox's Orange Pippin apple trees with annual daminozide spray treatments. Both trees were planted in November 1970 and photographed late in the summer of 1973. The tree on the right was treated with 1700 ppm daminozide in 1971 and 1972 and 1000 ppm in 1973. (From *East Malling Research Station Report for 1973*, 64, 1974. With permission.)

applied to individual buds or to all of the buds and bark in the desired trunk zone give comparable results. The result is more branches, wider branch angles, and increased total shoot growth. However, to date there has been very little use of this promising technique in commercial orcharding. Correct timing is important and may be difficult to achieve; or perhaps the treatment method is considered too costly and tedious.

Quite a different tree branching problem is encountered with some of the increasingly popular "spur type" cultivars of apple. These compact trees tend to form fruiting spurs rather than branches, particularly when tree vigor is low. It is not uncommon for trees of these cultivars to resemble a "fruiting pole", thus making very inefficient use of the allotted orchard space.

An early season application of Promalin® (Abbott Laboratories Inc.), a proprietary mixture of BA and gibberellins A_4 and A_7 (GA_{4+7}), is showing considerable promise in tests in North America and Belgium. It stimulates the development of lateral branches and increases the total length of shoot growth on treated trees (Table 2). The addition of a third gibberellin (GA_3) may further enhance treatment effectiveness.[13]

B. Promotion of Spur Development and Flower Initiation

Several important apple cultivars (e.g., Rome Beauty, Cox's Orange Pippin, Tydeman's Early Worcester) form long thin branches with a paucity of fruiting spurs. This condition can be partly corrected, and long-term tree productivity improved, by annual applications of butanedioic acid mono-(2,2-dimethylhydrazide) (daminozide or Alar-85®, Uniroyal Inc.). A program of 500 to 1000 ppm daminozide applied for several years within a month of bud break leads to shorter, sturdier branches supporting more spurs per unit of branch length (Figure 4). Flower initiation is also increased by these treatments but the more important aim is to modify growth habit.

Other apple cultivars, while displaying a suitable compact and potentially productive tree shape and growth habit, are nonetheless reluctant to commence full bearing. The growth regulator treatment prescribed in Washington State and elsewhere in the U.S. is a combination of daminozide and ethephon applied 4 to 5 weeks after bud break. Typical rates are 1000 to 1500 ppm daminozide and 300 ppm ethephon in a high volume spray. When the flowering problem is less severe, daminozide alone (applied somewhat earlier) is the more common treatment. These are likely to be one-time treatments occurring in the fourth or fifth orchard year.

A comparable reluctance to commence flowering and fruiting occurs in young pear trees. A chemical registered in several European countries to control vegetative growth and improve flowering of pear trees is (2-chloroethyl) trimethylammonium chloride (chlormequat, Cycocel®, Cyanamid Inc.). The Dutch recommendation for controlling growth and improving fruitfulness of Comice pear is to apply 1000 to 1600 ppm chlormequat when the new shoots have five to eight leaves and repeat in 2 to 3 weeks if necessary.

C. Suppression of Root Suckers

A troublesome characteristic of some otherwise desirable apple and pear rootstocks is their tendency to initiate strong growing shoots from adventitious buds near the soil surface (Figure 5). Root suckers are unproductive and unsightly; they hinder access to the tree and can harbor a wide range of orchard pests. Fireblight entry through root suckers is a particularly serious concern in some countries.

Procedures for the chemical control of root suckers are slowly being developed. A product originating in England and marketed in western Europe and the Southern Hemisphere combines 12.7% by weight of the free acid of NAA in an emulsion with decanol. Tipoff® (Midox Ltd., U.K.) is applied at a rate of about 4% in water when the suckers are about 10-cm long (Figure 5). At least 4 weeks should have elapsed since full bloom and spray drift must be avoided.

FIGURE 5. Control of sprout and sucker growth with NAA-based products. Top: Control of sprout growth on the trunks of 15-year-old Delicious apple trees with Trehold Inhibitor A-112® in 20% white latex paint applied in April. The active ingredient is the NAA ester at 10000 ppm. (Courtesy of Dr. T. J. Raese, USDA Tree Fruit Research Laboratory, Wenatchee, WA). Bottom: Control of root suckers on 14-year-old Beurré Hardy/Quince A pear trees with 4% Tipoff® applied in early summer. The active ingredients are alpha naphthaleneacetic acid (2240 ppm) and N-decanol (2.78%). (Courtesy of Dr. S. J. Wertheim, Research Station for Fruit Growing, Wilhelminadorp, The Netherlands). The treated tree (both top and bottom) is on the right.

In North America the material more commonly used for this purpose is Trehold Sprout Inhibitor A-112® (Union Carbide Inc.). This product contains 13.2% of the ethyl ester of NAA and is diluted to a concentration of 0.5 to 1.0% NAA with water or white latex paint, depending on use. For root sucker control it is usually diluted with water and used as described for Tipoff®. With both materials, best results are obtained when the sucker growth from previous seasons is cut back so that only current season growth is treated.

D. Suppression of Water Sprouts

Water sprouts are strong upright-growing shoots arising from the trunk and major scaffold limbs. They are particularly prevalent in vigorous trees subjected to containment pruning. On less vigorous trees they may be confined to regions near major pruning cuts. Water sprouts interfere with spray movement and coverage within the tree and seriously reduce light penetration. Removing them by hand significantly increases annual pruning costs.

Trehold Sprout Inhibitor A-112® diluted with 25 to 50% white latex paint in water to a final NAA concentration of 0.5 to 1.0% is used commercially in North America and parts of Europe to control water sprouts. It is applied as a paint to problem areas on the trunk and scaffold limbs. This may involve a complete coverage or spot applications to major pruning cuts. NAA applied to uncut surfaces is not translocated so thorough coverage is essential. However, applied to cut surfaces it will move about 30 cm into the tree and towards the root system, concentrating in the outer ring of xylem.[14] This NAA paint treatment is about equally effective on apple and pear trees.[15]

IV. REGULATION OF FLOWERING AND FRUIT SET

Apple and pear growers also use chemicals to regulate cropping by increasing "return" flowering of cultivars tending to bear biennially; by promoting fruit set on certain problem cultivars; and perhaps most importantly, by reducing set on trees that tend to overcrop in any given season. Interestingly, the use of one practice does not necessarily preclude the use of either of the others on the same tree and even in the same season.

A. Improving Return Bloom

Early fruit thinning is a powerful tool for increasing return flowering. However, the growth-retarding chemicals, daminozide, ethephon, and chlormequat, when applied during the flower initiation period provide additional help in this regard. Daminozide is the favored material for bearing apple trees because, unlike ethephon, early season use does not involve the risk of excessive fruit abscission. On cultivars tending to flower biennially, daminozide is often applied with one of the fruit thinning sprays. However, in England and parts of Europe a slightly different use pattern is developing. Annual mid-summer daminozide sprays, at considerably reduced rates, are said to improve flower strength or "quality" and thus improve cropping the following year.

With pears, increasing numbers of European growers are using multiple applications of chlormequat to improve the productivity of mature trees. Relatively low rates of chemical (generally less than 500 ppm) are applied several times at about 2-week intervals commencing shortly after petal fall. This treatment regime effectively controls shoot growth and improves return cropping without any appreciable reduction in fruit size. Increases in both flower numbers and fruit set explain the improved cropping.[16]

B. Increasing Fruit Set

Spring frosts and other climatic hazards periodically reduce the set of pome fruits in all major producing regions. This, coupled with the fact that growth regulator techniques have been developed to improve fruit set on crops as diverse as tomatoes and grapes, has encouraged the search for fruit setting aids applicable to apples and pears. Unfortunately, progress to date, especially with apples, has been rather disappointing with cultivars differing widely in their responsiveness to the various chemicals. Furthermore, treatments developed specifically for a given cultivar and region all too frequently fail to achieve the desired result in wider commercial practice.

Probably the most successful practice now in commercial use aims at improving the cropping of pears with blossom-time sprays of GA_3 or GA_{4+7}.[17] Rates of 15 to 25 ppm in a high volume spray are commonly suggested. Cultivars which normally set a high proportion of seedless fruit, such as Triomph de Vienne and Beurré Hardy, respond particularly well to gibberellin treatments and are treated annually by many orchardists in England, Holland, and Belgium. Some other cultivars (e.g., Conference) are treated only when spring frost has been severe enough to prevent the natural production of seeded fruits. This rescue operation can be highly cost effective even though some fruits may be misshapen, and blossom density the following season is likely to be reduced.

A second method for increasing fruit set of pears involves the use of 2,4,5-trichlorophenoxypropionic acid (fenoprop) as a postharvest (5 to 10 ppm) or a blossom-time (3 to 6 ppm) spray. This spray is safely applied to a small number of cultivars and dates back to early work in Washington State with Anjou pears.[18] This use of fenoprop no longer appears in North American spray guides, perhaps because positive responses often reveal other cultural ills, such as inadequate pollination, but it continues to find favor in some other countries. South African and Australian growers of Packhams Triumph pears continue to use fenoprop in this manner.

Chemical promotion of apple fruit set is even more problematic. The severity of the fruit set problem with Cox's Orange Pippin apples has led a group of English researchers to seek a growth regulator solution.[19,20] The present state-of-the-art fruit setting mixture for Cox combines GA_{4+7}, N^1N^1-diphenylurea, and 2-naphthoxyacetic acid. Applied at blossom time, it has increased the set of seedless fruits in a number of experiments. However, seedless fruits, or those with very few seeds, compete poorly with normally seeded fruits and the aim of increasing yields year in and year out by setting a proportion of such fruits is proving difficult to achieve. The key to success may be to grow Cox without cross-pollination and rely completely on hormone-induced set. This novel approach is now under test.

Clearly, the development of chemical techniques to reliably improve apple and pear fruit set is an important area for continued research. Early results with the ethylene biosynthesis inhibitor, aminoethoxyvinylglycine, have been outstanding in this regard.[21] This use of an inhibitory rather than promotive growth substance could be the conceptual breakthrough that will lead to some long-awaited practical solutions.

C. Chemical Thinning

Each year it becomes more difficult to sustain apple and pear production where fruit thinning is done entirely by hand. Hand thinning is costly, and because of labor availability problems, it is increasingly difficult to achieve at any price. However, chemical thinning is more than simply a cost-effective way of removing excess fruit. Its major advantage over hand thinning is that fruit removal is accomplished early in the growing season when the return bloom and fruit sizing benefits are greatest.

With the exception of the more recent introduction of ethephon, the chemicals now in commercial use for flower and fruit thinning have been widely available since the 1950s. Not surprisingly, some very localized usage patterns have evolved during these years of field experience. Table 3 illustrates the range of chemical thinning advice directed at growers of a single widely grown apple cultivar, Golden Delicious, in a number of countries and regions within countries. Unfortunately, a more complete presentation of the latest published advice concerning chemical thinning of apples and pears cannot be attempted here. The sheer volume of information indicates the importance placed on chemical thinning technology around the world and suggests an interest in, if not an absolute requirement for, a great deal of local advice.

However, some important regional differences are worthy of elaboration. For example, blossom thinning with sodium 4,6-dinitro-*ortho*-cresylate (DNOC) is an integral part of the apple thinning programs used by growers in western North America, yet its use is rare in other regions. Some interesting exceptions include Victoria, Australia, and Nova Scotia, Canada. Of the postbloom thinners, NAA is applied at petal fall to a range of apple cultivars in Queensland and Western Australia, although the common advice elsewhere is for a much

Table 3
CHEMICAL THINNING OF GOLDEN DELICIOUS APPLES: "SPRAY CALENDAR" ADVICE FROM 12 LOCATIONS

Location	Chemical	Timing	Concentration (g/100 ℓ)[a]	Remarks
Canada				
British Columbia	DNOC	Full bloom	40	For "concentrate" spraying apply 4.7 kg/ha in 1700 ℓ of water
	NAAm	8—12 days AFB[b]	1.7	Use a wetting agent; for "concentrate" spraying apply 118 g/ha as above
	NAA plus carbaryl	15 days AFB	0.5 50	Apply as a full volume spray
U.S.				
Washington	DNOC	See remarks	25—38	Apply when 3 blossoms per spur cluster are open on the N side of the tree
	NAAm	7—14 days AFB	1.7—3.4	Add a wetting agent
	NAAm plus ethephon	10—20 days AFB	1.7—3.4 30—40	Ethephon enhances thinning and return flowering
	NAA plus carbaryl	15—25 days AFB	0.3 45—60	
Ohio	NAA	14—21 days AFB	1.7—2.0	Reduce to 0.8 g if a wetting agent is added; NAAm not recommended
New York	NAA	Up to 21 days AFB or 18-mm fruit length	1.5—2.0	Without a wetting agent
	NAA plus carbaryl	As above	0.5—1.0 60	
Virginia	NAA	14—18 days AFB	0.8	Add awetting agent; carbaryl use can lead to russeting
England	Carbaryl	80% petal fall to 21 days AFB	50	Wet the tree thoroughly and apply under slow drying conditions
The Netherlands	NAAm	7 days AFB	6.0	Without wetting agent; apply as a full volume spray
	Carbaryl	12-mm fruit diameter; 19—28 days AFB	75	As above
France	NAAm	7—10 days AFB	5.0	Apply at least 1500 ℓ/ha
	NAA	9-mm fruit diameter	1.7	As above
	Carbaryl	15- to 16-mm fruit diameter; 18—25 days AFB	150	Apply 1500—2000 ℓ/ha
West Germany	NAAm	75% petal fall to 7 days AFB	5.0—5.8	NAA is not available for use in Germany
South Africa	NAAm	Full bloom to 90% petal fall	7.0	Reported to thin less than carbaryl
	Carbaryl	14—24 days AFB	75	May aggravate russeting

Table 3 (continued)
CHEMICAL THINNING OF GOLDEN DELICIOUS APPLES: "SPRAY CALENDAR" ADVICE FROM 12 LOCATIONS

Location	Chemical	Timing	Concentration (g/100 ℓ)[a]	Remarks
Australia				
Tasmania	NAA	Full bloom to 14 days AFB	1.0	Add a wetting agent
Western Australia	NAA	Late petal fall	1.0	Add a wetting agent; carbaryl aggravates russeting

[a] Values are grams of active chemical. These values multiplied by 10 yield ppm.
[b] After full bloom.

later spray. North Carolina growers are specifically advised that early NAA sprays may stick the fruit on rather than remove it. And in British Columbia, Canada successful thinning is achieved using low volume or "concentrate" spray equipment, whereas only 100 km away in Washington State the advice is to apply all chemical thinning sprays to run-off.

Some of these procedural differences relate directly to climate. Others relate to the orchard management systems used in various regions. Furthermore, it has been demonstrated repeatedly that cultivars of both species differ greatly in their responsiveness to various chemical thinning agents.

Undoubtedly, other explanations for this diversity could be listed, but suffice it to say that widely applicable recommendations for chemical thinning are seldom attempted. The following paragraphs describe in general terms the more important procedures.

1. Blossom Thinning of Apples

This is predicated on the following general observations:

1. Apple trees normally produce 5 to 20 times more blossoms than are required to produce a full crop.
2. Individual fruiting spurs produce 5 or 6 flowers yet should support no more than 2 fruits.
3. If a tree is to maintain annual cropping, a sizable proportion of these spurs should "rest" in any given season.
4. The flowers in each cluster open predictably, i.e., the central flower opens first.
5. With certain cultivars the position of the fruit within this cluster influences an important aspect of fruit quality, i.e., fruit shape.

Selective blossom thinning can be achieved with DNOC since: (1) fertilized flowers (fruits) are resistant to DNOC because the stigmatic tissues have achieved their function; (2) unopened flowers are protected from the spray; and (3) open but unfertilized flowers are damaged and therefore fail to set.

Obviously, timing of the spray is very important. Factors to consider include the stage of bloom and the conditions for pollen transfer and pollen tube growth that existed during the 24- to 48-hr period preceding the spray. If the spray is too late, very few flowers will be affected. If it is too early, the "king" or central flower may be thinned but too many lateral flowers will persist. This would be unwise for Delicious apples where the elongated "king-bloom" fruits are valued, but may be the correct procedure for McIntosh where this fruit is often misshapen. Weather conditions at the time of, and following spraying are also important in that wet weather extends the activity of DNOC. It is not surprising that DNOC is not widely used in maritime fruit growing regions.

Table 4
THINNING OF PEAR FRUITS WITH NAA AND NAAm: "SPRAY CALENDAR" ADVICE FROM 8 LOCATIONS

Location	Cultivar(s)	Chemical	Concentration (ppm)[a]	Timing[b]
Canada				
British Columbia	Bartlett	NAAm	16	13—21
Ontario	Bartlett and	NAAm or	10	2—3
	Anjou	NAA	10	5—9
	Keiffer and	NAAm and	15—20	2—3
	Winter Nelis	NAA	10	5—9
U.S.				
New Jersey	Bartlett and Bosc	NAAm	25—50[c]	7—10
New York	Bartlett	NAAm or	25	3—5
		NAA	2	5—10
Ohio	Bartlett	NAAm	25—35	3—7
Washington	Bartlett	NAAm or	10—15	15—21
		NAA	10—15	15—21
Denmark	Conference and Bonne Louise	NAA	15[c]	3—10
	Clapps Favorite	NAA	22[c]	3—10
Australia				
New South Wales	Williams	NAA	10—15	10—14

[a] Concentration of chemical in a full volume spray with added wetting agent.
[b] Days after full bloom.
[c] Wetting agent requirement not clearly specified.

Some suggested rates and times for DNOC application to thin Golden Delicious are listed in Table 3. Short-season cultivars and those prone to biennial bearing are most commonly treated with DNOC. Evidence to date suggests that pear trees cannot be safely thinned with this chemical.

2. *Postbloom Thinning with Hormone Materials*

Postbloom thinning with hormone materials, primarily NAA and naphthaleneacetamide (NAAm), is the most widely used approach to chemical fruit removal. These chemicals are sold under numerous trade names. NAA has greater auxinic activity than NAAm and this is usually reflected in lower application rates. Because of its attenuated auxinic activity NAAm is less likely to overthin, but because it appears to retain its auxinic activity for a longer period, NAAm usage can lead to some unexpected side effects. The thinning activity of both NAA and NAAm is usually enhanced when they are applied with wetting agents.

The mode of action of NAA and NAAm in inducing selective fruit abscission is far from clear. One would expect the same mode of action yet, as was suggested above, certain cultivars respond differently (e.g., NAAm is more likely than NAA to lead to "pygmy" Delicious apples). NAAm is typically, but not always, applied somewhat nearer blossom time than is NAA.

Most cultivars of apples appear to be responsive to NAA and NAAm as fruit thinning agents commencing at petal fall and continuing for 2 or 3 weeks.[22] Early sprays are suggested in some regions; later ones in others. Rather typical advice for thinning apples with NAA and NAAm on the eastern seaboard of North America is found in the 1981 *Virginia Spray Bulletin for Commercial Tree Fruit Growers*. For Grimes, Jonathan, Stayman, York Imperial, and Rome Beauty apples this bulletin suggests 50 ppm NAAm applied between 4 and 18 days after full bloom or 5 ppm NAA plus a wetting agent applied between 14 and 18 days. It suggests that early NAAm sprays (between 4 and 8 days) are particularly appropriate for short-season apples such as Yellow Transparent, but can also be used for the above-mentioned cultivars. NAAm is not recommended for Delicious or Golden Delicious and NAA is not recommended for the summer apples. The comparable bulletin from the neighboring state of North Carolina warns growers that NAA at rates higher than 5 ppm can lead to the retention of pygmy Delicious fruits.

Growers in New Jersey are advised that NAAm levels can be reduced to 25 ppm if Tween 20 is added to the spray mixture. This bulletin, *1981 Commercial Tree Fruit Production Recommendations for New Jersey,* also suggests that Golden Delicious apples can be thinned with 8 to 10 ppm NAA (plus Tween 20) applied 14 to 16 days after full bloom. This material is also suggested for petal-fall sprays for the early maturing apples and in the 7- to 14-day time period for mid-season apples like McIntosh, Spartan, and Empire.

Both NAA and NAAm are used to thin Bartlett or Williams pears in North America and parts of Australia and a number of other cultivars are similarly thinned in Europe and elsewhere (Table 4). And, as has been shown for apples, the specific advice concerning timing and application rates can vary rather widely, particularly within North America. It is interesting, for example, that in Washington State and Ontario, Canada NAA and NAAm are applied at the same rates (10 to 15 ppm), whereas in New York the suggested NAA rate is lower by a factor of 10. The recommended time of application of each of these chemicals ranges from petal fall in several eastern states to 15 to 21 days after full bloom in the west.

Generally speaking, the use of chemicals to thin pear fruits is less successful and less common than with apples. In some very important pear growing regions such as California and Victoria, Australia chemical thinning is either considered unnecessary or not reliable enough to recommend.

3. Postbloom Thinning of Apples with Carbaryl

Postbloom thinning of apples with carbaryl (1-naphthyl N-methylcarbamate), and more recently with carbaryl-containing combinations, is practiced worldwide. It dates back to the observation by Batjer and Westwood[23] that carbaryl, applied as an insecticide, reliably reduces fruit set. Advantages of carbaryl include the fact that it seldom causes overthinning yet is active well into the postbloom period. It is widely used on the Delicious cultivar because, unlike NAAm, it does not encourage the retention of pygmy apples. The insecticidal activity of carbaryl is a disadvantage when it leads to the destruction of beneficial insects and mites. However, careful monitoring of insect and mite populations can be used to avoid problems. For example, the excessive reduction of predaceous mite populations can lead to severe problems with spider mites, but mites resistant to carbaryl are becoming increasingly common. Consequently, this problem occurs less frequently.

The mechanism by which carbaryl causes fruit thinning, while not completely understood, appears to differ from that of NAA and NAAm.[22,24] Therefore, more effective and reliable thinning can be achieved by combining carbaryl with NAA (Table 5). This approach is rapidly gaining popularity.

Carbaryl is not a satisfactory pear fruit thinner.

4. Flower and Fruit Thinning with Ethephon

Applied as a blossom-time or postbloom spray, ethephon can be used to induce flower and fruit abscission. However, despite early promise as a thinning agent, ethephon is not widely used for this purpose in commercial fruit growing. It has proven more likely to overthin than the alternative materials, a problem that researchers hope to overcome by refining the procedures for its use. One of the very few published suggestions for its use as a fruit thinner in commercial orcharding appears in the *1981 Spray Guide for Tree Fruits in*

Table 5
CHEMICAL THINNING OF SPARTAN APPLES WITH CARBARYL AND/OR NAA, SUMMERLAND, BRITISH COLUMBIA 1979 TO 1981

	Final fruit set as a percent of the water controls[b]		
Treatment[a]	**1979**	**1980**	**1981**
Carbaryl (500 ppm)	86	95	74
NAA (10 ppm)	71	84	63
Carbaryl plus NAA	50	69	51

[a] Sprays applied 15 days after full bloom each year in the same orchard.

[b] The control fruit set values (fruit per 100 blossom clusters) were 88.9, 129.2, and 199.0 in 1979, 1980, and 1981, respectively.

Eastern Washington. A combination of NAAm (17 to 34 ppm) and ethephon (300 to 400 ppm) applied 10 to 20 days after full bloom is suggested for Golden Delicious to help counteract a biennial bearing tendency.

V. IMPROVING FRUIT CONDITION AND APPEARANCE

Modifying fruit condition and appearance is a relatively new but steadily expanding area of growth regulator usage in apple and pear production. Growth regulators can be used to improve the firmness, color, and shape of apples and to decrease the incidence or severity of several disorders of apple and pear fruits. Daminozide, introduced in the mid-1960s, is responsible for many of these effects, but some imaginative new uses for gibberellins and cytokinins are gaining in importance.

A. The Effects of Daminozide on Apple Fruits

Daminozide is now available in most fruit growing countries and, as will become increasingly apparent, is proving to be a very useful chemical. Its effects on fruit color and condition are still generally viewed as welcome side effects of sprays applied for more urgent reasons such as fruit drop control or the promotion of flowering. However, it is increasingly common for apple growers to apply daminozide in early to mid-summer with the primary aim of improving red color, increasing fruit firmness, or perhaps protecting the crop from a specific disorder such as "watercore" (Table 6).[25-33]

Daminozide is usually applied as a full volume spray containing 500 to 2000 ppm of active chemical. Low volume sprays have also proven effective in many regions. The amount of chemical applied is usually less with mature trees and is commonly reduced after several years of use on the same tree.

Some adverse carry-over effects of daminozide on fruit size, shape, and length of the pedicel have been noted and these are more likely to occur if the chemical is applied at high rates or too late in the growing season. Apple cultivars differ considerably in their susceptibility to these carry-over effects.

Another precaution is that a few storage disorders of apples can be advanced or intensified by daminozide treatment. These effects are often specific for a given cultivar and location,

Table 6
THE EFFECT OF DAMINOZIDE TREATMENT ON SOME APPEARANCE AND CONDITION PARAMETERS OF MCINTOSH AND DELICIOUS APPLES

	Control	Treated	Ref.
McIntosh			
Surface red color (%)	63	75[a]	25
	37	62	26
	62	80	27
Flesh firmness (kg) at harvest	8.1	8.7	28
	6.9	7.5	25
Flesh firmness (kg) after 3—4 months of 0°C storage	5.3	5.9	28
	7.9	7.7 n.s.	29
Flesh firmness (kg) after 5 months of CA storage	4.3	4.9	29
	4.5	5.7	25
Titrateable acidity (mℓ 0.1 N NaOH/25 mℓ of juice)	34.3	25.6	28
Juice soluble solids (%)	11.8	11.6 n.s.	28
Development of brown core in 0°C storage (% afflicted)	30	53	30
Delicious			
Flesh firmness (kg) at harvest	7.8	8.6	31
	8.7	9.1	28
	8.0	8.5	32
Flesh firmness (kg) after 3—6 months of 0°C storage	6.7	8.4	33
	7.3	7.5 n.s.	28
Titrateable acidity (mℓ 0.1 N NaOH/25 mℓ of juice)	11.4	11.4 n.s.	28
Juice soluble solids (%)	12.6	12.0	31
	11.1	10.9 n.s.	28
Incidence of water core at harvest (% afflicted)	56	8	31
Incidence of storage scald (% afflicted)	84	20	33

[a] Differences due to treatment are statistically significant at the 5% probability level unless indicated by n.s.

but nonetheless can be of serious concern. For example, daminozide sprays are reported to shorten the storage life of McIntosh apples in New York because they promote internal browning.[30] This effect has not been observed with McIntosh grown in British Columbia.

B. Preventing Premature Softening of Pear Fruits

A disorder of Bartlett pear fruits referred to as "pink-end" can be avoided or reduced by a daminozide spray (about 1000 ppm) applied 18 to 24 days before anticipated harvest. Pink-end is characterized by softening of the calyx end of fruits still on the tree and is induced by field temperatures below 10°C during the month preceding harvest.[34] This use of daminozide is presently confined to Oregon and Washington.

C. Reduction of Fruit Russeting

Most apple cultivars are susceptible to skin russeting when grown in humid climates. Severe fruit russeting frequently reduces the value of Golden Delicious apples, and an effective protectant would receive widespread acclaim. Australian work indicating that GA_{4+7} sprays reduce the severity of russet[35] is very promising and is being followed up at other

Table 7
REDUCTION OF APPLE FRUIT RUSSETING BY 4 SPRAYS OF GA_{4+7} APPLIED WITHIN 1 MONTH OF BLOOM

Orchard:	1	2	3	4	5	6
Cultivar:	Karmijn de Sonnaville				Golden Delicious	
No spray	82[a]	50	79	58	18	40
GA_{4+7} (10 ppm)	41	10	68	42	4	17

[a] Values are the percentage of moderately to severely russeted fruits in each treatment and orchard lot.

From Tromp, J. and Wertheim, S. J., *Proc. 15th Colloq. Int. Potash Inst.*, p. 137, 1980. With permission.

locations.[17,36] Relatively high concentrations of GA_{4+7} and multiple applications appear to be necessary for complete control,[36] but considerably lower rates can lead to a marked improvement in fruit finish (Table 7).[17] Assuming no adverse side effects, eventual registration and widespread commercial use can be anticipated. Interestingly, GA_3 is considerably less effective and very recent evidence suggests that it is the GA_4 in the GA_{4+7} mixture that reduces russeting. And since GA_7 is the mixture component most likely to inhibit flower initiation,[38] a pure GA_4 product for russet control would be ideal.

D. Increasing the Length to Diameter Ratio of Apples

A proprietary mixture of GA_{4+7} and 6-benzyladenine (6-BA) (Promalin®, Abbott Laboratories) is now registered in several countries and is used primarily to promote fruit elongation of Delicious apples. Historically, elongated or angular fruit of this cultivar have received a premium in the marketplace, and it has been clearly demonstrated that blossom-time sprays of Promalin® can lead to longer, more "typey" fruit.[39] It is also suggested that Promalin® treatment increases average fruit weight without noticeably reducing fruit numbers. However, other reports show quite clearly that this treatment enhances fruit thinning;[40] in fact, this effect is sometimes viewed as being beneficial and can be enhanced by adding DNOC to the spray mix.[41]

The time of Promalin® application has a great influence on the fruit elongation effect.[39,41,42] It is presently suggested that the spray be applied at the 80% bloom stage and that, unless fruit thinning is desired, high volume sprays and rainfall near the time of application are to be avoided. A medium volume spray containing 25 ppm of the active ingredients in Promalin® (approximately equal amounts by weight of 6-BA and GA_{4+7}) is used by Washington State apple growers.

VI. PREVENTING PREHARVEST FRUIT DROP

A major production problem with a number of important apple and pear cultivars is the tendency for fruits to loosen and drop before harvest can be completed. It is not uncommon for crop losses of 20 to 30% to be experienced unless some form of chemical protection is provided.

One of the earliest nonherbicidal uses of plant growth regulators was the use of synthetic auxins to control preharvest drop of apples and pears and these uses continue today (see Edgerton[43] for a review of this subject). More recently, daminozide has proven to be effective in this regard and, while more costly than the auxinic materials, is gaining in popularity.

A. Control of Fruit Drop with Synthetic Auxins

NAA and 2,4,5-trichlorophenoxypropionic acid (2,4,5-TP or fenoprop) are the synthetic

Table 8
SOME SELECTED "SPRAY CALENDAR" ADVICE FOR CONTROLLING PREHARVEST DROP OF APPLES AND PEARS WITH SYNTHETIC AUXINS

Location	Crop	Material	Specific advice
Canada			
British Columbia	Bartlett pear Early- to mid-season apples	NAA	20 ppm in a high volume spray or 1.4 kg of a 6.34% active NAA product per hectare; apply 2 days before expected drop
Ontario	Apples	NAA	10 ppm applied before drop commences; repeat in 7 days if necessary
England	Conference pear	NAA	10 ppm in a high volume spray
	Early apples	Fenoprop	10—15 ppm in a high volume spray
	Other apples	NAA	10 ppm 6—14 days before expected drop
The Netherlands	Apples and pears	NAA	5—7.5 ppm for early apples and pears — 10 ppm for all others; apply under fast-drying conditions to cultivars susceptible to skin damage (e.g., Boskoop); also, apply as a high volume spray 7—10 days before expected harvest
South Africa	Apples and pears (all cultivars)	NAA	20 ppm in a high volume spray at first sign of drop; normally 10—14 days before harvest
	Delicious and Granny Smith apples	Fenoprop	10 ppm as above
Israel	Apples	Fenoprop	10-20 ppm depending on cultivar; NAA added when quick action is desired
Japan	Japanese pear (all cultivars)	2,4-DB[a] plus MCPB[b]	10—15 ppm 7—14 days before harvest; 30—40 ppm just before harvest
U.S.			
North Carolina	Delicious apple	Fenoprop	10 ppm in a high volume spray 10—14 days before peak harvest date
	Delicious and other apples	NAA	10 ppm 4—6 days before expected fruit drop
Ohio	Early apples	NAA	5 ppm 4—5 days before heavy drop
	Apples later than McIntosh	NAA	10 ppm as above; repeat in 7 days if necessary
	Delicious and Grimes apples	Fenoprop	10 ppm in a high volume spray 10 days before peak harvest
	Late apples	Fenoprop	20 ppm as above
Washington	Bartlett pear	NAA	10 ppm in a high volume spray or 0.46—0.94 ℓ of a 5.6% active NAA product per hectare in a low volume spray
	Delicious and Winesap apples	Fenoprop	10 ppm in a high volume spray 7—12 days before expected drop; concentrate spraying can lead to blind wood
	Delicious apple	NAA	15—20 ppm in a high volume spray or 218 g actual NAA per hectare applied by aircraft

[a] 2,4-Dichlorophenoxybutyric acid.
[b] 2-Methyl-4-chlorophenoxybutyric acid.

auxins used most widely for this purpose (Table 8). Important advantages of these materials are their low cost and quick action. Relatively low levels of chemical (10 to 20 ppm), applied just before fruit drop commences, will delay abscission for several weeks. NAA provides fast but relatively short-lived protection whereas fenoprop, thought to be somewhat slower acting, gives excellent fruit drop control for up to 1 month.

Table 9
PERCENTAGE DROP OF MCINTOSH APPLES AS INFLUENCED BY NAA, FENOPROP, AND DICHLORPROP SPRAYS APPLIED 7 DAYS BEFORE NORMAL HARVEST, KELOWNA, BRITISH COLUMBIA — 1977

	Days after treatment		
Treatment	**7**	**11**	**16**
No spray	10.3	24.3	31.2
NAA (20 ppm)	3.6	7.3	10.9
Fenoprop (10 ppm)	1.9	3.6	6.0
Dichlorprop (20 ppm)	3.3	5.2	6.0

The loss or threatened loss of fenoprop as a registered agricultural chemical in a number of countries is of serious concern to apple and pear growers. This development relates to public concern about the trace levels of dioxins found in fenoprop, but considering the low rates of orchard application and the long history of user safety, it is highly unlikely that this concern is justified. This situation is mentioned because it is indicative of the stormy weather ahead for both the producers and users of agricultural chemicals.

This loss or potential loss of fenoprop has led researchers to reexamine some alternative chemicals. Dichlorprop (2,4-DP), at somewhat higher rates, gives comparable fruit drop control (Table 9) and may have a commercial future.[44] The most objectionable dioxin contaminant, TCDD, would not be expected to occur in dichlorprop. Interestingly, a similar chemical, 2,4-dichlorophenoxybutyric acid, is used to control fruit drop in Japan where neither NAA nor fenoprop is registered for this purpose.

The use of auxinic chemicals to delay fruit abscission is not without certain disadvantages. The ethylene generating properties of these chemicals can advance fruit softening and if harvest is unduly delayed, the fruit may be unmarketable. Furthermore, the herbicidal effects of these auxins can, under certain circumstances, cause bud and bark damage to the tree. And finally, the fact that the spray must be applied very near to harvest can be a serious disadvantage. In modern high density plantings it is often difficult to move a tractor and sprayer through the orchard rows just before harvest. The control of harvest drop with daminozide has none of these disadvantages although its use is limited to apples.

B. Control of Fruit Drop with Daminozide

Daminozide delays fruit abscission by a completely different mechanism to that of the synthetic auxins. It delays fruit ripening by suppressing the endogenous ethylene production associated with that event.[45] Since ethylene also triggers abscission, this physiological process is likewise delayed.

Daminozide is usually not applied solely to control preharvest fruit drop, but when it is applied, for whatever reason, additional fruit drop protection is unnecessary (except in conjunction with ethephon usage). The time of application can range from shortly after full bloom to about 60 days before expected harvest. The latter timing is common if drop protection is a primary aim, and sprays even closer to harvest may be appropriate for some early-maturing cultivars. Application rates usually range from 500 to 1000 ppm in a high volume spray or 2 to 4 kg of the 85% material per hectare in a reduced volume spray.

Since daminozide does not advance fruit softening, its use lends great flexibility to the harvesting operation. The fact that even the unripe fruits are firmer is particularly advantageous if these fruits are to be subsequently treated with ethephon and auxin to promote color development and early ripening.[26] This multiple use of growth regulators to program fruit harvest is discussed more thoroughly in the next section.

VII. REGULATION OF FRUIT RIPENING

A. Delay of Fruit Ripening with Daminozide

As noted above, daminozide treatment delays fruit ripening by suppressing autocatalytic ethylene production. In orchard practice this permits, but does not require, a harvest delay of several weeks. The limit to which this effect can be exploited is still not clear. The fruit may appear to be in excellent condition at harvest but deteriorates more rapidly in cold storage,[30] although this is certainly not always the case.[29,33] A harvest delay of up to 2 weeks appears to be safe for McIntosh apples in British Columbia, but would be strongly discouraged in New York State.

B. Advancing Fruit Ripening with Ethephon

Conversely, ethephon (in conjunction with fenoprop or NAA to control fruit drop) can be used to advance fruit coloring and ripening by several weeks.[27,46] The availability of highly colored apples with excellent out-of-hand eating quality, weeks earlier than normal, is advantageous to consumers as well as producers. However, the full potential of early fruit ripening with ethephon is yet to be recognized by the fruit growing industry. The main problem appears to be a reluctance to view the ethephon-treated fruit as a completely new commodity requiring a different strategy with regard to marketing and storage. Balancing the fruit quality improvements attributable to ethephon are (1) a greatly reduced storage life; (2) a shorter shelf life; and (3) a reduction in yield because the fruits are picked earlier than normal. Ethephon-treated apples are preferred by consumers when they are sold immediately following harvest. In addition to their attractive appearance, they are sweeter, juicier, and more aromatic than the nontreated fruits will be for several weeks. However, they soften rapidly in storage and cannot be expected to compete with the nontreated fruit a month or two later. Unfortunately, fruit growers have tended to use ethephon to improve fruit color and then attempt to dispose of these fruits through the traditional marketing channels. The results can be disastrous.

Successful use of the ethephon early-ripening technique requires a considerable knowledge about when and where the procedure is appropriately used, as well as careful attention to a number of procedural details. The need to couple this practice with a marketing strategy has already been discussed. Not only cultivars but strains within cultivars differ in their responsiveness to ethephon. Maintaining the fruit on the tree for the required period may require more than one application of NAA. The rate of fruit ripening is closely related to the prevailing air temperatures, and fruit coloring and fruit softening occur optimally at different temperatures. This list of technical details could continue.

C. Product Interactions

If trees to be treated with ethephon and auxin are previously treated with daminozide, several of the above-mentioned problems can be avoided.[27] Daminozide pretreatment increases initial fruit firmness and reduces the risk of excessive softening (Figure 6). It also promotes red coloration so that the combined effects are often superior, and imparts a degree of fruit drop control that is especially helpful where NAA is the only auxinic stop-drop material available (Table 10).

The extent to which the apple industry will use these tools for regulating fruit ripening

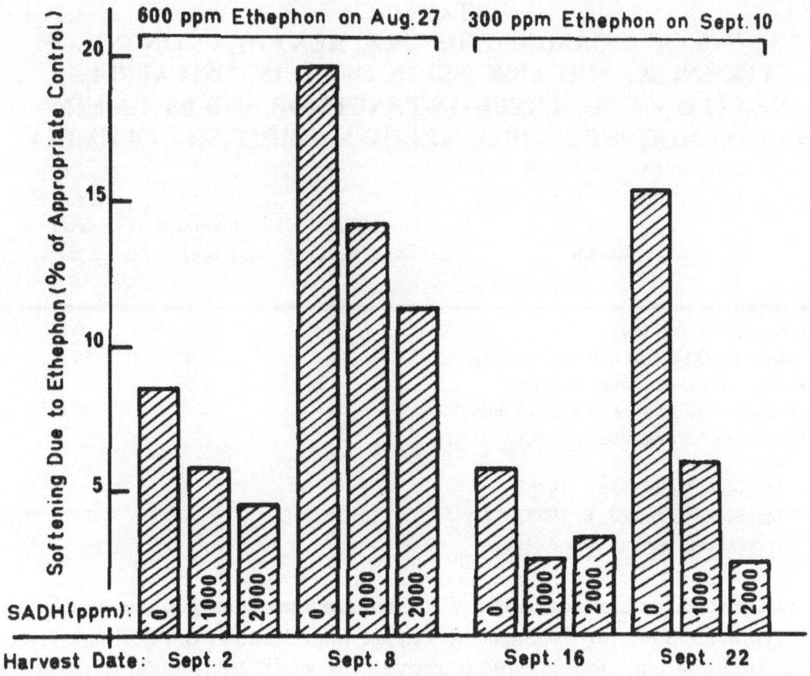

FIGURE 6. The effect of daminozide (SADH) pretreatment on ethephon-induced softening of McIntosh apples in 1973. Daminozide was applied July 7. (From Looney, N. E., *J. Am. Soc. Hortic. Sci.*, 100, 330, 1975. With permission.)

remains to be seen. With the available technology it is possible to harvest good quality McIntosh apples in British Columbia from mid-August to late September, but relatively few fruit growers are fully exploiting these techniques.

And finally, it is possible that the effects of chemicals like daminozide and aminoethoxyvinylglycine on fruit ethylene production will become increasingly important to fruit storage technologists. For example, English researchers hoping to extend the storage life of Cox's Orange Pippin apples by "scrubbing" ethylene from the storage environment have found that a daminozide pretreatment greatly facilitated this approach.[47]

VIII. FUTURE DEVELOPMENT OPPORTUNITIES

Clearly, the growers and shippers of deciduous tree fruits have contributed actively to the expansion of plant growth regulator technology. To an extent greater than is the case with other agricultural commodity groups they appreciate the potential of this technology, and this interest and support should be recognized as a resource by the scientific community. The extent to which they "sell" their growth regulator requirements and demonstrate the safety and utility of these chemicals to the general public and the various regulatory agencies could easily determine the future "beyond the lab" progress in this field.

At the time of this writing there are a number of fascinating and potentially important new plant growth regulators and plant growth regulator techniques of interest to deciduous fruit growers. Some have not been discussed because they are unavailable to commercial orchardists. It will be very interesting to see if these chemicals progress to commercial status. A powerful new growth retardant, 1-(4-chlorophenyl)4,4-dimethyl-2-(1,2,4-triazol-1-yl)pentan-3-ol (PP333, Plant Protection Division, I.C.I. Ltd., U.K.), promises to control shoot growth much more effectively than either daminozide or chlormequat and is being

Table 10
EFFECTS OF A DAMINOZIDE PRETREATMENT ON COLOR, FIRMNESS, AND ABSCISSION OF MCINTOSH APPLES TREATED WITH ETHEPHON-FENOPROP AND ETHEPHON-NAA ON AUGUST 27, 1973, KELOWNA, BRITISH COLUMBIA

Treatment	Surface red color (%)[a]	Firmness (kg)[a]	Fruit drop (%)[b]
No spray	54.5	8.5	1.5
Daminozide (1000 ppm)[c]	60.5	9.1	0.0
Ethephon (600 ppm) + fenoprop (20 ppm)	82.5	7.5	29.6
Daminozide + ethephon + fenoprop	88.5	8.7	7.2
Ethephon (600 ppm) + NAA (20 ppm)	—	—	64.6
Daminozide + ethephon + NAA	—	—	31.5

[a] Measured September 2, 1973.
[b] Measured September 6, 1973.
[c] Daminozide applied July 7, 1973.

studied intensively at several locations.[9,48,49] Aminoethoxyvinylglycine (AVG), a potent ethylene biosynthesis inhibitor, is known to release lateral buds from dormancy, delay fruit drop and fruit ripening, and dramatically improve fruit set.[21,40,50] If either of these chemicals achieves commercial status, our ability to control growth and productivity of fruit trees will be greatly extended.

However, this potential notwithstanding, there are still problem areas in fruit growing where satisfactory growth regulator aids have yet to evolve. A good example is the area of dealing with climatic hazards. By selecting appropriate cultivars, rootstocks, and management systems it is possible to grow apples and pears over a remarkably wide geographic range. Nonetheless, severe winter cold is a frequent hazard across a wide section of this range, and insufficient winter chilling limits tree productivity in warm-temperate regions. Interestingly, frost during the blossom period is a concern across most of this geographic range as is excessive heat and evapotranspiration during high summer.

It is possible, even likely, that we will eventually be able to deal effectively with the problem of insufficient chilling. Progress towards understanding the relationship of gibberellins and cytokinins to bud dormancy[51] bodes well for an eventual practical solution involving growth regulators. Likewise, it is likely that the spring frost hazard will eventually be dealt with by delaying anthesis. This has already been achieved with a limited number of species.[52]

On the other hand, the practical use of cryoprotectants to reduce the effects of severe winter freezes is still just a dream although interesting results have been reported.[53] Similarly, progress towards reducing water stress of orchard trees with growth regulators or antitranspirant films has been exceedingly slow; the challenge is to avoid a number of undesirable side effects on fruit, foliage, and tree productivity.[54]

Finally, a few other areas which are open for development deserve a mention. Growth regulators that influence the content or balance of mineral ions, especially calcium and other cations in fruit tissues, are being sought. A chemical to directly delay senescence in fruits could profoundly reduce both wastage and our dependency on expensive fruit storage technology. Still other chemicals may change our approach to dealing with orchard insects and diseases. For example, why not think in terms of exploiting the beneficial effects of a specific virus if the undesirable side effects (if any) could be treated with growth regulators?

In conclusion, the technology of using plant growth regulators in deciduous tree fruit production has developed to a high level in a relatively short period of time. Indeed, it is already imprudent if not impossible to compete in this industry without using plant growth regulators at several stages in the overall operation. Nonetheless, far more growth regulator techniques have been developed than are legally available to the fruit growers in most countries. The extent to which the use of plant growth regulators will continue to expand and contribute to efficient fruit production depends largely on the attitude of societies and governments regarding the "chemicals in agriculture" issue.

REFERENCES

1. **Howard, B. H.**, Propagation of leafless winter cuttings, *Plantsman*, 3, 99, 1981.
2. **Child, R. D. and Hughes, R. F.**, Factors influencing rooting in hardwood cuttings of apple cultivars, *Acta Hortic.*, 79, 43, 1978.
3. **Lane, W. D.**, Regeneration of apple plants from shoot meristem tips, *Plant Sci. Lett.*, 13, 281, 1978.
4. **Jones, O. P.**, Propagation *in vitro* of apple trees and other woody fruit plants: methods and applications, *Sci. Hortic.*, 30, 44, 1979.
5. **Knight, J. N.**, Chemical defoliation of nursery stock. III. Preliminary studies with chelated forms of copper and iron, *J. Hortic. Sci.*, 58, 1983.
6. **Larsen, F. E.**, Stimulation of leaf abscission of tree fruit nursery stock with ethephon-surfactant mixtures, *J. Am. Soc. Hortic. Sci.*, 98, 34, 1973.
7. **Larsen, F. E. and Lowell, G. D.**, Nursery stock defoliation, *Am. Nurseryman*, 148, 10, 1978.
8. **van Oosten, H. J.**, Effect of initial tree quality on yield, *Acta Hortic.*, 65, 123, 1978.
9. **Quinlan, J. D.**, New chemical approaches to the control of fruit tree form and size, *Acta Hortic.*, 120, 95, 1981.
10. **van Oosten, H. J.**, Effects of propyl 3-T-butylphenoxyacetate (M and B 25-105) on the branching of maiden apple trees, *Med. Fac. Landbouww. Rijksuniv. Gent.*, 46, 247, 1981.
11. **Duckworth, S. J., Abbas, M. F., and Quinlan, J. D.**, Influence of endogenous growth regulators on branching, *Rep. E. Malling Res. Stn. for 1978*, 39, 1979.
12. **Williams, M. W. and Billingsley, H. D.**, Increasing the number and crotch angles of primary branches of apple trees with cytokinins and gibberellic acid, *J. Am. Soc. Hortic. Sci.*, 95, 649, 1970.
13. **Costante, J. F.**, Update on growth regulators, *Proc. New Engl. Fruit Meet.*, 86, 65, 1980.
14. **Blanco Braná, A. and Jackson, J. E.**, Transport of NAA applied to the cut surfaces of pruned apple branches, *J. Hortic. Sci.*, 57, 31, 1982.
15. **Raese, J. T.**, Sprout control of apple and pear trees with NAA, *HortScience*, 10, 396, 1975.
16. **Reedijk, A.**, Resultaten van CCC-bespuitingen op Beurré Hardy, *Fruitteelt*, 62, 510, 1972.
17. **Tromp. J. and Wertheim, S. J.**, Synthetic growth regulators: mode of action and application in fruit production, *Proc. 15th Colloq. Int. Potash Inst.*, p. 137, 1980.
18. **Degman, E. S. and Batjer, L. P.**, Delayed effects of 2,4,5-trichlorophenoxypropionic acid sprays on Anjou pears, *Proc. Am. Soc. Hortic. Sci.*, 66, 84, 1955.
19. **Kotob, M. A. and Schwabe, W. W.**, Induction of parthenocarpic fruit in Cox's Orange Pippin apples, *J. Hortic. Sci.*, 46, 89, 1971.
20. **Goldwin, G. K.**, Hormone-induced setting of Cox apple, *Malus pumila*, as affected by time of application and flower type, *J. Hortic. Sci.*, 56, 345, 1981.
21. **Williams, M. W.**, Retention of fruit firmness and increase in vegetative growth and fruit set of apples with aminoethoxyvinylglycine, *HortScience*, 15, 76, 1980.
22. **Williams, M. W.**, Chemial thinning of apples, *Hortic. Rev.*, 1, 270, 1979.
23. **Batjer, L. P. and Westwood, M. N.**, 1-Naphthyl N-methylcarbamate, a new chemical for thinning apples, *Proc. Am. Soc. Hortic. Sci.*, 77, 1, 1960.
24. **Schneider, G. W.**, The mode of action of apple thinning agents, *Acta Hortic.*, 80, 225, 1978.
25. **Edgerton, L. J. and Blanpied, G. D.**, Interaction of succinic acid 2,2-dimethylhydrazide, 2-chloroethylphosphonic acid and auxins on maturity, quality and abscission of apples, *J. Am. Soc. Hortic. Sci.*, 95, 664, 1970.
26. **Looney, N. E.**, Interaction of ethylene, auxin and succinic acid 2,2-dimethylhydrazide in apple fruit ripening control, *J. Am. Soc. Hortic. Sci.*, 96, 350, 1971.
27. **Looney, N. E.**, Control of ripening in "McIntosh" apples. I. Growth regulator effects on preharvest drop and fruit quality at each of four harvest dates, *J. Am. Soc. Hortic. Sci.*, 100, 330, 1975.
28. **Fisher, D. V. and Looney, N. E.**, Growth, fruiting and storage response of five cultivars of bearing apple trees to N-dimethylaminosuccinamic acid (Alar), *Proc. Am. Soc. Hortic. Sci.*, 90, 9, 1967.

29. **Pollard, J. E.,** Effects of SADH, ethephon and 2,4,5-T on color and storage quality of "McIntosh" apples, *J. Am. Soc. Hortic. Sci.*, 99, 341, 1974.
30. **Blanpied, G. D., Smock, R. M., and Kollas, D. A.,** Effect of Alar on optimum harvest dates and keeping quality of apples, *Proc. Am. Soc. Hortic. Sci.*, 90, 467, 1967.
31. **Batjer, L. P. and Williams, M. W.,** Effects of N-dimethylamino succinamic acid (Alar) on water core and harvest drop of apples, *Proc. Am. Soc. Hortic. Sci.*, 88, 76, 1966.
32. **Looney, N. E., Fisher, D. V., and Parsons, J. E. W.,** Some effects of annual applications of N-dimethylamino succinamic acid (Alar) to apples, *Proc. Am. Soc. Hortic. Sci.*, 91, 18, 1968.
33. **Williams, M. W., Batjer, L. P., and Martin, G. C.,** Effects of N-dimethylamino succinamic acid (B-Nine) on apple quality, *Proc. Am. Soc. Hortic. Sci.*, 85, 17, 1964.
34. **Wang, C. Y., Mellenthin, W. M., and Hansen, E.,** Effect of temperature on development of premature ripening in "Bartlett" pears, *J. Am. Soc. Hortic. Sci.*, 96, 122, 1971.
35. **Taylor, B. K.,** Reduction of apple skin russeting by gibberellin A_{4+7}, *J. Hortic. Sci.*, 50, 169, 1975.
36. **Eccher, T. and Boffelli, G.,** Effects of dose and time of application of GA_{4+7} on russeting, fruit set and shape of Golden Delicious apples, *Sci. Hortic.*, 14, 307, 1981.
37. **Wertheim, S. J.,** Fruit russeting in apple as affected by various gibberellins, *J. Hortic. Sci.*, 57, 283, 1982.
38. **Tromp, J.,** Flower-bud formation in apple as affected by various types of gibberellins, *J. Hortic. Sci.*, 57, 277, 1982.
39. **Unrath, C. R.,** The commercial implications of gibberellins A_4A_7 plus benzyladenine for improving shape and yield of "Delicious" apples, *J. Am. Soc. Hortic. Sci.*, 99, 381, 1974.
40. **Green, D. W.,** Effect of silver nitrate, aminoethoxyvinylglycine, and gibberellins A_{4+7} plus 6-benzylamino purine on fruit set and development of "Delicious" apple, *J. Am. Soc. Hortic. Sci.*, 105, 717, 1980.
41. **Williams, M. W.,** Combinations of bio-regulants on apple trees, in *Tree Fruit Growth Regulators and Chemical Thinning*, Tukey, R. B. and Williams, M. W., Eds., Washington State University Press, Pullman, 1981, 213.
42. **Looney, N. E.,** Some effects of gibberellins A_{4+7} plus benzyladenine on fruit weight, shape, quality, Ca content, and storage behaviour of "Spartan" apple, *J. Am. Soc. Hortic. Sci.*, 104, 389, 1979.
43. **Edgerton, L. J.,** Control of abscission of apples with emphasis on thinning and pre-harvest drop, *Acta Hortic.*, 34, 333, 1973.
44. **Looney, N. E. and Cochrane, W. P.,** Relative effectiveness of, and residue declination values for dichlorprop, fenoprop and naphthaleneacetic acid used to control preharvest drop of McIntosh apples, *Can. J. Plant Sci.*, 61, 87, 1981.
45. **Looney, N. E.,** Control of apple fruit ripening by succinic acid 2,2-diamethylhydrazide, 2-chlororethyl-trimethylammonium chloride, and ethylene, *Plant Physiol.*, 44, 1127, 1969.
46. **Luckwill, L. C. and Child, R. D.,** Growth regulator effects on quality and pre-harvest drop of Worcester Pearmain apples, *J. Hortic. Sci.*, 47, 249, 1972.
47. **Knee, M., Hatfield, S., and Cockburn, J. T.,** Ethylene removal during storage of daminozide sprayed apples, *Ann. Rep. East Malling Res. Stn. for 1981*, 132, 1982.
48. **Tukey, L. D.,** personal communication, 1981.
49. **Williams, M. W.,** personal communication, 1981.
50. **Bangerth, F.,** The effect of a substituted amino acid on ethylene biosynthesis, respiration, ripening and preharvest drop of apple fruits, *J. Am. Soc. Hortic. Sci.*, 103, 401, 1978.
51. **Broome, O. C. and Zimmerman, R. H.,** Breaking dormancy in tea crabapple, *J. Am. Soc. Hortic. Sci.*, 101, 28, 1976.
52. **Howell, G. S., Jr. and Dennis, F. G., Jr.,** Cultural management of perennial plants to maximize resistance to cold stress, in *Analysis and Improvement of Plant Cold Hardiness*, Olien, C. R. and Smith, M. N., Eds., CRC Press, Boca Raton, Fla., 1981, 175.
53. **Ketchie, D. O. and Murren, C.,** Use of cryoprotectants on apple and pear trees, *J. Am. Soc. Hortic. Sci.*, 101, 57, 1976.
54. **Jones, H. G.,** PGRs and plant water relations, in *Aspects and Prospects of Plant Growth Regulators*, Monograph 6 of the British Plant Growth Regulator Group, Jeffcoat, B., Ed., ARC Letcombe Laboratory, Wantage, Oxfordshire, 1980, 91.

Chapter 2

GROWTH REGULATOR USE IN THE PRODUCTION OF *Prunus* SPECIES FRUITS

Norman E. Looney

TABLE OF CONTENTS

I. INTRODUCTION

With a few notable exceptions, growth regulator usage in the production of the various *Prunus* species fruits (often referred to as drupe or stone fruits) is less central to commercial success than is the case for apples and pears (*Malus* and *Pyrus* species, respectively). The reasons for this are not at all clear, but certainly the drupe fruits respond quite differently from pome fruits to a range of plant growth regulators. In general, the drupe fruits respond poorly to postbloom chemical thinning; there are many commercial cultivars of each species and these differ greatly in their responsiveness to growth regulators; and the drupe fruits are more likely to exhibit adverse side effects to plant growth regulators than are apples or pears.

But despite these limitations, there is a wide range of interesting plant growth regulator uses in drupe fruit production. Some are closely analogous to those already described for apples and pears and will not be detailed again. For example, the production of plum and cherry rootstocks by cuttings and by in vitro meristem culture involves growth regulator techniques comparable to those used for apples and pears. Likewise, nursery trees of various *Prunus* species can be chemically defoliated, and root suckers of plum and prune can be suppressed with the materials already discussed for apples and pears.

Furthermore, not all of the growth regulator practices applied to each specific crop will be discussed in detail. For example, the use of ethephon to loosen cherries for mechanical harvesting is applicable to both sweet and tart cherries but will only be discussed in the tart cherry section. Likewise, the plant growth regulator mixtures used in Europe to enhance fruit set on sweet cherries may also benefit plum and to a lesser extent sour cherry producers, but will be discussed only in the sweet cherry section.

In general, however, the growth regulator aids used in drupe fruit production are species, and even cultivar, specific. Furthermore, none of the practices discussed herein is important in all of the regions where these responsive cultivars are grown. Finally, it should be emphasized that in terms of world food production most of these practices are of minor importance.

In the discussion of each major crop, I will attempt to point out some of these differences and, in some cases, discuss the constraints to expanded growth regulator usage.

II. APRICOTS (*Prunus armeniaca* L.)

Apricots are used in a variety of ways. In addition to fresh consumption they are widely used for canning and drying and apricot juice and puree are important products of some regions. However, the total world production of about 1.5 million metric tons is small compared to that of apples (about 35 million metric tons) and pears (about 8 million metric tons).[1]

The U.S.S.R., Turkey, Spain, Italy, and U.S. all produce more than 100 thousand metric tons of apricots each year and 25 other countries (many categorized as "developing") each produce more than 20 thousand metric tons.[1]

At the present time, plant growth regulators play only a minor role in worldwide apricot production. Blossom thinning is practiced in a few regions and growers in a few scattered locations apply synthetic auxins to reduce fruit abscission and advance fruit maturity. Interestingly, a considerable amount is known about the role of plant growth substances in developing apricot fruits,[2,3] but this knowledge has yet to lead to a major plant growth regulator practice.

A. Blossom Thinning

Dinitro-*ortho*-cresylate (DNOC) and dinitro-*ortho*-butylphenol (DNBP or dinoseb) effectively thin apricot flowers when applied just before full bloom.[4-6] The techniques for using

Table 1
APRICOT BLOSSOM THINNING WITH 4,6-DINITRO-*ORTHO*-BUTYLPHENOL AT SUMMERLAND, BRITISH COLUMBIA

		Fruit		
Cultivar/year	Treatment	Set[a]	Weight (g)[b]	Remarks
Wenatchee (1977)	No spray	25.8	37.3	Hand thinning
	75 ppm at 90% bloom	18.0	43.7	reduced to half[c]
	150 ppm at 90% bloom	15.3	49.2	
Early Blenheim (1977)	No spray	9.5	39.3	150 ppm
	75 ppm at 80% bloom	5.6	41.2	overthinned
	150 ppm at 80% bloom	2.8	48.4	
Wenatchee (1978)	No spray	21.2	52.8	Hand thinning
	100 ppm at 90% bloom	10.7	61.4	reduced to half[c]
Early Blenheim (1978)	No spray	5.9	48.7	Treatment
	100 ppm at 95% bloom	2.0	53.3	overthinned
Tilton (1978)	No spray	43.3	32.9	Hand thinning
	100 ppm at 60—95% bloom	25.8	40.6	reduced to half[c]
Selection 8L-8-14 (1978)	No spray	26.6	33.4	Hand thinning
	100 ppm at 95% bloom	14.9	45.0	reduced to half[c]

[a] Number of fruits per cm^2 of branch cross-sectional area before hand thinning.
[b] Mean weight of individual fruits at harvest.
[c] The number of fruits removed by hand (to achieve a "normal" crop load) was reduced to half or less. Hand thinning was done between 30 and 40 days after full bloom.

After Looney, N. E. and Killick, R. G., *Can. J. Plant Sci.*, 59, 741, 1979.

the dinitro materials have been available for decades and the benefits of improved fruit size, reduced hand thinning, and more reliable flowering are widely recognized. However, apricots bloom early in the spring and are often grown at latitudes and locations where the frost hazard is high. Growers are understandably reluctant to thin before a crop is assured. Therefore, it is only in regions where the availability or cost of hand thinning labor makes this option increasingly problematic, that chemical thinning is gaining a foothold.

Two areas where chemical blossom thinning is superceding hand thinning are New Zealand and British Columbia, Canada. The blossom thinning practice used in these areas involves DNBP (about 100 ppm) applied when 60 to 90% of the blossoms has opened.[5,6] Most cultivars appear to respond satisfactorily to DNBP, but those that tend to set poorly (e.g., Early Blenheim) should not be treated (Table 1). The risk of overthinning with the dinitro materials does not appear to be as high as is commonly feared and can be reduced by reducing the amount of chemical or delaying the application.[4,6]

B. Reducing Preharvest Drop and Advancing Fruit Maturity

Apricots respond dramatically to synthetic auxins (especially 2,4-D, 2,4,5-T, and 2,4,5-TP) applied soon after the start of pit hardening.[7-11] The frequently encountered mid-season fruit drop, often of fruits that appear to be developing normally, can be virtually eliminated. Furthermore, the average weight of individual fruits is often increased and harvest maturity advanced by days or even weeks.

However, it appears that a number of adverse treatment side effects have combined to discourage product registration in most apricot producing countries. A Spanish report suggests that treated fruits are more susceptible to bruising during and after harvest,[11] and unpublished results from British Columbia show that treated fruits ripen unevenly and tend to be soft. High auxin concentrations can also cause shoot and bud damage and promote fruit cracking.[8]

Nonetheless, there appears to be great potential for improving the productivity of apricots with auxinic growth regulators and they are used successfully by growers in several countries (e.g., Israel and New Zealand). One is led to suspect that the chemical rates used by workers in the 1950s (50 to 200 ppm of materials like 2,4-D) were higher than necessary to control fruit drop and that they were aiming for other or perhaps a combination of benefits. Lower rates are likely to improve crop yields without the adverse effects on fruit quality.

III. TART CHERRIES (*Prunus cerasus* L.)

Tart or sour cherries are grown entirely for processing. They can be grown successfully in most temperate locations, but the availability of the necessary facilities and related infrastructure for processing characterizes the important production regions. The North American production of about 125 thousand metric tons is centered in Michigan, New York, and Pennsylvania,[12] but most of the tart cherries of the world are grown in Europe. East and West Germany, Hungary, Yugoslavia, Poland, and U.S.S.R. all produce significant quantities which total about 75% of the annual world production of about 450 thousand metric tons.[13]

A. Treatment of Tart Cherry Yellows Virus Disease

Trees infected with the tart cherry yellows virus produce too few vegetative buds to maintain an adequate leaf and fruiting area. The consequence is poor fruit size and a generally unproductive tree.

Gibberellic acid (GA_3) can be used to induce a proportion of vegetative buds to form in the leaf axils of current season growth. The following year these buds form lateral shoots and, with annual treatments, a near normal growth habit is achieved.

The treatment suggested to growers in New York State is 10 to 25 ppm GA_3 applied as a full volume spray about 2 weeks after petal fall.[14] The GA_3 rate is adjusted up or down depending upon the severity of the disease and the proportion of vegetative buds desired. The aim is to induce about three vegetative buds per terminal shoot. Fewer than three will depress the long-term cropping potential. More than three will depress cropping the year after treatment, but lead to excessive cropping in the subsequent year.

B. Fruit Loosening for Mechanical Harvesting

"Shake and catch" harvesting of tart cherries is now a common practice in North America and parts of Europe and is greatly facilitated by preharvest sprays of (2-chloroethyl)phosphonic acid (ethephon).[15,16] Ethephon treatment reliably reduces fruit removal force, and if used carefully, can be applied annually without deleterious effect to the crop or tree.

Reducing the fruit removal force means that more fruits can be removed mechanically; fewer fruits will have stems; the stem scar (on the fruit) is less likely to bleed; and less force needs to be applied with the trunk or limb shaker and, therefore, less tree damage is incurred.

The major hazard of this treatment is that the ethephon concentrations necessary for adequate fruit loosening can cause yellowing and abscission of older leaves and injury to shoot tips. Treated trees often exhibit slight to moderate gummosis around pruning wounds and lateral buds. But fortunately, these symptoms of phytotoxicity are usually minor and unlikely to reduce long-term productivity.

The amount of ethephon required for adequate loosening can be reduced by pretreating the trees, in mid-season, with butanedioic acid mono-(2,2-dimethylhydrazide) (daminozide).[16] Daminozide advances fruit coloring and loosening, and the crop matures more uniformly.

The procedure for using ethephon to loosen tart cherries is to apply a full volume spray containing 500 ppm ethephon between 7 and 10 days before harvest. The addition of a nonionic surfactant is sometimes advised.[14] Where a daminozide pretreatment is considered necessary, it is applied at 1000 to 2000 ppm about 2 months before the expected harvest.

IV. SWEET CHERRIES (*Prunus Avium* L.)

In 1975, the last year for which sweet and tart cherry production statistics were compiled separately, world production of sweet cherries totaled about 1.2 million metric tons.[13] There is substantial North American production (about 14% of the total), but most sweet cherries are grown in Europe and Asia. U.S.S.R., Italy, West Germany, France, and Turkey are all major producers.

Sweet cherries are used for both processing and local and distant fresh market sales. The price paid to producers is often quite high relative to annual production costs, but there are many risks in sweet cherry production and it is difficult to achieve consistent yields of high quality fruit. In regions considered marginal for this crop (in fact, much of the world production occurs in such regions) growers often speculate on sweet cherries, planting a portion of their acreage to a range of cultivars. It is only in regions where there is substantial production of a few sweet cherry cultivars that any of the following growth regulator aids are likely to be important.

A. Increasing Blossom Hardiness and Delaying Bloom

In common with other drupe fruits, sweet cherries bloom relatively early in the spring and are therefore highly susceptible to frost damage. Similarly, in common with most drupe fruits, susceptibility of the dormant buds to winter injury is a cause for concern in many production regions.

Pomologists have long sought a growth regulator treatment to reduce either or both of these risks, but it is only within the last decade that significant progress has been made. Proebsting (at Washington State University) and Dennis (of Michigan State University) have reported interesting results with ethephon, daminozide, and gibberellic acid applied to several of the drupe fruits, but it is the use of ethephon on sweet cherries that shows the most promise.[17-21]

Washington State sweet cherry growers are advised to apply ethephon in early autumn (mid-September) to increase blossom hardiness and achieve a modest delay in anthesis. Rates of 250 to 500 ppm are effective and relatively safe, but growers are warned that higher rates or concentrate spraying may lead to an excessive bloom delay, poor fruit set, and reduced fruit size.

B. Improving Fruit Set

Poor fruit set is another common limitation to sweet cherry productivity. A complete crop failure 1 year in 3 is not unusual in some countries in Northern Europe, yet interest in this crop remains high and researchers are exhorted to find ways to improve the regularity of cropping.

Research at the East Malling Research Station and at Wye College has shown that fruit set on a range of cultivars growing in the south of England can be improved with gibberellin-auxin mixtures applied during or shortly after anthesis (Table 2).[22] The addition of a cytokinin component, N,N′-diphenylurea, contributed little to treatment efficacy.

The amount of GA_3 required to increase set differs depending on cultivar. A 50-ppm treatment is effective on several important English cultivars,[23] whereas up to 200 ppm is required with others.[22] The auxin component can be either NAA (10 ppm), 2,4,5-TP (10 ppm), or 2-naphthoxyacetic acid (50 ppm).[22]

While the results of the commercial tests conducted to date have been quite promising,

Table 2

THE EFFECT OF PETAL-FALL SPRAYS OF 200 PPM GA_3 PLUS 300 PPM N,N′-DIPHENYLUREA AND EITHER 10 PPM NAA, 10 PPM 2,4,5-TP, OR 50 PPM NOXA ON FRUIT SET, FRUIT WEIGHT, AND FLOWERING OF 3 SWEET CHERRY CULTIVARS

Cultivar/treatment	Fruit set 1976 per 100 floral buds	Mean fruit weight 1976 (g)	No. of floral buds 1977 per 100 floral buds 1976	Fruit set 1977 per 100 floral buds	Mean fruit weight 1977 (g)	Mean total fruit no. 1976 + 1977 per 100 floral buds in 1976
Merton Glory						
Water control	2.2	5.58	124.6	1.0	8.30	3.5
Mixture with NAA	8.5[a]	6.46	103.2	6.0[b]	8.66	14.7[a]
Mixture with 2,4,5-TP	9.3[a]	5.95	74.1[a]	10.3[a]	8.39	16.3[a]
Mixture with NOXA	7.8[a]	5.66	89.1[a]	7.0[a]	8.94	13.9[a]
Merton Bigarreau						
Water control	40.1	4.48	134.7	2.2	—	42.9
Mixture with NAA	33.2	4.49	77.6[a]	7.4[a]	—	38.9
Mixture with 2,4,5-TP	53.3[b]	4.40	88.1[a]	4.9[b]	—	57.4[b]
Mixture with NOXA	40.6	4.49	86.1[a]	7.3[a]	—	46.1
Napoleon Bigarreau						
Water control	98.4	3.68	118.5	13.7	4.20	113.4
Mixture with NAA	112.4	3.71	115.0	34.7[a]	4.27	153.5[b]
Mixture with 2,4,5-TP	114.7	3.54	99.6	37.0[a]	4.17	149.3
Mixture with NOXA	125.8	3.37	110.8	40.8[a]	4.72	173.3[a]

[a] Values differ significantly from the control at $P = 0.05$.

[b] Values differ significantly from the control at $P = 0.01$.

From Webster, A. D., Goldwin, G. K., Schwabe, W. W., Dodd, P. B., and Pennell, D., *J. Hortic. Sci.*, 54, 27, 1979. With permission.

the commercial future of this procedure is far from established. The high rates of GA_3 required for some cultivars are not only costly, but are likely to suppress flower initiation.[24] Furthermore, the treatment appears to be primarily useful where natural set is low (Table 2) and cannot be used in a more general way to improve crop yields.

A related treatment using auxin (naphthyleneacetamide) alone is presently registered in West Germany. NAAm (70 ppm) is applied at full bloom and is said to reduce the June drop on a range of sweet cherry cultivars.

C. Advancing Fruit Ripening

As with tart cherries, daminozide treatment advances the ripening of sweet cherries by several days and leads to more uniform fruit ripening.[25-28] These effects are useful to growers interested in an early market or those simply wishing to program the harvest of a particular cultivar. However, these benefits are relatively modest considering treatment cost and, although registered for use on sweet cherries in a number of countries, daminozide is not widely used in commercial cherry growing.

The procedure is to apply between 1000 and 2000 ppm daminozide in a full volume spray 2 to 3 weeks after full bloom. A treatment of 2000 ppm daminozide applied 2 weeks after full bloom to Bing and Chinook cherries for 4 consecutive years advanced fruit coloring, sizing, soluble solids development, and softening by an average of 4 days.[28] There were no serious adverse effects on either the trees or the crop. Daminozide favors flower bud formation on cherry trees, but these buds tend to be less hardy and the bloom date is slightly advanced.[26,29] Some cultivars (e.g., Lambert) may respond to daminozide treatment with an excessive mid-season fruit drop.

D. Reducing Susceptibility to Rain Cracking

Rain cracking is a problem wherever sweet cherries are grown. Fruits approaching harvest maturity are particularly susceptible and it is not uncommon for a complete crop to be rendered unmarketable only days or hours before it would have been harvested. Unfortunately, the firmer, better quality sweet cherry cultivars are especially susceptible to rain cracking.[12]

A growth regulator treatment used to a limited extent in Western North America is 1 ppm NAA applied 30 to 35 days before harvest.[30,31] This treatment is unlikely to eliminate rain-induced cracking, but may reduce it to a tolerable level.

E. Delaying Fruit Ripening and Improving Fruit Quality

A practice of increasing interest to sweet cherry producers in Western North America involves the application of low rates of GA_3 a few weeks before harvest. Depending on the year and location, this treatment may produce one or more very beneficial effects.

Extensive trials in Washington State[26] and British Columbia[32] have shown that 10 to 20 ppm GA_3 applied to dark sweet cherries in late stage II (straw-yellow fruit color) delays red color development by 3 or 4 days, increases fruit size at a given color stage, and delays the period of maximum susceptibility to rain cracking. Furthermore, treated fruit tend to be firmer and are less likely to develop a postharvest condition, referred to as cherry pitting, that occurs after cold storage. White sweet cherries treated in a similar manner are a superior product for canning.[33] They are firmer after processing and less prone to discoloration in the can.

However, this treatment is not without certain hazards or drawbacks that may limit widespread acceptance. In the first place, delayed fruit ripening is seldom an advantage in terms of a fresh fruit marketing strategy. Secondly, fruit maturation or ripening as measured by red color development may be less uniform on treated trees. And finally, even low levels of GA_3 may inhibit flower initiation on particularly sensitive cultivars.

The sweet cherry growers in Europe most likely to benefit from this particular use of GA_3 are those (i.e., Italy and France) serving distant markets with fresh fruit. But while this procedure is mentioned by Cobianchi[34] in a recent discussion of growth regulator usage in Italian cherry production, there is no evidence of widespread commercial interest.

V. PEACHES AND NECTARINES (*Prunus persica* (L.) Batsch)

Peaches and nectarines for fresh sales and peaches for processing are important crops in a large number of countries and regions. Thirty-six countries produce more than 20 thousand metric tons annually and sixteen produce more than 100 thousand metric tons.[1] However, of the estimated world production of about 7.2 million metric tons, two countries, the U.S. and Italy, combine to produce nearly 3 million metric tons annually.[1]

Each of these countries has a history of strong support for horticultural research and development, and it is certainly not for lack of interest that plant growth regulators continue to play such a minor role in peach and nectarine production.

A. Chemical Thinning

Flower or fruit thinning early in the growing season is an essential operation in peach and nectarine production. Because thinning is largely accomplished with hand labor, it presently accounts for much of the cost of producing these crops. Since set on 10 to 15% of the flowers will usually result in a full crop, the number of flowers and fruits that must be removed is indeed substantial.

Hundreds of individual chemicals and chemical combinations have been screened for peach fruit thinning activity during the past 40 years. A few materials have shown enough promise to warrant commercial testing (Table 3),[35-49] but none has yet proved widely useful. One material, 3-chlorophenoxy-α-propionamide (3-CPA) is still commercially available and used to a limited extent.

Aside from the usual problems of unreliability, the peach fruit thinners have displayed an unfortunate tendency to suppress fruit sizing.[41,47] Thus, even when the crop load is reduced to the desired extent, the expected fruit sizing benefit may not materialize. Until this problem is solved it is unlikely that chemical thinning of peach fruits will be widely used.

Blossom thinning with materials like DNOC, however, does stimulate subsequent fruit sizing.[4,35,36] Thus, despite the risks involved in such early thinning, this practice is likely to become more important as labor costs escalate.

B. Advancing Fruit Ripening

Daminozide is used to a modest extent in commercial peach growing to advance and concentrate fruit harvest.[50-53] This effect is analogous to that already described for cherries and, as is the case for cherries, daminozide has been especially useful when used in conjunction with mechanical harvesting of peaches for processing.[52] Treated fruit develop both surface and internal red color 4 or 5 days earlier than normal, yet at that stage are as firm or firmer than untreated fruit. However, they tend to soften rapidly once they are harvested. This probably relates to the observation that daminozide enhances ethylene production by ripening peach fruits.[51,53]

The procedure is to apply 1000 to 2000 ppm daminozide between 2 and 4 weeks after full bloom. Late-season cultivars are usually treated somewhat later than early cultivars. This spray is normally applied at full volume and, if it is to be concentrated (i.e., the amount of spray volume reduced), the amount of chemical applied per hectare should also be reduced. Low volume sprays of daminozide can lead to a degree of foliar injury as evidenced by ''shot-holing''. Late-season peach cultivars have, in general, responded more satisfactorily to daminozide treatment than have early peaches.

Table 3
CHEMICALS WITH PEACH FLOWER AND FRUIT THINNING ACTIVITY

Chemical	Approximate rates[a] and timing	Ref.
DNOC	375 ppm; 80% bloom	4, 35, 36
N-1-naphthylphthalamic acid (NPA)	200—400 ppm; full bloom to 5 days postbloom	37, 38
3-Chlorophenoxy-α-propionamide (3-CPA)	150 ppm; 8-mm ovule length	39—41
Naphthaleneacetic acid (NAA)	30—40 ppm; at endosperm cytokinesis	42, 43
(2-Chloroethyl)phosphonic acid (ethephon)	100—300 ppm; full bloom to endosperm cytokinesis	44—47
1,1,5,5-Tetramethyl-3-dimethylaminodithiobiuret	100—300 ppm; at endosperm cytokinesis	46, 48
β-Chloroethyl-methyl-bis-benzyloxy-silane (CGA-15281)	250 ppm; 8- to 16-mm ovule length and warm weather	49

[a] Concentration of active chemical in a full volume spray.

VI. PLUMS (*Prunus domestica* L. and *Prunus salicina* Lindl.)

Prunus domestica L., the European plums, and *Prunus salicina* Lindl., the oriental plums, account for most of the prune and plum production of the world, but there are several other cultivated species. Thus, while plums as a general category rank second to peaches among the drupe fruits in total production (5.5 million metric tons in 1980),[1] their diversity is such that even the most important cultivars are generally considered minor crops.

The European plums include the various prune cultivars (plums with sufficient sugar to permit drying without seed removal) such as d'Agen (French) and Italian, and several types of processing and fresh market plums. These include Green Gage, Victoria, Yellow Egg, Bradshaw, and scores more.

The oriental plums are adapted to warmer climates, are usually consumed fresh, and are often shipped to distant markets. For example, California and Mexico supply much of North America with cultivars like Santa Rosa, Burmosa, and Duarte. Both China and Japan produce substantial quantities of *Prunus salicina* plums, mostly for domestic use, and a number of other countries have small but important oriental plum shipping industries. These include Israel, Argentina, and South Africa to name but a few.

Combining all plum species, the U.S. and U.S.S.R. are the leading producers, but the bulk of the world production occurs in Europe outside of the U.S.S.R. Romania, West Germany, and Yugoslavia all produce in excess of 500 thousand metric tons annually.[1]

A. Chemical Thinning

Various cultivars of European and oriental plums have been thinned satisfactorily with blossom thinners such as DNOC, DNBP or, in parts of Northern Europe, lime sulfur.[54,55] Postbloom thinning with materials like 3-CPA and ethephon has been attempted but the results have been excessively variable.[47,55,56]

Plums bloom very early in the spring, and this is a major deterent to blossom thinning. Cold or damp weather during the bloom period is a frequent occurrence, and fruit set can be erratic. Nonetheless, there are regions where blossom thinning is an established practice. Queensland, Australia growers apply 150 ppm DNBP at full bloom to thin Wilson and Santa Rosa plums but are advised against spraying European plums. However, a blossom-time spray of 380 ppm DNOC is suggested to Ontario, Canada growers of European plums experiencing problems with biennial bearing or inadequate fruit size, and plum growers in parts of Europe have traditionally used blossom-time sprays of lime sulfur to reduce fruit set.

A recent development is that lime sulfur is being replaced with newer fungicides and is increasingly difficult to obtain. Dutch growers are now advised to use wettable sulfur as a substitute treatment. A rate of 5 kg/100 ℓ is applied at full bloom and is said to be most effective when applied during warm, sunny weather.

B. Reducing Fruit Drop

Several cultivars of prunes (*P. domestica* L.) are treated with fenoprop (2,4,5-TP) to reduce mid- to late-season fruit drop.[57,58] This problem, referred to as the "blue drop" in the Pacific Northwest, can dramatically reduce yields and is particularly serious because it occurs so late in the season. There is little time for compensatory growth to occur in the remaining fruit.

The practice used successfully in Washington, Oregon, and Idaho with Italian and Early Italian prunes is to apply 20 ppm fenoprop (or as little as 5 ppm with added wetting agent) about 2 weeks after the start of pit hardening. Neither fruit size nor fruit quality is reduced by this treatment and yields are increased by 25 to 100%. Fenoprop is used in a similar manner by South African growers of French (d'Agen) prunes.

C. Advancing Fruit Maturity and Abscission

Santa Rosa (*P. salicina*) plums can be treated with ethephon to advance and condense fruit harvest.[59] The treatment suggested to South African growers is 100 ppm ethephon applied about 2 weeks before the start of harvest. A full volume spray with added wetting agent is advised.

Ethephon is also used to advantage in several regions where prunes are harvested mechanically and used for drying. A representative rate and timing is 200 ppm of ethephon applied 5 to 7 days before harvest.

D. Improving Fruit Quality

Early Italian prunes grown in Western North America are prone to an internal browning disorder that can seriously reduce their acceptability as a fresh market product.[60] The incidence of internal browning can be reduced and fruit firmness and shelf life improved with 50 ppm of GA_3 applied 4 weeks before harvest.[61]

VII. CONCLUSIONS

Fifteen plant growth regulator techniques have been discussed and several others briefly mentioned. These encompass all six of the important *Prunus* species fruits, but, despite this rather impressive number, it is fair to say that plant growth regulators have not yet "arrived" on the drupe fruit production scene. In general, the practices discussed can be described as being diverse in nature, often complicated in their execution, and very often tenuous with regard to their future. We appear to be grasping for answers to the various production problems and using plant growth regulators (always borrowed from other crops) as our grappling hook.

It is well worth noting that the few practices which are firmly established are those where sound physiological and morphological research have paved the way.

A good example is the use of ethephon to loosen tart cherries for mechanical harvesting. Researchers at Michigan State University systematically studied the abscission process in tart cherries and described and discussed the interaction of the ethephon treatment with this natural phenomenon.[63-65] Others have described the reaction of the tree.[66] The result has been that ethephon is widely used and is deemed a safe and reliable practice.

A similar approach to understanding the role of endogenous plant growth substances in the growth and development of tart cherry fruits[67,68] bodes well for the development of other successful plant growth regulator techniques for this crop.

The validity and utility of this approach to developing new plant growth regulator practices are too obvious to be ignored. The arrival of the long-awaited peach fruit thinner will, almost certainly, not be a "lucky catch".

REFERENCES

1. **Anon.**, FAO Production Yearbook, Vol. 34, Rome, 1980.
2. **Jackson, D. I. and Coombe, B. G.**, Gibberellin-like substances in the developing apricot fruit, *Science*, 154, 277, 1966.
3. **Nitsch, J. P.**, Hormonal factors in growth and development, in *The Biochemistry of Fruits and Their Products*, Vol. 1, Hulme, A. C., Ed., Academic Press, London, 1970, 427.
4. **Westwood, M. N.**, Chemical thinning of peaches and apricots, *Proc. Wash. State Hortic. Assoc.*, 52, 100, 1956.
5. **Lewthwaite, J. R.**, Apricot Thinning with Dinoseb and Carbaryl Sprays, N.Z. Minist. Agric. Food, Roxburgh, 1974.
6. **Looney, N. E. and Killick, R. G.**, Apricot blossom thinning with dinitro-ortho-butylphenol, *Can. J. Plant Sci.*, 59, 741, 1979.
7. **Crane, J. C. and Brooks, R. M.**, Growth of apricot fruits as influenced by 2,4,5-trichlorophenoxyacetic acid application, *Proc. Am. Soc. Hortic. Sci.*, 59, 218, 1952.
8. **Crane, J. C.**, The comparative effectiveness of several growth regulators for controlling preharvest drop, increasing size and hastening maturity of Stewart apricots, *Proc. Am. Soc. Hortic. Sci.*, 67, 153, 1956.
9. **Rogers, B. L. and Batjer, L. P.**, Effects of 2,4,5-T on size, maturity, and drop of apricots, *Proc. Wash. State Hortic. Assoc.*, 50, 81, 1954.
10. **Chadra, K. L. and Bajwa, M. S.**, Effect of 2,4,5-T and 2,4,5-TP on the control of fruit drop, increase in yield and fruit quality of New Castle apricot, *Acta Hortic.*, 11(3), 573, 1968.
11. **Albert Bernal, A. and Martínez Javéga, J. M.**, Influencia del 2,4,5-T en la maduración y comercialización posterior del albaricoque "Canino", *An. Inst. Nac Invest. Agron.*, 18, 87, 1969.
12. **Westwood, M. N.**, *Temperate Zone Pomology*, W. H. Freeman & Co., San Francisco, 1978.
13. **Anon.**, FAO Production Yearbook, Vol. 29, Rome, 1975.
14. **Burr, T. J., Leeper, J. R., and Stiles, W. C.**, *1981 Tree Fruit Production Recommendations*, Cornell University, Ithaca, N.Y., 1981.
15. **Bukovac, M. J., Zucconi, F., Larsen, R. P., and Kesner, C. D.**, Chemical promotion of fruit abscission in cherries and plums with special reference to 2-chloroethylphosphonic acid, *J. Am. Soc. Hortic. Sci.*, 94, 226, 1969.
16. **Looney, N. E. and McMechan, A. D.**, The use of 2-chloroethylphosphonic acid and succinic acid 2,2-dimethylhydrazide to aid in mechanical shaking of sour cherries, *J. Am. Soc. Hortic. Sci.*, 95, 452, 1970.
17. **Proebsting, E. L., Jr. and Mills, H. H.**, Bloom delay and frost survival in ethephon-treated sweet cherry, *HortScience*, 8, 46, 1973.
18. **Proebsting, E. L., Jr. and Mills, H. H.**, Time of gibberellin application determines hardiness responses of "Bing" cherry buds and wood, *J. Am. Soc. Hortic. Sci.*, 99, 464, 1974.
19. **Proebsting, E. L., Jr. and Mills, H. H.**, Ethephon increases cold hardiness of sweet cherry, *J. Am. Soc. Hortic. Sci.*, 101, 31, 1976.
20. **Dennis, F. G., Jr.**, Trials of ethephon and other growth regulators for delaying bloom in tree fruits, *J. Am. Soc. Hortic. Sci.*, 101, 241, 1976.
21. **Howell, G. S., Jr. and Dennis, F. G., Jr.**, Cultural management of perennial plants to maximize resistance to cold stress, in *Analysis and Improvement of Plant Cold Hardiness*, Olein, C. R. and Smith, M. N., Eds., CRC Press, Boca Raton, Fla., 1981, 175.
22. **Webster, A. D., Goldwin, G. K., Schwabe, W. W., Dodd, P. B., and Pennell, D.**, Improved setting of sweet cherry cultivars, *Prunus avium* L., with hormone mixtures containing NOXA, NAA or 2,4,5-TP, *J. Hortic. Sci.*, 54, 27, 1979.
23. **Modlibowska, I. and Wickenden, M. F.**, Effects of chemical growth regulators on fruit production of cherries. I. Effects of fruit-setting hormone sprays on the cropping of Merton Glory and Van cherry trees, *J. Hortic. Sci.*, 57, 413, 1982.

24. **Crane, J. C. and Hicks, J. R.,** Further studies on growth-regulator-induced parthenocarpy in the Bing cherry, *Proc. Am. Soc. Hortic. Sci.*, 92, 113, 1968.
25. **Looney, N. E.,** Regulation of sweet cherry maturity with succinic acid 2,2-dimethyl hydrazide (Alar) and 2-chloroethylphosphonic acid (Ethrel), *Can. J. Plant Sci.*, 49, 625, 1969.
26. **Proebsting, E. L., Jr.,** Chemical Sprays to Extend Sweet Cherry Harvest, *Wash. State Univ. Coop. Ext. Serv., Multilith*, 3520, 1972.
27. **Schumaker, R.,** The influence of growth regulators on fruit development of sweet cherry, *Acta Hortic.*, 34, 317, 1973.
28. **Proebsting, E. L., Jr. and Mills, H. H.,** Effect of daminozide on growth, maturity, quality and yield of sweet cherries, *J. Am. Soc. Hortic. Sci.*, 101, 175, 1976.
29. **Proebsting, E. L., Jr. and Mills, H. H.,** Effects of growth regulators on fruit bud hardiness in *Prunus*, *HortScience*, 4, 254, 1969.
30. **Bullock, R. M.,** A study of some inorganic compounds and growth promoting chemicals in relation to fruit cracking of Bing cherries at maturity, *Proc. Am. Soc. Hortic. Sci.*, 59, 243, 1952.
31. **Westwood, M. N.,** Use of growth regulators in stone fruit production, *Ore. State Hortic. Soc. Annu. Rep.*, 67, 27, 1976.
32. **Looney, N. E. and Lidster, P. D.,** Some growth regulator effects on fruit quality, mesocarp composition, and susceptibility to postharvest surface marking of sweet cherries, *J. Am. Soc. Hortic. Sci.*, 105, 130, 1980.
33. **Proebsting, E. L., Jr., Carter, G. H., and Mills, H. H.,** Quality improvement in canned "Rainier" cherries (*P. avium* L.) with gibberellic acid, *J. Am. Soc. Hortic. Sci.*, 98, 334, 1973.
34. **Cobianchi, D.,** Possibilita applicative di fitoregulatori nella coltivazione del ciliegio, in *La Coltura del Ciliegio Dolce; Indirizzi e Prospettive*, "Centro Reg. Sper. Agraria-Friuli-Venezia", Ed., Guilia, Udine, Italy, 1980, 123.
35. **Hoffman, M. B. and Van Doren, A.,** Some results in thinning peaches with a blossom removal spray, *Proc. Am. Soc. Hortic. Sci.*, 46, 173, 1945.
36. **Batjer, L. P. and Rogers, B. L.,** Chemical thinning of stone fruits, *Proc. Wash. State Hortic. Assoc.*, 47, 115, 1951.
37. **Edgerton, L. J. and Hoffman, M. B.,** Effect of N-1-naphthylphthalamic acid on fruit set of peaches, *Science*, 121, 467, 1955.
38. **Aitken, J. B., Buchanan, D. W., and Sauls, J. W.,** Thinning short-cycle Florida peaches with N-1-naphthylphthalamic acid, *HortScience*, 7, 255, 1972.
39. **Beutel, J., Gerdts, M., Larue, J., and Carlson, C.,** Chemical thinning for shipping peaches, nectarines and plums, *Calif. Agric.*, 23, 6, 1969.
40. **Stembridge, G. E. and Gambrell, C. E., Jr.,** Thinning peaches with 3-chlorophenoxy-γ-propionamide, *J. Am. Soc. Hortic. Sci.*, 94, 570, 1969.
41. **Morini, S., Vitagliano, C., and Xiloyannis, C.,** Effects of 3-CPA and ethephon on growth and abscission of "Cardinal" peach, *J. Am. Soc. Hortic. Sci.*, 101, 640, 1976.
42. **Kelley, V. W.,** Time of application of naphthaleneacetic acid for fruit thinning of peach in relation to June drop, *Proc. Am. Soc. Hortic. Sci.*, 66, 70, 1955.
43. **Leuty, S. J. and Bukovac, M. J.,** The effect of naphthaleneacetic acid on abscission of peach fruits in relation to endosperm development, *Proc. Am. Soc. Hortic. Sci.*, 92, 124, 1968.
44. **Stembridge, G. E. and Gambrell, C. E., Jr.,** Thinning peaches with bloom and postbloom application of 2-chloroethylphosphonic acid, *J. Am. Soc. Hortic. Sci.*, 96, 7, 1971.
45. **Thompson, A. H. and Rogers, B. L.,** Three years results with chemical thinning of peaches with (2-chloroethyl)phosphonic acid, *J. Am. Soc. Hortic. Sci.*, 97, 644, 1972.
46. **Martin, G. C., Nelson, M. M., and Nishijima, C.,** 2-Chloroethylphosphonic acid and 1,1,5,5-tetramethyl-3-dimethylaminodithiobiuret as chemical thinners for peach, *HortScience*, 6, 169, 1971.
47. **Weinbaum, S. A., Guilivo, C., and Ramina, A.,** Chemical thinning. Ethylene and pre-treatment fruit size influence enlargement, auxin transport, and apparent sink strength of French prune and "Andross" peach, *J. Am. Soc. Hortic. Sci.*, 102, 781, 1977.
48. **Kiel, H. L. and Fogle, H. W.,** 1,1,5,5-Tetramethyl-3-dimethylaminodithiobiuret, a promising new peach thinner, *HortScience*, 6, 403, 1971.
49. **Byers, R. E,,** Chemical thinning of peach fruit with CGA 15281 and CGA 17856, *J. Am. Soc. Hortic. Sci.*, 103, 232, 1978.
50. **Sansavini, S., Martin, J. M., and Ryugo, K.,** The effect of succinic acid 2,2-dimethyl hydrazide on the uniform maturity of peaches and nectarines, *J. Am. Soc. Hortic. Sci.*, 95, 708, 1970.
51. **Looney, N. E.,** Effects of succinic acid 2,2-dimethylhydrazide, 2-chloroethylphosphonic acid, and ethylene on respiration, ethylene production, and ripening of "Redhaven" peaches, *Can. J. Plant Sci.*, 52, 73, 1972.
52. **Baumgardner, R. A., Stembridge, G. E., Van Blaricom, L. O., and Gambrell, C. E., Jr.,** Effects of succinic acid-2,2-dimethylhydrazide on the color, firmness, and uniformity of processing peaches, *J. Am. Soc. Hortic. Sci.*, 97, 485, 1972.

53. **Looney, N. E., McGlasson, W. B., and Coombe, B. G.,** Control of fruit maturation in peach, *Prunus persica* (L. Batsch): action of succinic acid-2,2-dimethylhydrazide and (2-chloroethyl)phosphonic acid, *Aust. J. Plant Physiol.*, 1, 77, 1974.
54. **Dodd, B. C.,** Thinning Wilson plums with chemicals, *Queensland Agric. J.*, 93, 476, 1967.
55. **Webster, A. D.,** Flower and fruitlet thinning of the plum (*Prunus domestica* L.) cv Victoria, *J. Hortic. Sci.*, 55, 19, 1980.
56. **Kvale, A.,** Ethephon (2-chloroethylphosphonic acid) a potential thinning agent for plums, *Acta Agric. Scand.*, 28, 279, 1978.
57. **Verner, L., Kochan, W., Braun, R., and Kamal, A.,** Sprays to reduce the dropping of prunes, *Proc. Wash. State Hortic. Assoc.*, 53, 19, 1957.
58. **Proebsting, E. L., Jr. and Mills, H. H.,** Response of Richards Early Italian prunes to 2,4,5-TP with varied time of application, *Proc. Wash. State Hortic. Assoc.*, 57, 164, 1961.
59. **Blommaert, K. L. J., Hanekom, A. N., and Steenkamp, J.,** Earlier and more uniform ripening of Santa Rosa plums using ethephon, *Decid. Fruit Grower*, 25, 267, 1975.
60. **Proebsting, E. L., Jr. and Mills, H. H.,** Effect of gibberellic acid and other growth regulators on quality of Early Italian prunes (*Prunus domestica* L.), *Proc. Am. Soc. Hortic. Sci.*, 89, 135, 1966.
61. **Proebsting, E. L., Jr., Carter, G. H., and Mills, H. H.,** Interaction of low temperature storage and maturity on quality of "Early Italian" prunes, *J. Am. Soc. Hortic. Sci.*, 99, 117, 1974.
62. **Proebsting, E. L., Jr. and Tukey, R. B.,** Gibberellic Acid to Reduce Internal Browning of Early Italian Prune, *Wash. State Univ. Coop. Ext. Serv., Ext. M.*, 3967, 1975.
63. **Wittenbach, V. A. and Bukovac, M. J.,** Cherry fruit abscission: a role for ethylene in mechanically induced abscission of immature fruits, *J. Am. Soc. Hortic. Sci.*, 100, 302, 1975.
64. **Wittenbach, V. A. and Bukovac, M. J.,** Cherry fruit abscission: peroxidase activity in the abscission zone in relation to separation, *J. Am. Soc. Hortic. Sci.*, 100, 387, 1975.
65. **Olein, W. C. and Bukovac, M. J.,** The effect of temperature on rate of ethylene evolution from ethephon and from ethephon-treated leaves of sour cherry, *J. Am. Soc. Hortic. Sci.*, 103, 199, 1978.
66. **Wilde, M. H. and Edgerton, L. J.,** Histology of ethephon injury on "Montmorency" cherry branches, *HortScience*, 10, 79, 1975.
67. **Hopping, M. E. and Bukovac, M. J.,** Endogenous plant growth substances in developing fruit of *Prunus cerasus* L. III. Isolation of indole-3-acetic acid from the seed, *J. Am. Soc. Hortic. Sci.*, 100, 384, 1975.
68. **Bukovac, M. J. and Yuda, E.,** Endogenous plant growth substances in developing fruit of *Prunus cerasus* L. VII. Isolation of gibberellin A_{32}, *Plant Physiol.*, 63, 129, 1979.

Chapter 3

PLANT GROWTH REGULATOR USE IN NATURAL RUBBER (*Hevea brasiliensis*)

Kenneth Bridge

TABLE OF CONTENTS

I. INTRODUCTION

The natural rubber (NR) industry of today is more than 100 years old, but as observed by Dr. B. C. Sekhar, who in 1973 was the Director of the Rubber Research Institute of Malaysia, "Natural rubber has been considered by many as the grandfather of all elastomers, but unlike many others it shows no visible or discernible symptoms of aging or weakening." The fact that NR production is an industry based on a renewable resource and essentially is nonpolluting is almost unique in the world of today. The last three decades have seen many developments to increase production efficiency and among these have been the use of plant growth regulators. Present opportunities can be categorized as follows and will be discussed in detail:

1. Stimulation of latex flow, or the use of plant growth regulators to allow for a longer release of latex from the rubber tree than considered normal, following the tapping operation
2. Stimulation of root development of propagation material such as budded stumps, stump buddings, and seedling stumps
3. Defoliation as a management tool in the control of leaf diseases

II. STIMULATION OF LATEX FLOW

Today use of plant growth regulators has the same physiological objectives that were present during the early years of the NR industry, when small holders applied mixtures of cowdung and clay to the tapping panels. It has been suggested that such mixtures contained small amounts of indole acetic acid, known to stimulate the production of ethylene in some plants.

To understand the purpose of *Hevea* latex stimulation and the biological effects of plant growth regulators which have been developed since the early 1950s, a description of modern-day tapping techniques and the arrangement of the latex bearing vessels are necessary.

The latex is contained in latex vessels which are arranged in concentric cylinders in the bark. These latex vessels are not precisely vertical, but run in a counterclockwise spiral up the trunk. The latex is formed in the latex vessels from assimilates produced by photosynthetic processes. Latex vessels or laticifers in plants are systems of tubes filled with liquid, for which no function is known to physiologists. Like the lymphatic systems in animals, the liquid in laticifer tubes is apparently static. Movement probably occurs only when the laticifer system is ruptured. Morphologically, the laticifer has rather thin walls, and lacks crosswalls in *Hevea*. It has been suggested that cellulase, present in latex sap, is involved in dissolving crosswalls between new laticifer cells, resulting in a series of connected tubes. Inside the wall, one finds multiple nuclei, mitochondria, and some inclusions including rather large lutoids, protein bodies, and oval bodies containing the rubber itself. Very soon after the laticifer matures the nuclei and mitochondria become scarce or absent. Hence, the latex sap from almost all laticifers involved in tapping consists of lutoids (like vacuoles), protein bodies, and rubber bodies suspended in a serum which contains many soluble enzymes.

Upon tapping, flow of the latex is driven by the hydrostatic or turgor pressure within the latex vessels. This is at its highest during the night and early morning and lowest following heavy transpiration during the heat of midday; thus tapping is regularly carried out in the morning, and collection of the latex is usually completed by 10:30 to 11:00 a.m. The latex ceases to flow after a period of time because a plug of coagulated rubber seals off the end of the laticifers. To direct the flow of latex toward a cup or container for collection, the tapping cut is made in an oblique direction. Since the latex vessels run in a counterclockwise spiral upward, a cut running from left to right of any given length and slope will open up more latex vessels than a cut running from right to left. This is why in modern-day tapping where only a half-spiral cut is made, it is from left to right at a 45° angle with the horizontal.

A. Tapping Systems

Since the late 1960s, a high level of success has been achieved in the stimulation of latex flow, and while the development of new plant regulators has played the most important role, changes in tapping systems have occurred at the same time. A brief description of what constitutes any given tapping system and examples of some of these in use today will be helpful.

Two major factors constitute a tapping system and determine its intensity. These are the length of the tapping cut and the interval between successive tappings. A standard international notation of tapping was agreed upon by the major rubber producing countries in 1940 and has been used ever since. The notation has up to three separate terms, followed by a percentage. The first term indicates the length of the tapping cut. In the second term, the denominator indicates the frequency of tapping, and in the third term (where this applies) the form of panel changing and the periodicity of tappings are detailed. The percentage is that which the "intensity" of tapping forms of a basic standard: this is taken as an average of a "quarter cut" (or a cut compassing one quarter of the circumference) per day. Therefore, the "half-spiral" alternate daily tapping system, which is perhaps still the most widely used system today, averages out at this standard and is thus characterized fully as S/2, d/2, 100%. Similarly, "full spiral fourth daily" tapping is characterized as S/1, d/4, 100%. A standard description of the panel being tapped has also been employed although not quite so generally as have tapping notations. The description essentially relates to a half-spiral cut. Thus, the first panel of virgin bark which extends from its opening at a height of 125 to 150 cm (in buddings) to the base of the tree is termed "A". The second panel of virgin bark, on the opposite side of the tree, is termed "B". The subsequent panels on first and second renewed bark are termed "C" and "E" and "D" and "F", respectively.

In some parts of the world, the trees are allowed to rest completely, either on a rotational basis or at fixed times of the year, the latter often being determined by climatic conditions (monsoons or severe drought).

B. *Hevea* — Nature's Own Factory

Templeton[1] observed that the efficiency of solar radiation utilization in a stand of *Hevea* trees with a closed canopy and having a rate of dry matter production of 35.5 tons/ha/year was about 2.8%, which in itself is not a low value compared to many other species. A monoculture of *Hevea*, therefore, appears to be a relatively efficient converter of solar energy to dry matter production. Bonner[2] observed that, when rubber trees are tapped, a portion of the dry weight that they would otherwise have made is lost. This loss of photosynthate varies between 10 and 80%, depending on the age of the tree, the particular clone, etc. The energy required to make rubber from the sugar produced in photosynthesis is considerable, and it is to be expected, therefore, that there should be a great loss in dry weight as the result of tapping. Nevertheless, Bonner hypothesized that the modern-day *Hevea* clones or varieties probably were capable of producing in the region of 7000 kilos dry rubber per hectare per year, which is nearly six times the average from high yielding varieties in Malaysia today.

Since the artificial stimulation of rubber trees is responsible for a longer flow of latex (although this is also influenced by the tapping system used as well as clonal characteristics), it is perhaps important to note that Wycherley[3] observed that a long flow of latex seems to lead to an adverse partition of assimilates, with the rubber being harvested at a greater cost in accumulation of dry weight by the tree than the calorific equivalent of the rubber. Wycherley also concluded that the comparative yield between clones is correlated with the vigor of the immature tree, the number of latex vessel rows, and the rate at which these vessels become plugged. Therefore, maximum yields might be obtained by simultaneous positive selection for vigor, numerous latex vessels, and long flow. However, the correlation of long flow with adverse partition, retarded growth, and trunk snap indicates that such a procedure is unlikely to produce the highest potential yields over the whole economic life of the trees. Partition of assimilates has been shown to be heritable. Since this partition also seems to be associated with the latex flow pattern, Wycherley suggested that improvement programs might continue to select for vigor and numerous latex vessels, but could be redirected towards high plugging, short flow, and, therefore, advantageous partition of assimilates. Since it has also been demonstrated that ethylene producing stimulants are effective on several clones to prolong flow and increase yield when desired, it does appear that a reorientated selection program could be coordinated with modern methods of exploitation such as stimulation to give a higher degree of control over yields, as well as maintaining the security of the trees themselves.

C. Latex Stimulant Development — Historical

Since the early years of the NR industry there was an interest in latex yield stimulants, beginning with the application of mixtures containing cowdung and clay by small holders to scraped bark below the cut and to renewing bark above it. Early trials showed that the yield was increased slightly, and that the renewal of bark was improved. Even in the early 1970s, such mixtures were still being used quite widely on small holdings in Sri Lanka. It was found later that palm oil had similar effects and this was applied to older trees on plantations as well as small holdings. It was postulated that this oil contained auxins although later, petrolatum grease which contains no auxins was evaluated and was also shown to stimulate latex yields and improve bark renewal.

By the early 1950s, the preoccupation was with the synthetic auxins and the Rubber Research Institute of Malaysia (RRIM) set up a series of experiments to evaluate IBA, NAA, 4-CPA, and others applied in paraffin oil emulsions; varying degrees of yield increases were obtained. At about the same time, Chapman[4] reported large yield increases by applying 2,4-D to scraped bark below the tapping cut, and a number of different esters of 2,4-D were included in the test as well as the sodium salt. Subsequently, the n-butyl ester of 2,4-D formulated in a palm oil-based paste was offered commercially as a latex stimulant. Later it was recommended that this palm oil-based carrier be thickened with a petrolatum grease to prevent the downward flow of 2,4-D from the treated area which was causing some swelling and splitting of the bark. Having noticed that 2,4-D with two chlorine substitutions in the benzene ring was more effective than 4-CPA with one chlorine substitution, this led to the testing of 2,4,5-T with three chlorine substitutions in the benzene ring, and, formulated with palm oil and petrolatum grease, this synthetic auxin along with 2,4-D became a commercial latex stimulant which was only partially discontinued following the development of ethylene releasing compounds in the late 1960s.

D. The Role of Ethylene in *Hevea* Latex Flow Stimulation

The effect of the "plugging" on the length of the latex flow period has already been referred to. This process was found to be partly inhibited in trees treated with 2,4,5-T and the prolonged latex flow largely accounts for the increase in latex yields, although probably not entirely. 2,4-D and 2,4,5-T are analogues of the natural auxin indoleacetic acid (IAA), which is itself a yield stimulant. Screening of other auxin analogues for yield stimulant activity, which included 2,4-dichloro-5-fluorophenoxy acetic acid and the herbicide 4-amino-3,5,6-trichloropicolinic acid, led to Abraham's remark that "it may be significant that all the active compounds are to some degree phytotoxic....the common factor may be a selective and limited toxicity."

As early as 1961, Taysum[5] reported the first use of a gas, ethylene oxide, which is toxic, to increase latex flow. Subsequently acetylene was found to be a yield stimulant, and in 1968 Abraham et al.[6] reported yield stimulation effects by treating several halogenoparaffins with ethylene gas and [(2-chloroethyl)phosphonic acid] (ethephon). It was noted that all effective compounds increased the period of latex flow and the volume of latex harvested consistently with the supposition that they act, at least in large measure, by inhibiting plugging. In 1969 D'Auzac and Ribailler[7] also reported the yield stimulant action of acetylene, ethylene, and [(2-chloroethyl)phosphonic acid] and similar activity from the application of b-hydroxyethyl hydrazine, another compound which can decompose releasing ethylene. It became evident that the results of all previous research indicated that all active compounds have the capacity to produce ethylene. Bonner had already suggested this possibility from the mounting evidence of its involvement in the effects produced by the phenoxy acetic acids and other hormone-like substances now generally known as "ethylene inducers" on the growth of many types of plants. The confirmation of the importance of ethylene in yield stimulation was obtained experimentally by its direct application as a gas under polythene sheet surrounding the tapping panel, or in chambers cemented to the tapping panel area. Archer and Audley in 1970[7] confirmed the correlation by demonstrating that all yield stimulants of the hormone type tested induced the formation of significantly increased quantities of ethylene from both excised leaflets and stem segments of *Hevea*.

It is now believed that ethylene is the fundamental active substance involved in all effective artificial means of yield stimulation of *Hevea brasiliensis*. It has long been known that ethylene occurs in traces in plant tissues and that it plays an important role in plant physiology and biochemistry affecting growth and development, flowering, fruiting, fruit ripening, and seed dormancy and germination. Ethylene is now fully recognized as a plant hormone and among this class of substances it is unique in that it is a gas. Applying ethylene gas to *Hevea* trees is much too cumbersome for commercial use, and it was fortunate that in 1968 classes of chemicals were available which under certain conditions decomposed to produce ethylene among other products. [(2-Chloroethyl)phosphonic acid] (ethephon) is such a substance which under slightly alkaline conditions as exist in plant tissues, decomposes to produce mainly ethylene, phosphate, and chlorine ions. This has been described as a base-catalyzed reaction of the following order:

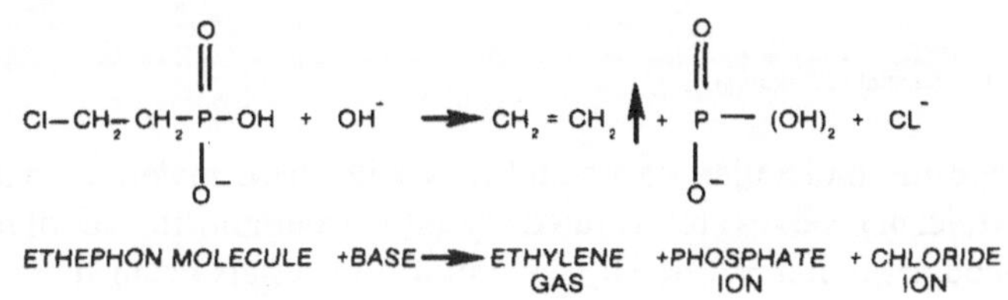

(Stable Below pH 3.5)

In addition to ethephon and other than b-hydroxyethyl hydrazine (the latter compound being withdrawn from further evaluation presumably because of commercial and economic reasons), a wide range of organosilicon compounds, also classified as ethylene-generating substances, were screened for stimulant activity. These derivatives of 2-chloroethyltrialkoxysilane were developed and synthesized at the Malaysian Rubber Producers Research Association (MRPRA) laboratories in the U.K. and were readily hydrolyzed by water to release ethylene and hydrogen chloride with the rate of release of ethylene governed by the size of the alkyl group. The compounds tested are listed in Table 1.

While some of the organosilicon compounds have shown promise as yield stimulants, the responses obtained in general have been slightly lower than those of ethephon, which became a commercially available latex stimulant in 1970. No further developments of the organosilicon compounds as latex stimulants have taken place since their initial testing.

1. Ethylene-Adsorbed Substances

Ethylene gas adsorbed onto activated charcoal and applied as a slurry is used commercially by the pineapple industry to induce flowering. During the early evaluation of ethylene gas to stimulate latex flow, application difficulties have been observed. However, research has continued in this area and although it is only very recently (1981) that new formulations have been suggested as being commercially acceptable, studies have continued.

Activated charcoal was the first of a number of potential adsorbent materials examined which included among others, Kieselguhr, bentonite, Fuller's earth, brick dust, silica gel, and particularly artificial zeolites otherwise known as molecular sieves. Of these, by far the

Table 1
RANGE OF ORGANOSILICON COMPOUNDS SCREENED FOR STIMULANT ACTIVITY

2-Chloroethyltrimethoxysilane
2-Chloroethyltrisopropoxysilane
2-Chlororethyltrilauroxysilane
2-Trimethoxysilylethyl ethyl methsulphonium iodide
Poly-2-chloroethylmethoxysilane
2-Chloroethylsilane
2-Chlororethyldimethoxyacetoxysilane
2-Chloroethyldimethoxychlorosilane
2-Chloroethyldimethoxyethylsilane
2-Bromoethyltriphenylsilane
Diphenyl-2(trimethylsilyl)ethyl phosphate
2-Chlororethyldimethylbutoxysilane
2-Chloroethyltrimethylsilane

Note: In comparison with the decomposition of ethephon, the organosilicon compounds hydrolyze as follows:

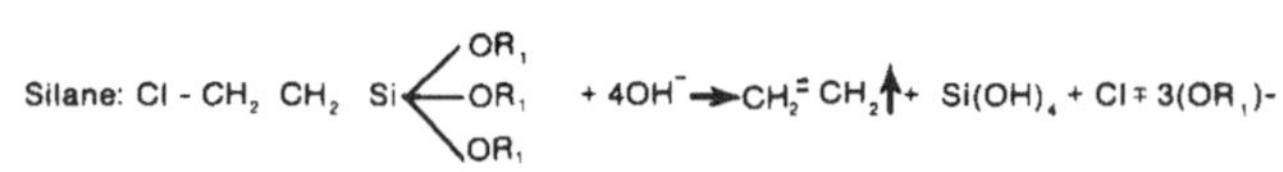

R_1 = Alcohol substitution

most effective were the molecular sieves and it is with these materials that developments have continued. Molecular sieves are hydrated crystalline aluminosilicates of metals produced synthetically, although similar but much less uniform materials occur naturally as zeolites.

Molecular sieves are activated for adsorption by removing the water of hydration, usually by heating, when in contrast to the behavior of most other hydrated crystalline material, the crystal lattice structure does not collapse but remains virtually unchanged. This produces highly porous adsorbents with a strong affinity for water and certain other gaseous or liquid molecules of suitable size. The pores in any particular type of molecular sieve are extremely uniform in size and hence confer the sieve-like capacity to accept or reject molecules whose effective diameter is less or greater than that of the individual pores.

Activated charcoal and 4a molecular sieve become saturated with ethylene at the rates of approximately 5 and 8 g/100 g of adsorbent, respectively, but 80% of the adsorbent gas is dissipated within a few minutes of exposure to the atmosphere. This latter disadvantage for developing a practical field application material has apparently been overcome by enshrouding the ethylene-saturated adsorbent in a suitable viscous carrier, usually a grease or mixtures of greases and oil, thus trapping semipermanently by occlusion the major portion of the adsorbed ethylene. At the time of writing, little is known about the ethylene-adsorbed formulations being introduced commercially, particularly in Malaysia, but it is believed that the sieve content is between 20 and 45%. Lower sieve levels may possibly be used when applying the materials by the groove method, which will be discussed under "Methods of Application for Latex Stimulants", Section II.E.3.

2. *The Development of Ethephon as a Latex Stimulant*

Ethephon has been used commercially as a latex stimulant since early 1970 and is available under the trade name ETHREL® Latex Stimulant. Following the discovery that ethephon decomposed to release ethylene, practical uses were developed in a wide variety of crops which included evaluations for maturity enhancement and fruit loosening in sweet and sour cherries. It was observed during this trial program that when high concentrations of ethephon were applied to cherries, particularly if the trees were under stress conditions, gummosis could occur. This observation led to the postulation that ethephon would be a candidate *Hevea* latex stimulant, and in 1968 samples of ethephon as various formulations were sent to the RRIM for screening.

Ethephon is very soluble in water unlike the esters of the phenoxy derivatives and while this physical characteristic was eminently suitable for applying ethephon to most other crops, special formulation techniques were needed for the practical application of ethephon to rubber trees. Initially following experiences with earlier latex stimulants, palm oil was used as the carrier for ethephon, but was soon discarded because of poor storage qualities involving stratification, errors in mixing the correct concentration, and inadequate rain fastness. Formulation objectives included long shelf life, physical and chemical stability, adequate painting characteristics, rain fastness, and an inexpensive carrier. These objectives were achieved in the commercially available ethephon product by combining the chemical with a series of synthetic gum materials and including a dye to facilitate field supervision. The initial formulation contained 10% wt/wt ethephon acid equivalent since early trials using the first of many application methods indicated an optimum treatment of 300 to 200 mg ethephon per treated panel per application. Subsequently, as tapping systems changed with stimulation and as more data became available, the amount of ethephon using the initial application method was reduced to approximately 150 mg ethephon per panel per treatment. Ethephon formulations are now available which contain 5 and 2.5% ethephon on a wt/wt basis, in addition to the original 10% material.

E. Methods of Application for Latex Stimulants

While ethephon formulated as a latex stimulant is the plant regulator of choice as of this writing, the application methods described will apply equally to other stimulants where these are used.

1. Bark Application

In normally practiced downward directional tapping, this method involves the scraping of the bark directly below the tapping cut for a depth usually equal to the amount of bark consumed over a 2-month tapping period. A tool similar to a paint scraper is used for this operation. In older rubber areas where tapping of the lower bark has become difficult, upward tapping techniques have now been developed. When tapping upward, the bark above the tapping cut is scraped, again equal to an area of bark which will be consumed over a 2-month period. Scraping of the outer bark allows for better adsorption of the stimulant into the phloem and is used today mainly in upward tapping systems where the bark has not been tapped previously. In virgin bark of this type, scraping removes the outer corky layers and the greenish layer of collenchyma cells until the whitish layer of hard tissue (the sclerenchyma ring) is visible. Scraping of the bark should not be deep enough to result in latex bleeding. The bark application method as described is labor intensive.

2. Panel Application

Currently, this is probably the most widely used application method for ethephon formulated as a latex stimulant since the stimulant is painted directly onto the thin layer of renewing bark left after the preceding 1 or 2 months of tapping. Both monthly and twice monthly applications are practiced. For monthly applications, a band approximately 1.5-cm deep is painted on the renewing bark with an artist's brush immediately above the tapping cut for the more conventional downward tapping and immediately below the tapping cut when the tapping direction is upward. Preferably the application should be made on a nontapping day without removing the coagulated strip of rubber on the tapping cut (known in the industry as tree lace) in order to prevent the mixing of the latex with the stimulant.

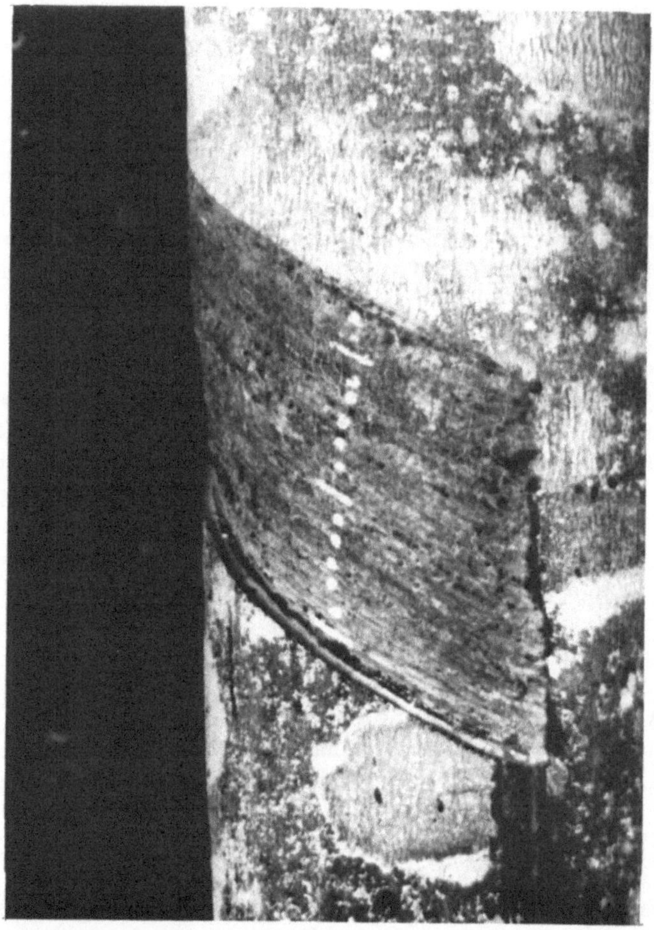

FIGURE 1. Modified groove method of applying ethephon-based latex stimulant to young mature rubber tree. Stimulant applied on coagulated strip on tapping cut, slightly overlapping onto freshly tapped bark.

This application method is also preferred when trees are being tapped on poorly renewed bark which would be difficult to scrape without exposing the latex vessels. The method is relatively nonlabor intensive.

3. Groove Application

This application method was developed after both the bark and panel application methods although it is itself being replaced by the "modified" or "new" groove method. The groove application method involves the removal of the coagulated strip of latex left behind following normal tapping. A thin layer of the stimulant is then applied directly to the tapping groove with a small brush; application preferably should be made 1 or 2 days before tapping is due. Generally, approximately 0.5 g of the stimulant formulation per tree is made every month — compared to approximately 1.5 g stimulant per tree when using the bark application method.

4. "Modified" or "New" Groove Application

This method of application (Figure 1) is extremely popular at the present time since it does not require the removal of the coagulated strip of rubber on the tapping cut (the tree lace). The stimulant is painted directly onto the tree lace overlapping slightly onto the renewing bark.

5. Vertical Groove Application

This application method is only used in conjunction with the experimental puncture tapping system which will be discussed further in the proceeding section. The groove is usually made with a gouge type tapping knife and while different strip lengths are being evaluated, 80 to 100 cm will be average. The stimulant is painted along the whole length of the vertical groove with new grooves being made usually on a monthly basis. An alternative to the vertical groove is a vertical strip approximately 1-cm wide (although a 2-cm strip also is satisfactory) and whether a strip or groove is prepared, the tissue should be removed to a depth of 1 to 2 mm. The vertical groove made with the tapping tool has the advantage of guiding the latex to the spout and cup following the puncture tapping operation.

F. Tapping Systems in Relation to Stimulation

There has been a great deal of research involving the economics of the stimulation of rubber in relation to different tapping systems, but primarily because of variabilities in the world price of rubber, the labor payment systems in the individual rubber producing countries, and the different rates of agronomic progress especially in terms of improved clonal introductions, it is impossible to generalize on this subject. However, in all rubber producing countries where stimulation has now become a regular management tool, certain observations will apply in most if not all cases.

1. A tapper will harvest more total rubber (including latex and coagulated cup lump from late dripping) per day when tapping the trees every third (or more) day than then tapping every second day.
2. Even with varying rates of incentive pay for the yield harvested, the cost of the tapping and collection operations per kilo of rubber produced will be less when the trees are tapped every third vs. every second day.
3. When all production costs are considered, the break-even yield is greater when tapping every second day than it is when tapping every third day.
4. Even with the more rapid introduction of newer higher yielding clones, the economic life of a rubber tree is influenced by the amount of bark excised over a given period of time. Assuming the same level of tapping skill, the bark consumed from third or fourth daily tapping is commensurately less than when the trees are tapped every second day. Similarly, less bark is consumed on a short tapping cut (such as a one quarter spiral) over a given period of time in comparison to a longer tapping cut (one half spiral or greater).

The intensity of tapping was discussed earlier. Allowing for generalization, it is a fact that tapping intensities have been reduced side by side with the introduction of stimulation.

1. The S/2 System

Where stimulation has not been introduced, probably the most common tapping system used today in high yielding clonal rubber both on plantations and small holdings is the half-spiral alternate daily or S/2, d/2, 100%. With the introduction of stimulation (Figure 2), the most common change is to a S/2, d/3 or 4 system, 67 or 50%. Either the bark, panel, or groove method of applying the stimulant can be used. The objectives of these tapping system changes are to increase the yield obtained on the day of tapping, thus reducing the cost of the tapping operation itself; to conserve bark consumption increasing the economic life of the most productive part of the tree; and by lessening the intensity of tapping the physiological

FIGURE 2. Typical area of rubber, stimulated with an ethephon-based latex stimulant. Note bark application method with stimulant applied to a band of scraped bark equal in depth to 2 months tapping and bark consumption.

stress on the tree is reduced which is considered to be beneficial. The increase in the interval between tappings from d/2 to d/3 or 4 significantly reduces the number of tappers needed to cover a given area, releasing workers for other plantation tasks such as replanting older rubber areas. This aspect is particularly important where there is a shortage of skilled labor and on small holdings where family labor tends to migrate to urban employment, often leaving the older family members to tend to the rubber areas.

Under the S/2, d/3 or 4 systems, the amount of stimulant used per tapping panel (when ethephon is the active material) will vary depending on the application method and the interval between applications. With the bark application (Figure 3), about 150 mg ethephon active ingredient or 1.5 g of 10% ethephon product is used per panel if applied every 2 months. With the panel application method to the renewing bark, 50 to 80 mg is applied every month. With the groove or "modified" groove method, approximately 50 mg ethephon is applied again on a monthly basis.

2. The S/4, d/4 System

While seemingly a simple quarter spiral, fourth daily system, this does in fact have many variations, one of which is to open a half spiral cut, tapping one half of this every fourth

FIGURE 3. Close-up of bark application of ethephon-based latex stimulant applied every 2 months. Tapping system used is S/2, d/3, 67%. Note double cups for extra latex collection.

day and the other half every fourth day so that each cut is only tapped once every 8 days. This is a very low intensity tapping system, being rated at 25%. This system can only be used effectively when a stimulant is applied to counteract the rapid plugging which occurs following the tapping of a short cut. In general terms, there is an inverse relationship between cut length and response to stimulants just as there is a relationship between the length of the tapping cut and the rapidity of plugging. This relationship no doubt is due to the natural production of ethylene as a response to the degree of physical damage inflicted on the tree. Thus, a full spiral tapping cut results in slow plugging and late dripping, whereas a one quarter spiral tapping cut plugs rapidly, with cessation of latex flow being relatively rapid. A full spiral tapping cut while producing high yields of latex also severely interferes with the growth of the tree (especially the rate of girth) and imposes considerable physiological strain, often causing trees to ''go dry'' (Figure 4).

The S/4, d/4 system is very conservative in terms of the consumption of bark and thus is often used in young mature rubber in the early tapping years when maximum girth development is required. Under this system, the stimulant is applied to the full half spiral length of the tapping cut whether the bark, panel, or groove application method is used.

FIGURE 4. "Thy cup runneth over". The ultimate in latex stimulation — but not recommended for young high yielding rubber! A PBIG seedling tree, Malaysia, 1970.

3. Puncture Tapping and the Micro-X Tapping Systems

While both the puncture and micro-X tapping systems are considered to be experimental and under development, both are important in relation to the use of stimulation because they cannot be used without assistance from an antiplugging agent. Both systems essentially are designed to alleviate the growing problem of a shortage of skilled tappers, but also bring to the rubber industry the possibility of mechanizing the tapping operation. Additional benefits include a reduction in the normal consumption of bark and accompanying physiological stress, and the possibility of bringing trees into production at a younger age than is now considered advisable with conventional tapping systems.

Puncture tapping (Figure 5) has been mentioned briefly but consists of a vertical strip being prepared on the tree, 80 to 100 cm in length. The width of the strip has varied from 1 to 2 cm and while the 2-cm wide strip has yielded somewhat better than the 1-cm strip, the width chosen will depend on the length of period the full circumference of the tree is to be exploited. As mentioned earlier, the vertical strip can either be prepared by scraping or grooving, removing the tissue to a depth of 1 to 2 mm. The stimulant is applied to the whole length of the strip with the amount and concentration of the stimulant having varied between 1 g to 2.5 g per application. A new vertical strip is prepared on a monthly basis and, therefore, this is the frequency of the stimulant application.

FIGURE 5. Puncture tapping of young mature rubber tree. Vertical strip treated with ethephon-based latex stimulant.

The strip is divided equally into sections depending on the number of punctures to be made on each day of tapping; although this has varied between five and ten punctures per tapping, presently the general consensus is that good results can be expected from six punctures per tapping. For the most part, the yield response to puncture tapping has been better on a third daily tapping frequency (d/3) than on alternate daily tapping (d/2), although satisfactory yields have also been obtained on fourth daily (d/4). To date, there have not been sufficient long-term data collected to establish a clear pattern of yield, particularly with regard to establishing a decline in response to puncture tapping with time.

A number of puncture tapping tools have been developed with the most commonly used needle size being 1 mm in diameter. The tools used are designed to allow for setting the needle to penetrate a specific depth. The penetration depth is calculated by measuring a large enough sample of bark thickness along the prepared vertical strip using a bark tester; having made this calculation the puncture tapping tool is set to penetrate this thickness minus 1 mm. Following the puncture tapping operation, the latex is led down the vertical strip to a conventional spout and latex cup for collection.

A system of high level puncture tapping (Figure 6) is being developed particularly in older rubber areas where previous conventional tapping has exhausted the lower bark areas and when the recovery or renewal of the lower bark has been insufficient for further tapping

FIGURE 6. Result of high level puncture tapping following low level puncture tapping. Conventional tapping (bark shaving) can be introduced following puncture tapping without waiting for the normal bark renewal process.

to be continued. This system known as HLPT uses the same methods of strip preparation and stimulant application as on the lower level.

The micro-X system is a combination of both puncture and conventional tapping systems and it can be introduced either when the trees are ''opened'' for the first time or while the trees are on conventional tapping. Micro-X uses the half spiral tapping cut which is either opened at or corrected to a 40° slope to enhance the latex flow into the cup. The stimulant ethephon is applied by the groove method on a monthly interval, generally using 1 g of stimulant per tree per application. The concentration of the ethephon stimulant used depends on the age of the panel being tapped and in general is 2.5% for panel A, 5% for panel B, and 10% for panels C and above.

The punctures are made on the tapping groove itself and currently available experience indicates that three punctures equidistant on the half spiral groove are made on 9 tapping days with the alternate daily frequency the most commonly used. This is followed by 3 alternate days of conventional tapping by shaving the bark which removes the puncture marks and the traces of stimulant. On the d/2 system, only approximately 1 cm of bark is consumed per month, or roughly one third of the bark which would be used on a normal S/2, d/2 tapping system.

Micro-X tapping can also be used on an upward tapping exploitation program, and even quarter spiral tapping upward with this alternate puncture/shaving technique has given satisfactory yield results.

The ratio of puncture tapping to conventional tapping (shaving) using the micro-X technique can depend on the availability of skilled tappers and, therefore, is determined by the tapping frequency. With the different combinations possible with the micro-X system more trees may be tapped with a less number of skilled labor. The puncture tapping operation whether using the total puncture tapping or the micro-X system can utilize casual labor for the puncture tapping operation while skilled labor is needed for the conventional tapping (shaving). There would appear to be a distinct future for the further development of these somewhat revolutionary exploitation methods of rubber trees.

G. Further Opportunities for Plant Regulators in *Hevea* Latex Production

Notwithstanding the gains already made in the development of plant regulators, particularly in the area of stimulation, there are many opportunities which still are available in the plant regulator field. Some of these are described in the following sections.

1. Stimulation of Rubber Biosynthesis

While it is an established fact that following stimulation (e.g., with ethephon) the rubber content of the latex sap does decrease because of increased flow time, this does not fall off more than about 20%, suggesting that the act of stimulation may additionally, stimulate rubber synthesis. Leopold[8] has suggested the dependency of yields and flow upon the latex turgor, the duration of "bleeding", and the rubber content of the latex sap as indicated in the following diagram:

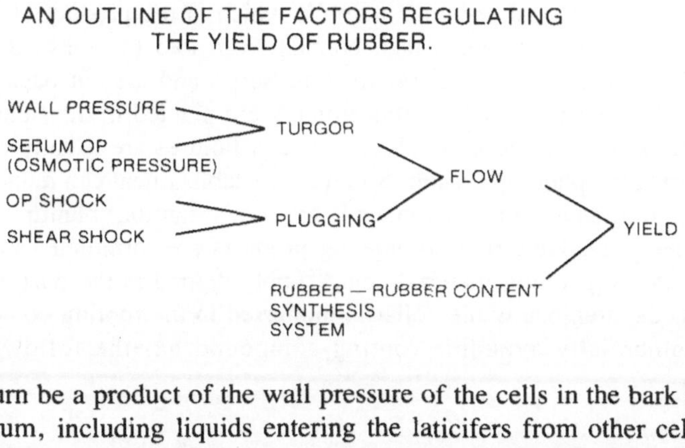

Turgor will in turn be a product of the wall pressure of the cells in the bark and the osmotic value of the serum, including liquids entering the laticifers from other cells. "Bleeding" time will depend upon the effectiveness of coagulation, which appears to be principally regulated by osmotic shock or shearing forces on the lutoids. However, rubber content will depend ultimately upon the biosynthetic system. Plant regulator treatment could be envisioned which would alter rubber yields, then, through an alteration of turgor, serum levels or osmolarity, the coagulation system, or the rubber biosynthesis system. The darkness surrounding the biosynthesis of rubber is a serious handicap in the selection of plant regulators which may have the capability of influencing this biosynthetic process.

2. Regulation of Ethylene Release

The mechanism of the degradation of ethephon and organosilicon compounds has been outlined. While there is no evidence available as to the persistence of applied ethephon to rubber trees, experience with other species suggests that the breakdown would probably be completed in 1 or 2 weeks. Since yield stimulation continues well beyond the time of assumed ethephon persistence, the stimulated state may be a condition of general ethylene biosynthesis in the bark area, since it is known that ethylene applications can autocatalytically generate ethylene production. Nevertheless, it would appear useful to devise some method of controlling the release of ethylene from compounds known to degrade to this end molecule.

III. ROOT STIMULATION IN *Hevea* RUBBER PLANTING

Plant growth regulators in the form of synthetic auxins have been used to promote rooting and aid in the establishment of rubber planting materials for more than three decades. The most commonly used active substances are derivatives of naphthalene acetic acid and will include 1-naphthalene acetamide, 2-methyl-1-naphthalene acetic acid, and 2-methyl-1-naphthalene acetamide. In addition, indole-3-butyric acid is often included as a component in commercially available rooting products. In recent years, fungicides have been formulated with the synthetic auxins as a measure to prevent fungal infection, particularly on the roots of rubber planting material, which are often pruned prior to planting to minimize ''drying out''. One such fungicide would be tetramethylthiuram disulfide.

In preserving the characteristics of clonal selections in the rubber breeding program, multiplication of planting material is carried out by using bud grafting techniques. Today the rubber industry usually produces the material for planting in the field in special nurseries by green budding selected seedlings. After the bud patch has ''taken'', one technique is to cut back the stem above the bud and prune the tap and lateral roots after lifting. This ''budded stump'' is then transferred to the field for planting.

An alternative to this technique is to cut back the stem above the bud and allow the scion to develop in the nursery for a year or more prior to cutting back and planting in the field. This material is referred to as a ''stumped budding''. Yet another variation in this type of planting material is the ''clonal seedling stump'', which is the product of seeds collected from special isolation seed gardens which contain from two to many carefully selected clones. The seeds are germinated and grown in nurseries and are cut back after at least 1 year of growth and after tap and lateral root pruning are planted in the field.

Failures after transplanting are expensive, and such failures are usually caused by inadequate rainfall during the planting season. Rapid root establishment can reduce transplanting failures to a minimum, and rooting compounds are now a familiar planting practice.

Most commercially available root stimulating products are formulated as dusts for ease of application, with the planting material being simply dipped in the dust and, in the case of rubber stumps, the area below the collar is immersed in the rooting compound.

A typical commercially available rooting compound has the following ingredient components.

	% wt/wt
1-Naphthalene acetamide	0.067
2-Methyl-1-naphthaleneacetic acid	0.033
2-Methyl-1-naphthaleneacetamide	0.013
Indole-3-butyric acid	0.057
Tetramethylthiuram disulfide	4.000
Inert ingredients	95.830

New application methods have been developed for the specific application of rooting compounds to rubber stump planting material, and one such method is to prepare a slurry using a 1:2 or 1:3 ratio of rooting compound/water by weight. The rubber stump, after pruning, is simply dipped in the slurry so that the whole of the root area is treated.

While research continues in an effort to find more effective root stimulating plant regulators, the naphthalene acetic acid derivatives and indole butyric acid have contributed significantly to the planting efficiency of rubber stump material.

IV. DEFOLIATION AS A MANAGEMENT TOOL IN THE CONTROL OF LEAF DISEASES

The technique of large-scale artificial defoliation of rubber with plant regulators was first developed in Malaysia as a standby emergency measure in the event of the accidental introduction into that country of South American Leaf Blight (SALB) from the continent for which it was named. This most destructive of all of the leaf diseases of rubber is caused by the organism *Microcyclus ulei* which fortunately is still confined to the rubber areas of Latin America, and primarily Brazil. Assuming that an outbreak of SALB was detected early enough in Malaysia, it was reasoned that immediate defoliation of a large enough area surrounding the infestation would prevent the spread of this disease, which because of the susceptibility of the major commercial clones, would have constituted a national disaster.

Of more importance in the major rubber producing areas of Southeast Asia is secondary leaf fall (SLF) with the causal fungi being *Oidium heveae* or *Colletotrichum gloesporioides*; although SLF has rarely required attention in the past, the position is somewhat different today because of the increased planting of some modern clones selected for their good yield, but which are of high disease susceptibility. Infected areas suffer severe leaf fall during the refoliating season, resulting in a poor canopy for the remainder of the year.

The severity of SLF in susceptible clones is largely determined by weather conditions during the wintering (natural leaf drop) and refoliating season which in Malaysia falls between February and April, and the most important factors are rainfall and the frequency of the precipitation. Dry weather during January and February helps to bring on wintering with subsequent refoliation ideally commencing before the rains that normally follow in March. Infrequent showers of short duration, high humidity, and low temperatures favor *Oidium heveae*, while continuous periods of rain are required for the development of *Colletotrichum gloeosporioides*. SLF therefore assumes importance when wintering is gradual or slow, followed similarly by slow and late refoliation occurring during the wetter weather in the latter part of March and April (under Malaysian conditions).

The process of annual wintering is set in motion by a series of physiological changes in the foliage commencing much earlier than the first leaf drop symptoms are observed, which suggests that application of a plant regulator about 2 weeks in advance of wintering may hasten this process and bring about a rapid and complete defoliation.

2,4,5-Trichlorophenoxy acetic acid was the original auxin used as a standby measure to defoliate rubber in the event of an SALB infestation, and the n-butyl ester was determined to be the most economic form of 2,4,5-T. Application was made in diesoline at a volume of 35 ℓ/ha. The concentration of 5% by volume was used. While this defoliation treatment was adequate as an emergency measure to counteract an invasion of SALB, it is not suitable to obtain controlled wintering, as 2,4,5-T has the disadvantage of moving beyond the leaves to the buds and shoots, causing extensive dieback.

Since the early experiments with 2,4,5-T a number of contact herbicides or dessicants have been tested to induce early and uniform wintering as a means of avoiding annual attack of SLF on susceptible clones. Of these, two organoarsenates, dimethylarsinic acid (cacodylic acid) together with its sodium salt and monosodium methane arsonate (MSMA), were found to give the optimum defoliation effect without any shoot dieback. Between 1974 and 1976 large-scale commercial applications using this technique for avoiding SLF were made with some 8000 ha treated. In general the speed and uniformity of defoliation have been satisfactory in combating SLF and most sprayed areas have derived the benefit expected. However it was noted that where results were poor, or not completely satisfactory, one of the reasons was the unavoidable interference by rain during or soon after the aerial spray application since the organoarsenates are water soluble and wash-out resulted. Repeat spraying, apart from being expensive, is difficult to fit into the short-season spraying schedule.

By 1977 a number of other candidate defoliants were tested in Malaysia which included ammonium sulfate, endothall, merphos, bromacil, diuron, MSMA, and linuron. Testing was carried out on nursery seedlings and among the seven defoliants tested only merphos proved to be as good as MSMA in bringing about discoloration and complete defoliation of the mature leaves, respectively, in 1 or 2 weeks.

Merphos subsequently was shown to be readily soluble in diesoline and the effect of merphos in this oil carrier was evaluated on mature leaves of seedlings in the nursery. The results showed that there was no difference in the effectiveness of merphos in water or diesoline, the chemical performing satisfactorily even at the lowest concentration used of 5% by volume.

Rainfastness tests were then performed to compare the merphos in oil with MSMA, with the plants drenched with running water for 20 min to simulate heavy rain wash. Results indicated almost complete defoliation after 2 weeks with the merphos/diesoline carrier whereas MSMA was essentially inactivated following the simulated rain treatment.

There are obviously further opportunities for developing new techniques in the use of plant regulators to effectively and safely defoliate *Hevea* rubber as part of a management system in the control of leaf diseases.

REFERENCES

1. **Templeton, J. K.,** Partition of assimilates, *J. Rubb. Res. Inst. Malaya,* 21, 3, 1969.
2. **Bonner, J.,** personal communication, 1973.
3. **Wycherley, R. P.,** unpublished work, 1973.
4. **Chapman, G. W.,** Plant hormones and yield in *Hevea brasiliensis, J. Rubb. Res. Inst. Malaya,* 13, 167, 1951.
5. **Taysum, D. H.,** Effect of ethylene oxide on the tapping of *Hevea brasiliensis, Nature (London),* 191(4795), 1319, 1961.
6. **Abraham, P. D.,** Stimulation of the Yield of *Hevea brasiliensis* by Ethylene Releasing Substances, Doctoral thesis, University of Reading, U.K., 1977.
7. **D'Auzac, J. and Ribailler, D.,** L' ethylene, nouvel agent stimulant de la production de latex chez l' *Hevea brasiliensis, Rev. Gen. Caoutch. Plastg.,* 46(7-8), 857, 1969.
8. **Leopold, A. C.,** personal communication, 1970.

Chapter 4

BIOREGULATION OF RUBBER SYNTHESIS IN GUAYULE PLANT

H. Yokoyama, W. J. Hsu, E. Hayman, and S. M. Poling

TABLE OF CONTENTS

I. INTRODUCTION

Rubber is distributed extensively in the plant kingdom. Some 2000 species of plants are known to contain rubber, but only a few have ever produced rubber in substantial quantities for commercial use. Two of these, the rubber tree *Hevea brasiliensis* and the guayule shrub *Parthenium argentatum*, have been continuing sources of natural rubber. The two plants have contrasting climatic requirements. *Hevea* is native to equatorial lowland rain forest regions in the Amazon basin; guayule flourishes wild in the upland plateaus in Mexico and Texas with subtropical-temperate climates and low rainfall. Despite these differences, the two plants produce a similar rubber. Studies have shown that physicochemical and mechanical properties of guayule rubber and *Hevea* rubber are similar.[1]

II. GUAYULE PLANT

Guayule is a member of the sunflower family *Compositae* and belongs to the genus *Parthenium*. There are 16 species of *Parthenium*, and guayule is *Parthenium argentatum*, so designated because of a silvery sheen on its gray-green leaves. It is the only *Parthenium* species known to produce rubber in any quantity.[2] Unlike the rubber in *Hevea* and other latex producing plants, guayule rubber is not contained in ducts but in single thin-walled cells. These rubber-filled cells are mainly in the outer layers. Curtis reported larger accumulation of rubber in older plants and that commercially important quantities are found only in the bark and roots.[3] Artschwager showed that in plants older than 1 year the vascular rays of phloem and xylem carry the major portion of rubber.[4] Other parenchymatous tissues, the pith, the primary cortex, epithelial cells of the resin canals, and xylem parenchyma contain less amounts of rubber. In younger plants these latter tissues contribute a relatively larger part of the total rubber. Rubber is evident in the leaf parenchyma, but only in trace amounts. The rubber is suspended in cell sap to form a latex as in the other rubber producing plants.

A bushy perennial shrub, guayule could be cultivated in its native habitat and in the warmer regions in California, Arizona, and New Mexico. It could also be cultivated in the subtropical-temperate regions in other continents, particularly Australia and Africa.

III. PLANT BIOREGULATION

The single major obstacle to commercial production of guayule rubber is the low yield of the guayule plant. Without increased yield to make it cost competitive, the guayule plant will not become a commercially viable source of natural rubber. Traditional breeding programs and variety of more exotic hybridization techniques may not be the only ways to improve the content of rubber in guayule. A promising approach to improving the yield characteristics is through the bioregulation of the synthesis of polyisoprenoids to cause an accumulation of increased amounts of rubber. This approach is based on the discovery of bioregulators that appear to derepress specific genes, thereby inducing the production of additional quantities of constituents in plants.[5]

IV. BIOREGULATORS

A series of bioregulators was developed for possible use in influencing rubber formation in guayule. Studies on structural activity relationships indicate that onium type structural features are necessary for biological activity relative to isoprenoid biosynthesis whether it be at the lower terpenoid or higher polyisoprenoid level.[6] Like *Hevea* rubber, guayule rubber is a polymer of the simple 5-carbon molecule isoprene and has the *cis*-1,4 shape (Figure

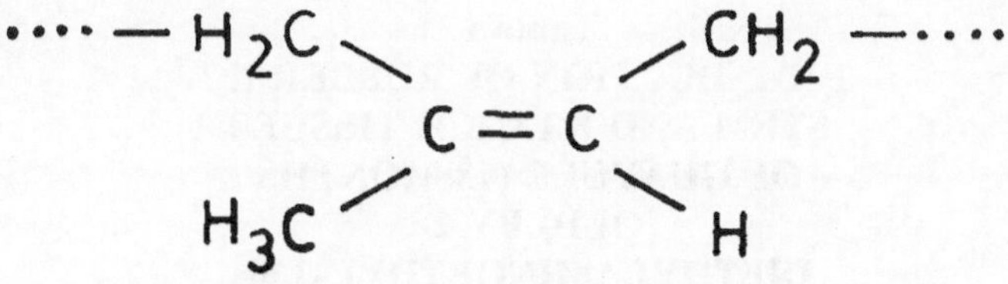

FIGURE 1. Simple 5-carbon isoprene unit of rubber.

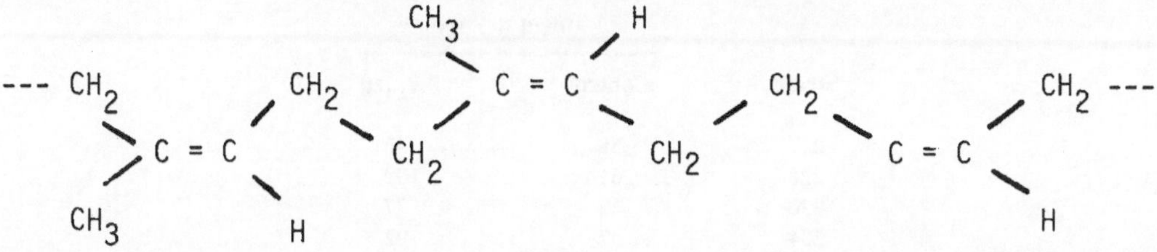

FIGURE 2. Structure of guayule rubber. The polymer consists of thousands of isoprene units joined end-to-end and all the double bonds have *cis* stereochemistry.

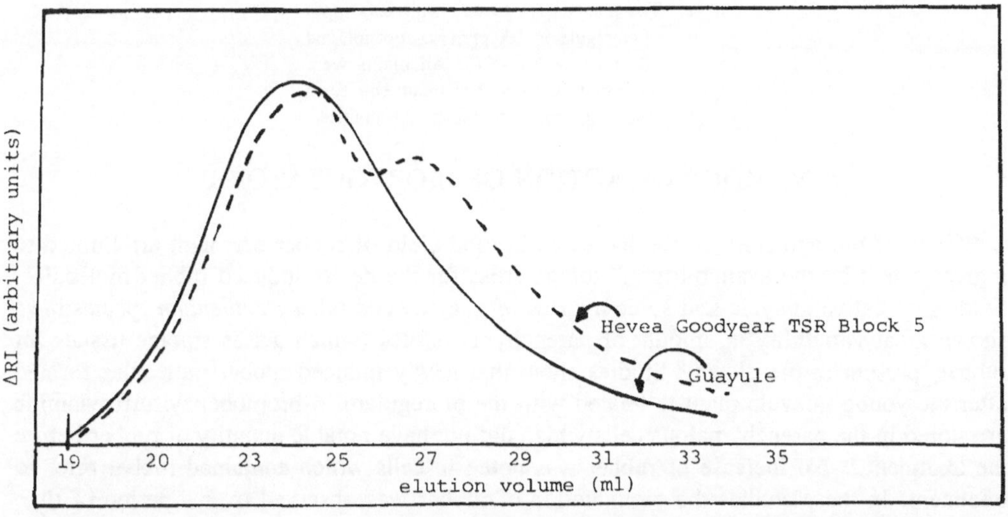

FIGURE 3. Comparison of gel permeation chromatography of guayule rubber and *Hevea* rubber.

1). The isoprene units are joined together end-to-end to form a giant molecule containing tens of thousands of carbon atoms in a linear chain identical to that of *Hevea* rubber and with similar molecular weight (Figures 2 and 3).

The initial bioregulator which was reported to affect rubber biosynthesis in guayule is the onium type compound 2-diethylaminoethyl-3,4-dichlorophenylether (Table 1).[7] Subsequently, a number of benzylfurfurylamines and benzylalkylamines were synthesized (Figure 4).[8] *p*-Bromobenzylfurfurylamine and N-methylbenzylfurfurylamine caused marked increases in the content of rubber in guayule. N-methylbenzylhexylamine exhibits nearly the activity as *p*-bromobenzylfurfurylamine (Tables 2 and 3). The former compound is more stable than the latter and would be cheaper to manufacture.

Table 1
BIOINDUCTION OF RUBBER IN STEM AND BRANCH TISSUES OF GUAYULE (18 MONTHS OLD) BY 2-DIETHYLAMINOETHYL-3,4-DICHLOROPHENYLETHER

	Rubber content (mg/g dry wt)	
Strain	**Control**	**Treated**
212	58	146
228	61	100
230	42	77
234	52	92
239	57	122
241	48	122
242	51	108

Note: The plants were treated with 2000 ppm of bioregulator, 100 ppm isopropanol, and 500 ppm Ortho®-X77. All plants were harvested 40 days after treatment. Each result represents the mean of 6 plants.

V. MODE OF ACTION OF BIOREGULATORS

Effects of bioregulators on the increases in total yield of rubber per plant are limited to a great extent by the availability of storage areas for the newly induced rubber molecules. Studies on native guayule and some hybrids of guayule and other *Parthenium* species have shown great variability in amount of parenchyma tissues (which act as storage tissues for rubber) present in the plants.[9] Studies show that newly induced rubber molecules formed after the young guayule plant is treated with the bioregulator *p*-bromobenzylfurfurylamine are stored in the parenchymatous cells which did not have notable quantity of rubber before the treatment.[10] No increase of rubber was noted in cells which contained rubber prior to treatment. In these studies the bioinduction of rubber was observed over a period of time (55 days) by taking tissue slices of the stems at intervals of 0, 13, 24, 34, and 55 days after treatment. Micrographs taken at 24 and 34 days show significant increases in rubber content (dark spots) (Figures 5 to 7). No increases in rubber content were noted between 0 and 13 days and between 34 and 55 days. Bioinduction appears to cease after all the cells which are capable of forming and storing rubber have been filled with rubber. Approximately a threefold increase in the number of rubber particles was seen.

Findings thus far indicate that the guayule plant has a certain capacity dependent on the particular strain. Also, bioinduction will not be feasible beyond the biological potential unless this potential is increased by cell differentiation and enlargement.

VI. RUBBER QUALITY

Any improvement of yield in rubber in guayule must be accomplished without diminishing the physicochemical characteristics of rubber. The precise stereochemistry (*cis*-1,4 isoprene unit) of natural rubber must be maintained unaltered. These are essential requirements in

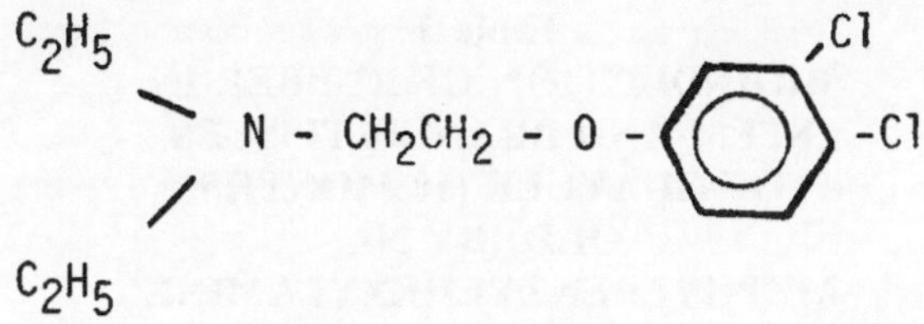

1

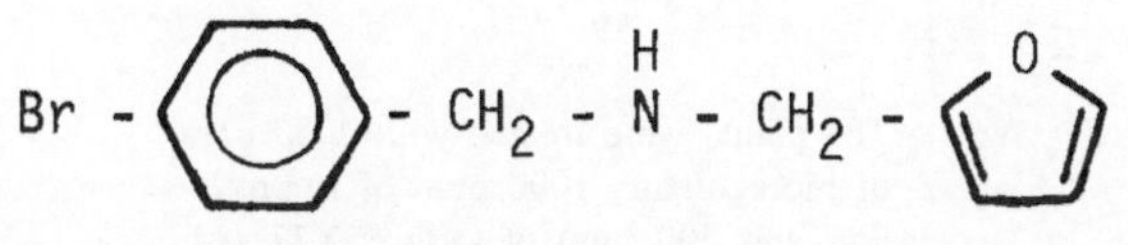

2

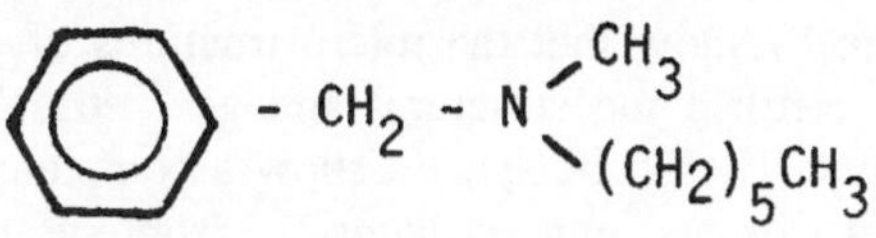

3

FIGURE 4. Structures of bioregulators. (1) 2-Diethylamino-ethyl-3,4-dichlorophenylether; (2) *p*-bromobenzylfurfurylamine; (3) N-methylbenzylhexylamine.

Table 2
BIOINDUCTION OF RUBBER IN STEM AND BRANCH TISSUES OF GUAYULE (18 MONTHS OLD) BY *p*-BROMOBENZYLFURFURYLAMINE

	Rubber content (mg/g dry wt)	
Strain	**Control**	**Treated**
228	58	146
239	52	121
234	54	162
89	72	207

Note: The plants were treated with 2000 ppm of bioregulator, 1000 ppm of isopropanol, and 500 ppm of Ortho®-X77. All plants were harvested 40 days after treatment. Each result represents the mean of 6 plants.

Table 3
BIOINDUCTION OF RUBBER IN STEM AND BRANCH TISSUES OF GUAYULE (10 MONTHS OLD) BY N-METHYLBENZYLHEXYLAMINE

	Rubber content (mg/g dry wt)	
Strain	**Control**	**Treated**
593	39	131

Note: The plants were treated with 2000 ppm of bioregulator, 1000 ppm of isopropanol, and 500 ppm of Ortho®-X77. All plants were harvested 40 days after treatment. Each result represents the mean of 5 plants.

the bioinduction of rubber formation. Rubber samples isolated from guayule plants treated with 2-diethylaminoethyl-3,4-dichlorophenylether and *p*-bromobenzylfurfurylamine were compared to those from untreated plants as well as *Hevea* rubber. Carbon-13 nuclear magnetic resonance (NMR) spectra (Figure 8) show that the microstructures of guayule rubber and *Hevea* rubber are identical and confirm the structural and geometrical purity of guayule rubber isolated from treated plants. The NMR spectra show a complete absence of signals attributable to stereo or structural isomers, namely *trans*-1,4 isoprene units, demonstrating that guayule rubber from treated plants is highly stereospecific polymer composed entirely of *cis*-1,4 isoprene units. The NMR spectra confirm that improvement of rubber yield is accomplished without altering the microstructure of the rubber.

The molecular weight distribution of guayule rubber and *Hevea* rubber was examined by gel permeation chromatography (GPC) (Figure 9). The distribution of molecular weights of the polyisoprene chain in rubber from untreated plants is identical to that from treated plants and both are unimodal. The GPC results (Figure 10) suggest that the bioregulators induce the formation of new rubber molecules rather than a chain extension of rubber molecules at the surface of existing rubber particles.

ACKNOWLEDGMENT

Work reported here supported in part by National Science Foundation Grant #PFR-7807567.

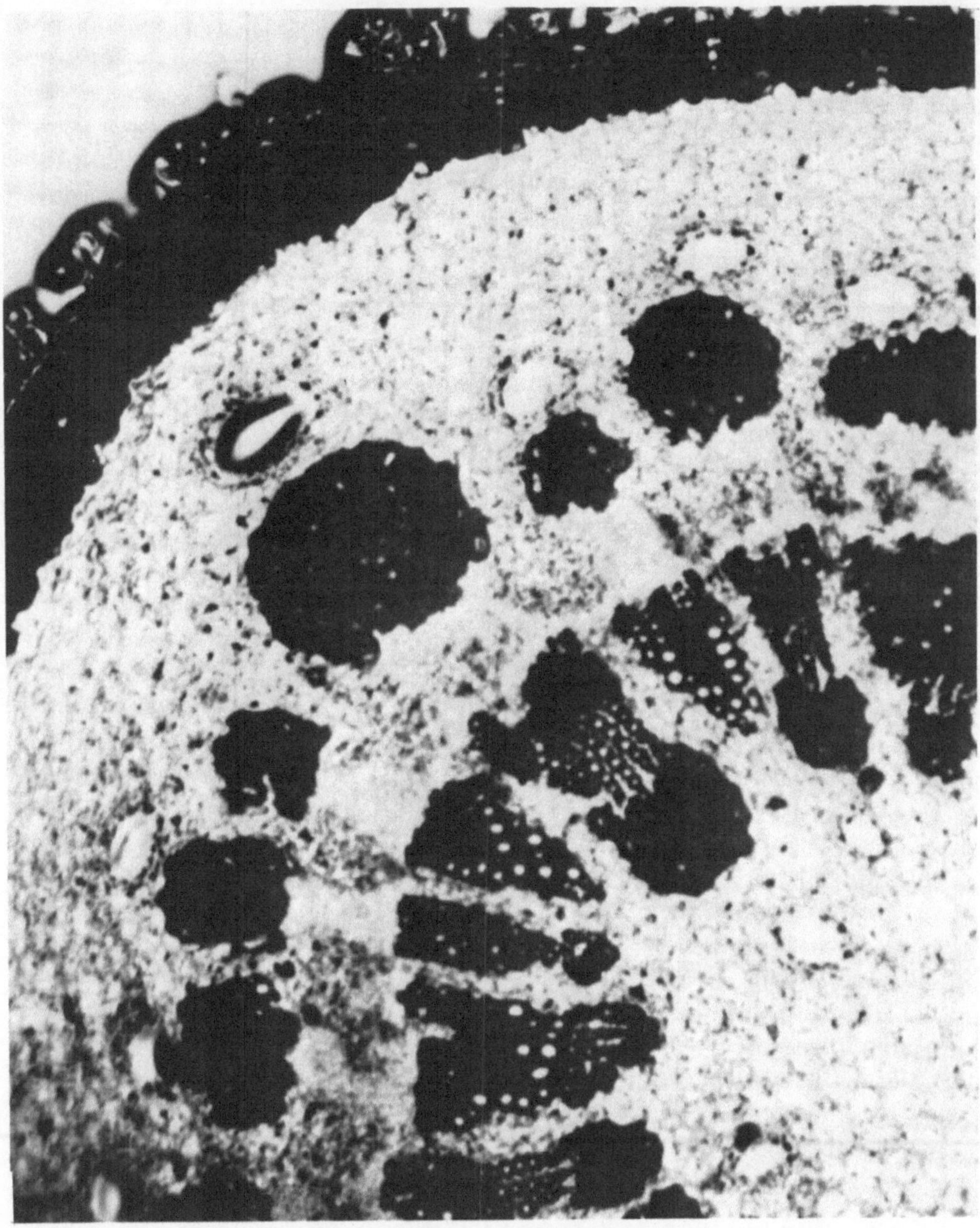

FIGURE 5. Cross-section of a young guayule stem 13 days after treatment with *p*-bromobenzylfurfurylamine. No increase in rubber content (dark spots) is observed when compared to cross section at 0 day.

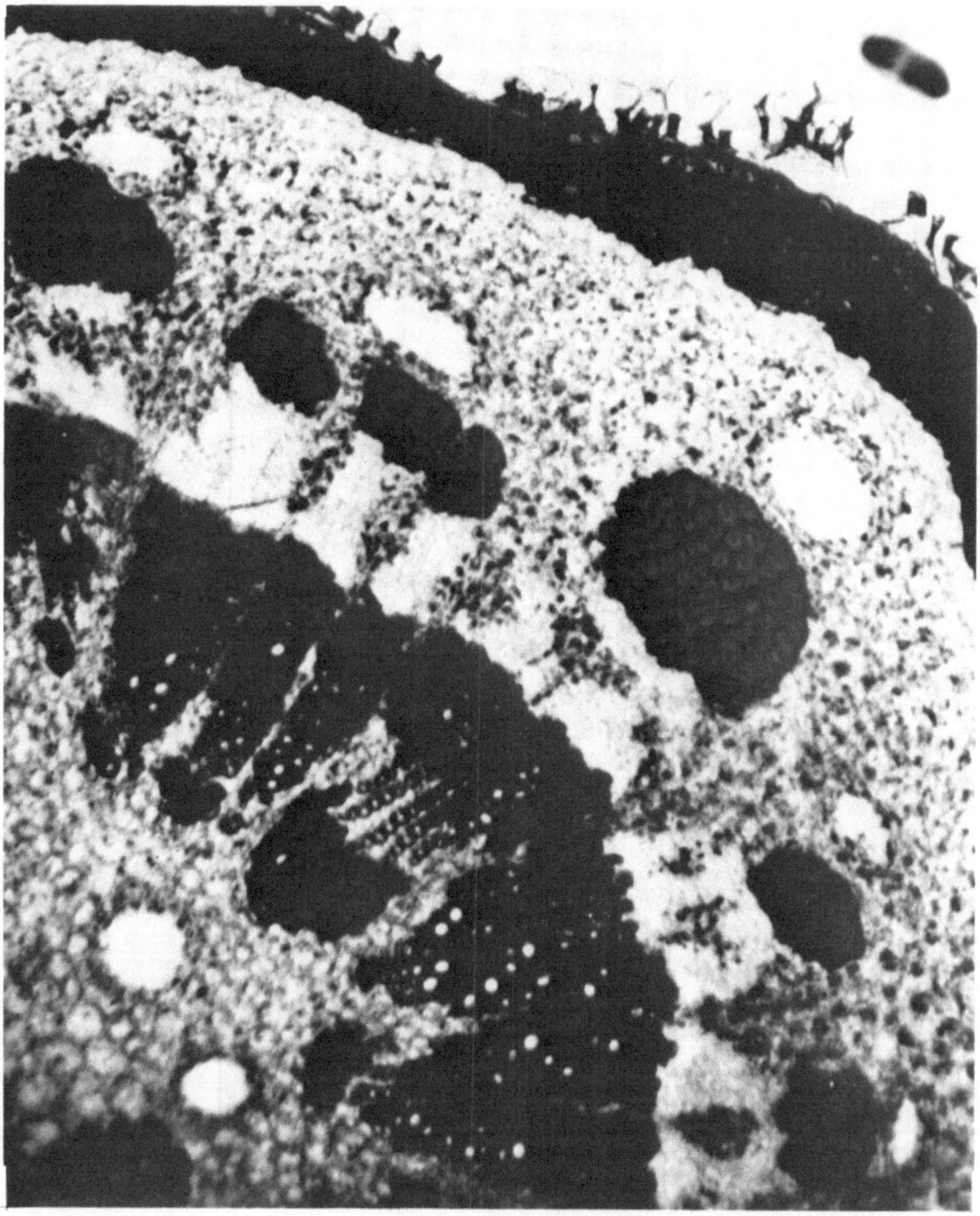

FIGURE 6. Rubber content (dark spots) 24 days after treatment. Cross-section of young stem shows rubber content increasing in response to application of bioregulator *p*-bromobenzylfurfurylamine.

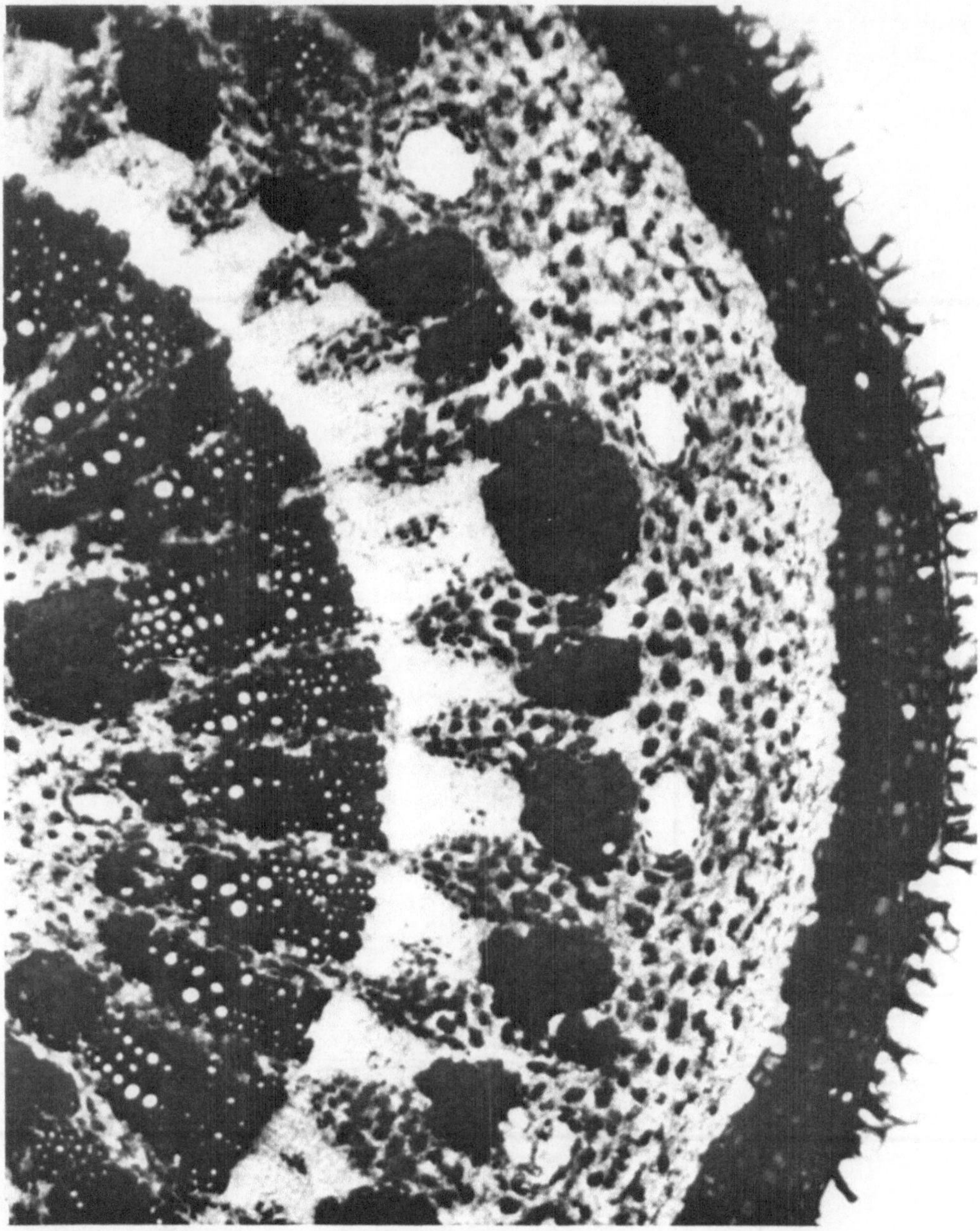

FIGURE 7. Rubber content (dark spots) 34 days after treatment with bioregulator *p*-bromobenzylfurfurylamine. Cross-section of young stem at 55 days after treatment shows no increase in rubber content.

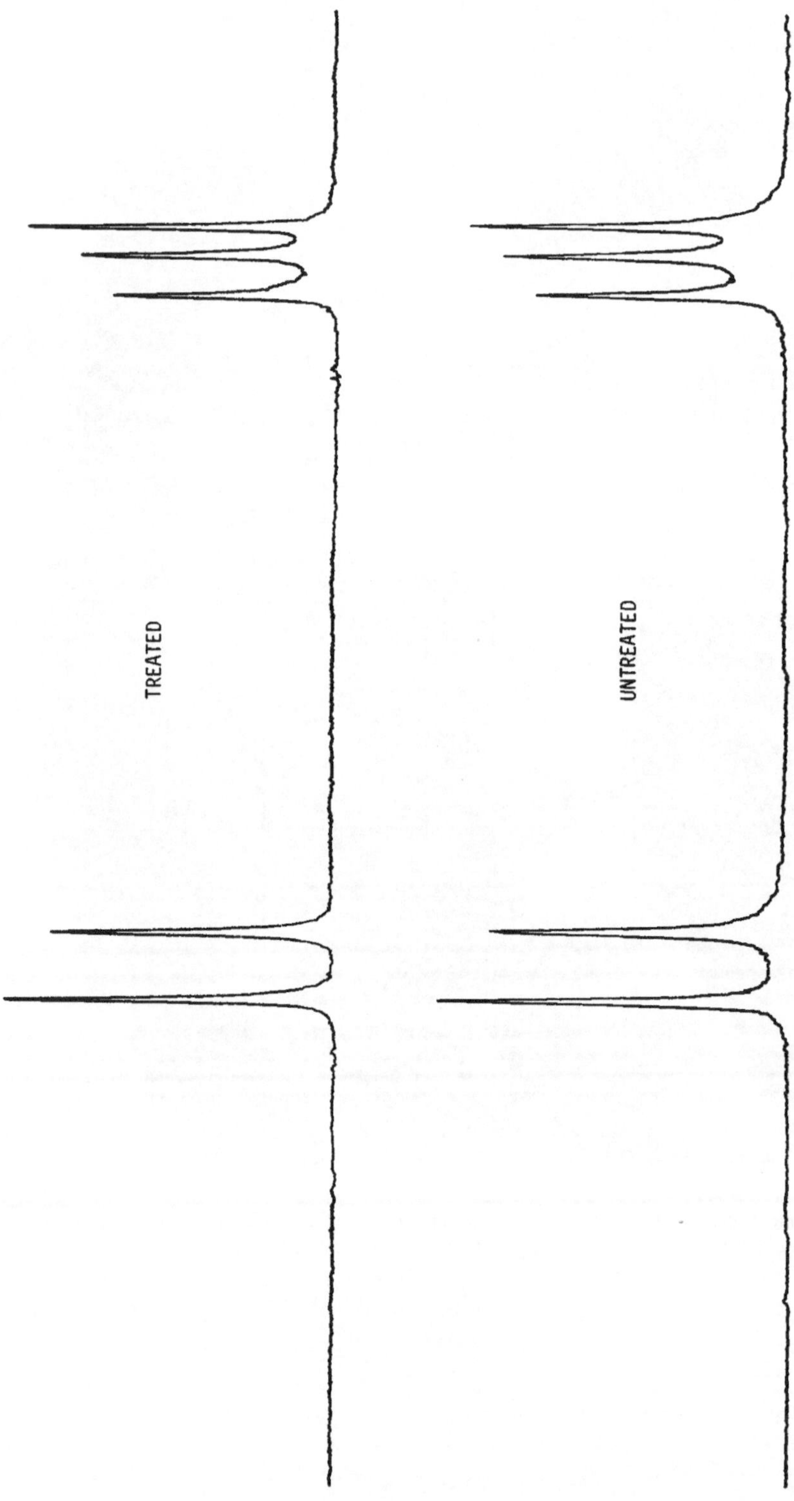

FIGURE 8. Comparison of C-13 NMR spectra of guayule rubber from treated and untreated plants.

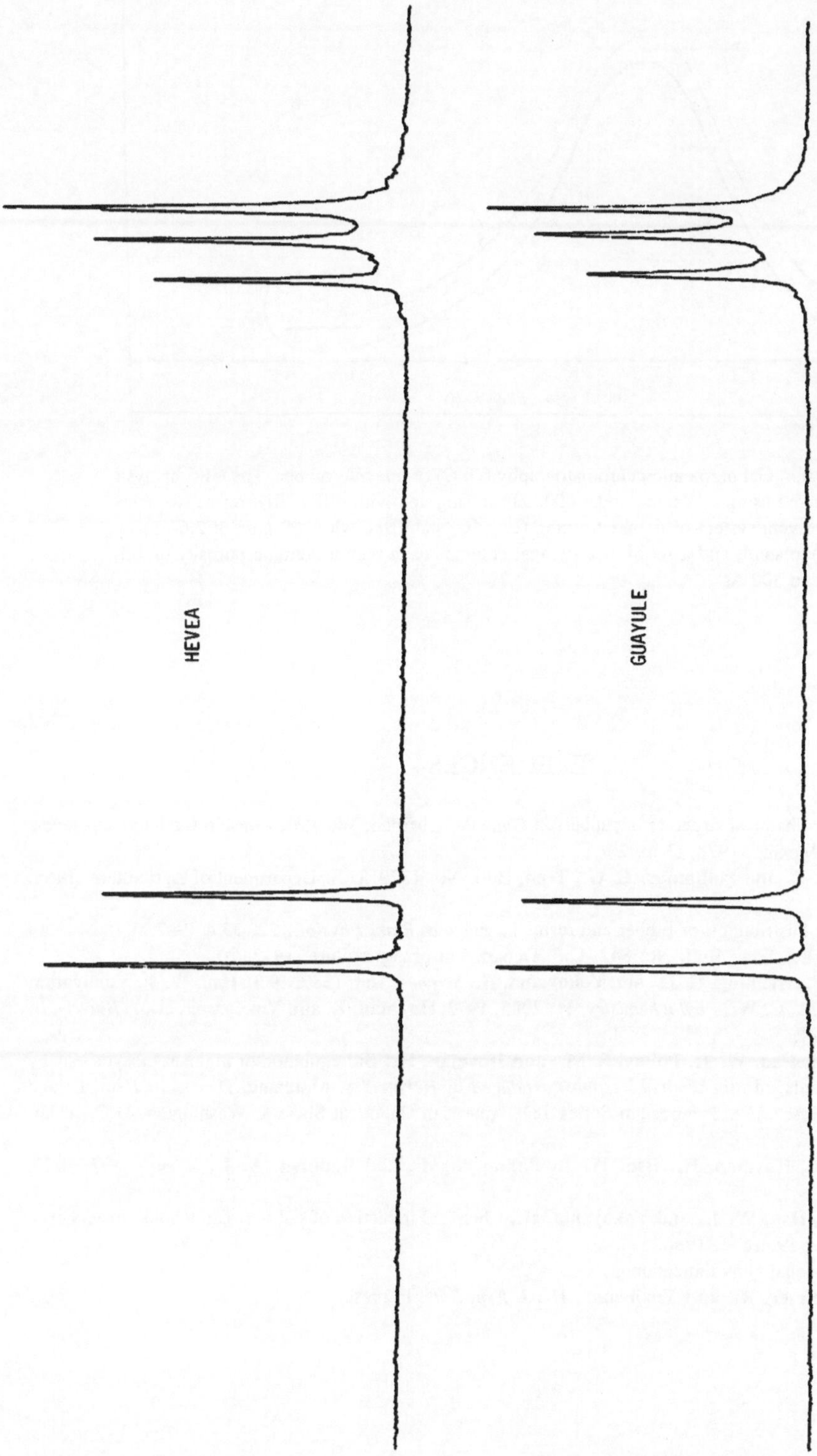

FIGURE 9. Comparison of C-13 NMR spectra of guayule rubber and *Hevea* rubber.

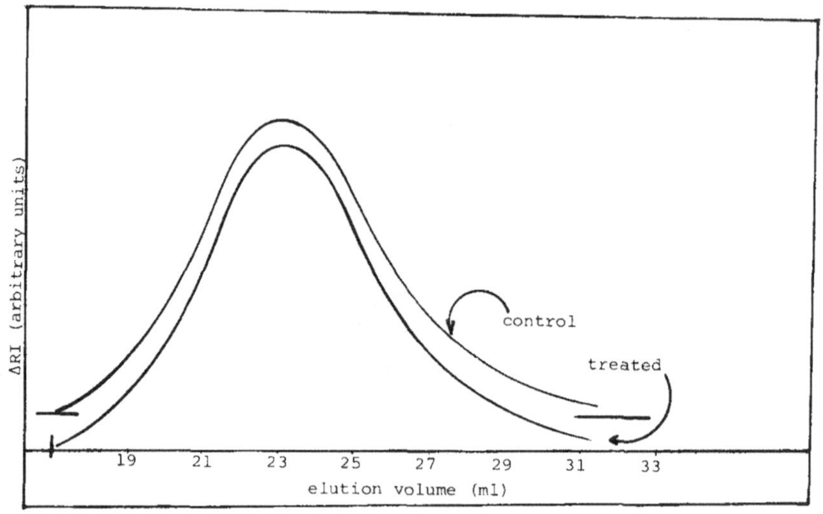

FIGURE 10. Gel permeation chromatography (GPC) of guayule rubber. The GPC analysis was conducted using a Waters model 6000 chromatograph with differential refractive index detector, solvent system of tetrahydrofuran (at 28°C) stabilized with 200 ppm of 2,6-di-*tert*-butyl-4-ethyphenol, and a set of 5-μ styragel columns each with a nominal porosity of 10^6, 10^5, 10^3, and 500 A.

REFERENCES

1. **McIntyre, D.,** Chemical structure of rubber, in *Guayule*, Guzman, W., Ed., Consejo Nacional de Ciencia y Tecnologia, Mexico, 1978, chap. 2.
2. **Hammond, B. L. and Polhamus, L. G.,** Tech. Bull. No. 1329, U.S. Department of Agriculture, 1965, 5.
3. **Curtis, O. F.,** Distribution of rubber and resins in guayule, *Plant Physiol.*, 22, 333, 1947.
4. **Artschwager, E.,** Tech. Bull. No. 842, U.S. Department of Agriculture, 1943, 20.
5. **Coggins, C. W., Henning, G. L., and Yokoyama, H.,** *Science*, 168, 1589, 1970; **Hsu, W. J., Yokoyama, H., and Coggins, C. W.,** *Phytochemistry*, 11, 2985, 1972; **Hayman, E. and Yokoyama, H.,** *J. Bacteriol.*, 120, 1339, 1974.
6. **Yokoyama, H., Hsu, W. J., Poling, S. M., and Hayman, H.,** Bioregulation of pigment biosynthesis by onium compounds, in *Biochemical Responses Induced by Herbicides*, Moreland, D. E., St. John, J., and Hess, F. D., Eds., ACS Symposium Series 181, American Chemical Society, Washington, D.C., chap. 9, 1982.
7. **Yokoyama, H., Hayman, E., Hsu, W. J., Poling, S. M., and Bauman, A. J.,** *Science*, 197, 1076, 1977.
8. **Poling, S. M., Hsu, W. J., and Yokoyama, H.,** Chemical induction of poly-*cis* carotenoid biosynthesis, *Phytochemistry*, 19, 1677, 1980.
9. **Mehta, I.,** personal communication.
10. **Bauer, T., Glaeser, R., and Yokoyama, H.,** *J. Exp. Bot.*, in press.

Chapter 5

TOBACCO

George L. Steffens

TABLE OF CONTENTS

I. INTRODUCTION

Tobacco producers over the years have found that removal of terminal buds (topping) along with removal or control of axillary buds (suckers) improve cured leaf yield and quality. Topping stimulates the growth and development of suckers so that many of the economic gains and physiological effects of topping are lost if suckers are allowed to grow and develop. In the past, suckers were removed by hand, but three decades have now elapsed since chemicals to control sucker growth were first introduced in the U.S. With the exception of several specialized tobacco types, almost all U.S.-grown tobacco is treated with one or more sucker control agents (Figure 1). Maleic hydrazide (MH), a systemic inhibitor, is the most widely used growth regulator on tobacco. The fatty alcohol (FA) type contact agents are usually used in the U.S. and several other countries on flue-cured tobacco along with MH, but the contacts are also used alone in some countries. Sucker control chemicals may modify cured leaf quality and remain as chemical residues on cured leaf so they may affect smoke properties as well. During the last several years chemical ripening or yellowing agents have also been used in some flue-cured tobacco producing areas of the U.S. and Canada. Their physiological effects are related to the well-known phytohormonal activity of ethylene.

Several recent reviews[1-3] have considered some of the specific aspects of the use of plant growth regulators for the production of tobacco and much of the information contained in them will be utilized for this present review.

II. GROWTH AND CONTROL OF AXILLARY BUDS (SUCKERS)

Growth in length of *Nicotiana tabacum* L. takes place at the apex of the main axis of the plant. Even though lateral buds are present in the axils of every leaf, the side shoots develop only slowly while the terminal bud is vegetative. When the terminal bud is allowed to develop flowers and seed, or if the terminal bud is removed, the upper lateral buds develop rapidly. The lateral buds — called suckers by tobacco producers — are vegetative initially, but they also become reproductive organs if allowed to remain on the plant.

Removal of terminal buds — called topping — causes an interruption in apical dominance which allows rapid development of lateral or axillary buds. Early work with the auxins[4] showed that this plant hormone was associated with apical dominance. Indole-3-acetic acid or other auxin type growth regulators can inhibit lateral bud development if they are applied to the cut stem in place of the terminal bud. It is thought that upon removal of the terminal bud, the supply of auxin and perhaps other hormones is interrupted which allows the formerly quiescent axillary buds to grow and develop. Buds in the upper three or four leaf axils on topped commercially grown tobacco cultivars usually develop most rapidly. If these upper suckers are allowed to grow they will reestablish partial apical dominance so that growth of axillary buds in lower leaf axils is suppressed. If the upper three or four suckers are then removed, the axillary buds lower down the stalk develop, as well as secondary suckers in the upper three or four leaf axils.[5] Seltmann and Kim[6] found that there are potentially three suckers in each leaf axil, but usually two at most develop under commercial tobacco production conditions. Hand suckering in many tobacco producing areas is therefore a recurring process.

Modern management systems for the production of most tobacco classes require that tobacco be topped. Topping is done by hand in most cases but machines for topping tobacco are used sometimes. One such machine is attached to a tractor and has rotating blades which decapitate the plants. Presently, no plant growth regulator has been developed for commercially topping tobacco, although chemicals have been evaluated for this purpose.[7,8] Because the reproductive bud or "sink" is removed by topping, leaf yield is increased but, in addition, cured leaf quality is usually improved.[9,10] Topping plus the control of sucker

FIGURE 1. (Left) An untreated tobacco plant which was topped showing the growth of suckers in the upper leaf axils. (Right) A similar tobacco plant treated with a sucker control agent which controlled growth of suckers in the leaf axils.

growth, either by hand or with chemicals, increases yield to a greater extent than topping alone. Time of topping and degree or effectiveness of sucker control also affect cured leaf chemical and physical properties.[11-13] Depending upon tobacco class, modifications may occur in cured leaf total nitrogen, soluble nitrogen, nicotine, reducing sugars, total volatile bases minus nicotine, alpha amino nitrogen, water-soluble acids, total ash, potassium, filling value, equilibrium moisture content, and other leaf quality factors as well as smoke characteristics.[14-18] The removal of metabolic sinks usually results in an increased growth of the root system which is associated with increases in cured leaf nicotine and leaf weight per unit area. A larger root system is associated with reduced lodging. The elimination of sucker growth also reduces populations of both tobacco horn- and budworms due to decreased food supplies.

The number and size of suckers which grow and develop on commercially produced plants depend upon management and cultural practices such as tobacco class and cultivar grown, amount of fertilizer applied, rainfall or irrigation received, plant spacing, and other factors which normally affect plant growth and development. Early topping before florets begin to open will maintain high yields if sucker growth is also controlled. Marshall and Seltmann[9] showed that leaf yield of flue-cured tobacco may be reduced about 1%/day when topping is delayed until after the button stage or florets begin to open. However, early topping allows for increased sucker growth over an extended period of time until leaf maturity. To maintain their yield potential, plants topped early will therefore require several hand suckerings or effective chemicals to control sucker growth. Topping height will also influence sucker growth. Low topping enhances the growth of suckers and usually they will be more numerous in the lower portions of the plant.

To maintain cured leaf yield and quality, it is necessary to control sucker growth on topped plants. If suckers are to be removed manually, and if topping is done at the early- to mid-flower stage of development, some suckers most likely will have already developed and will need to be removed during the topping operation. About 7 to 10 days after topping, suckers about 4- to 8-in. long will have to be removed from the upper leaf axils. Later hand suckering will require that suckers be removed from the middle and lower leaf axils. Depending upon tobacco class and growing season, as many as 5 to 15 suckers per plant may have to be removed by the manual method to maintain a high degree of control. Because of the difficulty of removing these suckers, many times leaves are lost through breakage or bruising.

Chemical methods of sucker control have therefore been enthusiastically received by tobacco producers around the world. Effective chemical control usually increases yield and is much more efficient than hand suckering. Hand suckering is very disagreeable, costly, and must be performed at a time when labor demands for other activities are at their peak. Mechanized harvesting of flue-cured tobacco requires that control of sucker growth be nearly 100% effective. Efficient and complete sucker control cannot be achieved manually, so chemical control is considered to be an essential prerequisite for mechanized harvesting. The historical development and use of sucker control chemicals in the U.S. through 1970 have been reviewed by Seltmann.[10] Since that time with increased mechanical harvesting, especially of flue-cured tobacco, the use pattern of sucker control chemicals in the U.S. has changed. While the systemic growth inhibitor MH is still widely used, there has been a marked increase in the amount of the C_8 and C_{10} FA contact agents used. Almost all U.S.-grown flue-cured tobacco is treated with one or two applications of a "contact" before MH is applied. The use of both MH and FA has also markedly increased in other tobacco producing countries of the world. Several other promising chemicals and formulations are in various stages of development,[19] but their commercialization is as much dependent upon their toxicological characteristics as upon their efficacy for controlling sucker growth.

A. Sucker Control with Maleic Hydrazide (MH)

The structure of MH (1,2-dihydro-3,6-pyridazinedione) is shown in Figure 2. In the early 1950s Petersen[20] recognized the effectiveness of MH as a systemic inhibitor of sucker growth on tobacco. Almost all U.S.-grown tobacco is now treated with MH and its use is increasing in many other tobacco producing countries of the world. However, in some countries MH is not used because it has been associated with undesirable effects on leaf quality.[21,22] Until a few years ago, MH was available only as the diethanolamine (DEA) salt. The potassium (K) salt became available and its use by U.S. tobacco producers is increasing. The U.S. Environmental Protection Agency in September 1981 suspended further production of the DEA salt because of toxicological-related implications, but existing supplies of the DEA salt may be used until exhausted. The KMH formulation, which is considered to be somewhat more effective than the DEA salt on an equivalent active ingredient (a.i.) basis, will be available to U.S. tobacco producers.

MH can be applied to plants with a wide variety of spray equipment. In the U.S. it is applied using hand-held compressed air sprayers, tractor drawn or tractor mounted equipment, and in some cases, by airplane, depending upon tobacco class and acreage grown. With tractor mounted or drawn equipment, for example, 6 pints of the 3 lb a.i. per gallon DEA-MH formulation or 12 pints of the 1.5 lb. a.i. per gallon KMH formulation (170 mg MH as the acid equivalent/plant) is diluted with 20 to 50 gal of water for treating flue-cured tobacco.[3] This spray solution is usually applied by use of a conventional three-nozzle boom arrangement above the plants. The nozzles move perpendicular to the row which allows for

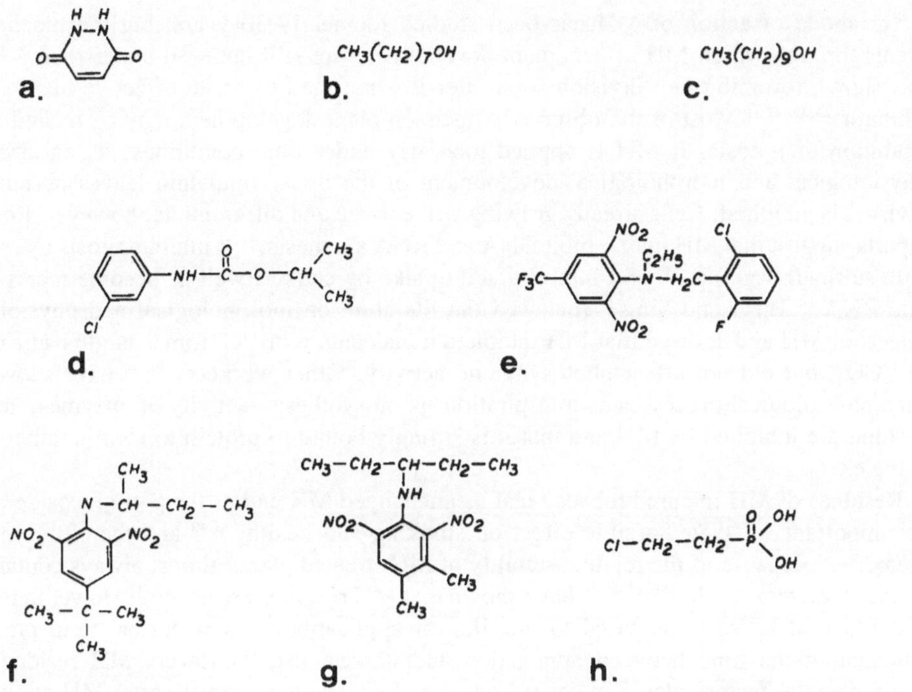

FIGURE 2. Plant growth regulators used in tobacco production. (a) Maleic hydrazide — 1,2-dihydro-3,6-pyridazinedione; (b) 1-octanol; (c) 1-decanol; (d) chlorprophan — isopropyl *m*-chlorocarbanilate; (e) butralin — 4(1,1-dimethylethyl)-N-(1-methylpropyl)-2,6-dinitrobenzenamine; (f) pendimethalin — N-(1-ethylpropyl)-3-4-dimethyl-2,6-dinitrobenzenamine; (g) CGA 41065 — *N*-ethyl-*N*-(2-chloro-6-fluorobenzyl)-2′,6′-dinitro-4′-trifluoromethylaniline; (h) ethephon — (2-chloroethyl)phosphonic acid.

thorough wetting of all upper exposed leaf surfaces. For effective sucker control, the diluted MH solution should be applied as a fine spray so that it thoroughly wets the upper one third of the tobacco plant. MH must be taken up by the leaves to be effective, but absorption efficiency is related to environmental conditions. Uptake is usually increased when plants are rapidly growing and under high humidity conditions.[23] MH applications made in the morning and at midday are more effective than those made in the evening.[24] Applications of less than the recommended amount of MH under some conditions may provide effective sucker control, but because of the wide variety of environmental conditions encountered in the tobacco producing areas of the U.S., the full recommended amount of MH is usually required for effective sucker control. In addition to controlling sucker growth, MH-treated plants wilt less readily than untreated plants under dry weather conditions and they are less susceptible to brown spot disease.[10]

After MH enters the plant, it is readily translocated in the xylem and phloem as shown by the early work of Crafts and others.[23,25] Labeling studies by Frear and Swanson[26] show that MH rapidly moves without conversion within the plant and is translocated from source to sink. This result suggests that the absence of an active apical meristem (terminal bud) causes a larger portion of MH to be translocated to the roots. The major methanol metabolite was the β-D-glucoside of MH and the methanol-soluble residue was mostly unchanged MH. Their work also suggests that residues in green and cured leaf are similar and little change takes place during curing. Cheng and Steffens[27] showed that there was approximately an 80% reduction in level of MH residues from time of treatment until after air-curing Maryland tobacco, but MH losses occurred only during the time plants remained in the field, with no losses occurring during the curing period.

The mode of action of MH has been studied for nearly 30 years, but the biochemical events through which MH affects plant development are still not well understood.[28-30] MH was shown to inhibit cell division soon after it was found to be an effective plant growth inhibitor.[23,25,30,31] Most of the observed effects on plant development may be related to this inhibition of mitosis. If MH is applied too early under field conditions, it can affect the physiological and morphological development of the upper immature leaves because cell division is inhibited. Cells already growing will enlarge and differentiate, however. Research reports suggest that MH may inhibit DNA and RNA synthesis,[31,32] inhibit mitosis by reaction with sulfhydryl groups,[33] and inhibit uracil uptake by cells as well as become incorporated into RNA.[29] Bush and Sims[28] reviewed the literature on morphological and physiological effects of MH and showed that MH inhibited translocation of ^{14}C from a single leaf exposed to $^{14}CO_2$, but did not affect photosynthetic activity. Other workers[23,25,34] have shown that such physiological processes as transpiration, photosynthesis, activity of enzymes, and respiration are inhibited by MH and that it is strongly bound to protein to form a rather stable complex.

Residues of MH in cured tobacco leaf as unchanged MH and as the β-D-glucoside of MH are important for their possible effect on smoking and health[3,35-40] and therefore on commerce.[41-44] Because of the relative stability of MH, treated plants almost always contain MH residues. Several studies[26,27,45,46] have shown that MH residues are generally lower in tobacco leaf tissue at harvest compared to the time of application. Under normal field practices, lengthening the time between application and harvest usually lowers MH residues, but environmental factors play a major role. Rainfall or irrigation shortly after MH application can lower residue levels as well as effectiveness of sucker control.[23,47,110] Wide variation in MH residue levels occur due to position of leaves on the stalk and environmental conditions. Residue levels many times are highest in lower leaves of flue-cured tobacco, probably because they are harvested nearer the time of MH application than leaves farther up the stalk.

The application of MH at recommended rates to flue-cured tobacco (170 mg a.i. per plant) resulted in average residues ranging from 31 to 156 ppm.[46] Residues from different stalk positions ranged from a low of 36 ppm to a high of 368 ppm over three locations and five harvests in 1974. Sheets et al.[48] found that average MH residue levels of flue-cured market samples tended to increase and then leveled off as follows: 1972, 87 ppm; 1974, 101 ppm; 1976, 129 ppm; 1978, 117 ppm; and 1980, 127 ppm. Storage of flue-cured tobacco also had little effect on MH residues.[49]

In the 1975 crop for Kentucky-grown burley tobacco, MH residue levels averaged 71 ppm,[50] but the actual average was 48 ppm when the samples which contained no MH were included (about 50%). Residue levels ranged from 0 to 270 ppm. Such a wide range is the result of environmental factors, leaf stalk positions, amounts of MH applied, and harvest date.[45] In cured leaf of Maryland tobacco, 50 ppm of MH residue was found from the field application of 170 mg a.i. MH per plant. Average MH residue levels of some of the minor tobacco classes grown in the U.S. may be less than those found in flue-cured or burley tobacco because a lower percentage of the tobacco may be treated. Sheets et al.[48] evaluated a number of U.S. tobacco products obtained in 1977 and found MH residues of 47 ppm in cigarettes, 8 ppm in cigars, 10 ppm in little cigars, 23 ppm in smoking tobacco, 16 ppm in chewing tobacco, and 22 ppm in snuff.

To affect the user of tobacco products, MH per se, a metabolite of MH, or a pyrolytic product would first have to be transferred to the user or be contained in the mainstream smoke. Almost all MH residues in tobacco leaf can be accounted for as intact MH or the glucoside,[26] and reports show transfer percentages of MH to mainstream smoke ranging from 0.3 to 10%.[3,23,51,52,111] Further information about possible MH-related smoke constituents may be found in the review by Steffens[1] and a recent paper by Iwasaki et al.[53]

A number of practical ways to reduce MH residue have been suggested.[54] Among these are to use only the dosage recommended by the label instructions; apply MH at the most appropriate time to achieve effective sucker control in order to avoid the need for additional application; and allow as much time between MH applications and harvest as possible. Because of problems associated with relatively high residue levels in cured tobacco leaf, several changes have been made recently in the label instructions for use of MH to control suckers on U.S.-grown tobacco. The label now states that only one application of MH is to be made per season at the stated application rate. Retreatment is permissible only if rain washes off the K-MH salt formulation within 6 hr of application or the DEA-MH salt formulation within 12 hr. An interval of 7 days between MH application and harvest is also now specified. Because of the MH label changes and because of the importance in keeping MH residue levels at a minimum, the Tobacco Extension Specialists of the North Carolina Agricultural Extension Service have developed a six-step flue-cured tobacco fertilization, chemical application, and topping program[54] for use by growers. The program is designed to allow the growers to control sucker growth effectively by using only recommended amounts of MH.

Research has been underway for several years in an effort to reduce levels of MH residues on cured leaf. Seltmann[1,55] has reported on research involving directed MH sprays which provided encouraging results for reducing MH residues on cured leaf. Steffens[56] investigated the influence of such growth regulators as gibberellic acid on the effectiveness of MH for the control of tobacco sucker growth. If pretreatment with a PGR or a PGR mixed with MH increases the effectiveness of MH, it may be possible to reduce amounts of MH necessary for effective sucker control. GA mixed with MH or GA applied prior to MH resulted in some increase in effectiveness over MH alone, but the increase was not great enough to be considered practical. Additional evaluations combining MH with other PGR[1] have also been conducted to determine if such mixtures would show additive or synergistic effects for inhibiting tobacco sucker growth. Preliminary results show some increases in activity over MH alone and may provide a starting point for the development of effective "combination" sucker control agents.

Shortly after MH began to be used by U.S. tobacco producers in the late 1950s, a number of tobacco manufacturers reported that undesirable changes in the chemical and physical properties of the cured leaf resulted.[57,58] Jeffery and Cox[59] reported in 1962 that a number of changes occurred from use of MH. Opinions were mixed, however, as to whether or not the changes adversely affected the suitability of MH-treated tobacco for cigarette manufacture. Changes were found in reducing sugars, petroleum ether extractables, alkaloids, and ash. The filling value of cured leaf treated with MH was reduced which meant that fewer cigarettes could be manufactured per unit leaf weight. Questions that remained included which and to what degree were cured leaf chemical and physical property changes caused by physiological activity of MH per se and which were caused by the almost complete inhibition of sucker growth? Simply reducing the number of active meristematic regions (suckers) has a profound effect on the physiology of the whole plant. Although the use of MH caused some changes in all tobacco classes examined, flue-cured tobacco showed the greatest number of changes. Chaplin[60] reported in 1967 that cured leaf from flue-cured plants which were closely hand suckered (suckers removed every 2 or 3 days) tended to be similar to that from MH-treated plants in some respects but not in others. Filling value was reduced and moisture equilibrium and sugar content increased in both MH-treated and close-suckered tobacco. However, close suckering resulted in a smaller reducing sugar increase, and total alkaloids were considerably lower in MH-treated leaves. Chaplin concluded that lower alkaloids were more closely associated with MH use than with the degree of sucker control. Peedin[61] also found that MH-treated leaves had a lower nicotine content than closely suckered tobacco, but nicotine content increased as the degree of control via hand suckering increased. The differences in alkaloid content may be associated with changes in the growth of roots since nicotine is synthesized in the roots and translocated to shoots and leaves.[62] Seltmann's results[63] strongly suggest that amount of sucker growth on flue-cured plants influenced cured leaf characteristics more than sucker control chemicals per se. His experiments include MH as well as several contact type chemicals and show that good sucker control (compared to poor control) resulted in higher yield, total alkaloids, equilibrium moisture content, and lower ash and filling value. A number of other reports indicate differences in chemical composition of cured leaf treated with MH as compared to cured leaf from hand-suckered plants.[14,15,23,64,65] Even though some of the changes in cured leaf characteristics may be ascribed to MH per se, it is difficult to separate effects of the chemical from effects due to inhibited sucker growth.

Early reports[59,66] also gave indications that members of expert smoking panels preferred cigarettes made from hand-suckered tobacco over those made from MH-treated tobacco. The results were not very clear-cut because panel members did not always discriminate in favor of cigarettes made of hand-suckered tobacco and against those made of MH-treated tobacco.

B. Sucker Control with Fatty Alcohols (FA)

Fatty acid derivatives act as contact sucker growth inhibitors when emulsified in water with surfactants.[67,68] Of the fatty acid derivatives evaluated as sucker control agents, the C_8 and C_{10} FA (Figure 2) are among the most effective and are now used as contact sucker control agents in the U.S. as well as in other tobacco producing countries of the world.

FA (C_8 + C_{10} or C_{10} alone) emulsions are applied as relatively coarse sprays and must drain down the stalk to *contact* the immature suckers to kill them. The emulsions are phytotoxic to young meristematic tissue, but cause little or no visible injury to more mature tissue. FAs with the aid of surfactants appear to penetrate the surface cuticular layer of the immature leaves and meristematic region of the axillary buds. Within about an hour the treated axillary buds turn brown and are killed by cell disruption and desiccation. The type and amount of surfactant are important for controlling the selectivity of the FA emulsions.[69] If the surfactant is not suitable or if the concentration of the FA is too great, the spray solution may injure tissue adjacent to the leaf axils as well as the leaves. It is also important that the spray be applied to the plants at a relatively low pressure to avoid leaf injury. When injury to leaf tissue does occur, the injured areas desiccate and fall out.

Either 2 gal of the 85% active C_8 + C_{10} formulation or 2 gal of the approximately 78% active C_{10} formulation is usually mixed with 48 gal of water for application with tractor mounted or drawn spray equipment to each acre of flue-cured tobacco.[3,54] It is suggested that a three-nozzle spray boom arrangement be used over each row and that the equipment be operated slowly enough to apply at least 50 gal of spray per acre. Plants may be treated before topping, but care in application must be exercised in order to avoid injury to young, tender leaves. Concentrations greater than the recommended amount may cause excessive injury, and less concentrated spray solutions may result in poor control. Low concentration sprays do not effectively control secondary suckers which are usually more difficult to control or they may only suppress sucker growth so that further control measures are less effective. It is also necessary to apply a sufficient amount of spray solution to each plant so it runs down the entire length of the stalk to contact each immature sucker. At the proper concentration, the FA spray solution will destroy the visible first sucker in each leaf axil and in some cases the second sucker also. To insure that all first suckers are killed, especially in fields where plant growth is irregular, a second application is usually made 3 to 5 days after the first. The two applications also usually control some second suckers. It is difficult to contact each sucker with the spray solution, especially on plants which are not upright. A delay in applying the second application of a contact will allow these missed suckers or "escapes" to grow too large to be controlled with a contact. Because secondary suckers develop before the flue-cured tobacco crop is harvested, the contact sprays are followed by a final application of MH[54,70] (see Section II.C). Because MH is more effective and easier to apply than contact-type sucker control agents,[14,15] FAs are usually used in the U.S. only on flue-cured and dark tobacco classes where the sequential method (contact followed by systemic) significantly improves effectiveness of control. In countries where MH is not used, the FA contact-type agents are used on those tobacco classes produced.

A major deterrent to the widespread acceptability of the contact type agents is that their effectiveness is not consistent. Any suckers not contacted by the spray emulsion will grow rapidly. Even on upright plants, many times one or two suckers in the upper part of the plant are not controlled. Seltmann[1,71] therefore studied over a 3-year period the effectiveness of FA application number with and without hand suckering for controlling sucker growth on flue-cured tobacco. The data show that one, two, and three applications of FA did not provide a high enough degree of control for efficient mechanical leaf harvesting. Suckers present on plants treated with FA were in the middle and upper portions of the plant which, because of their position, tend to interfere with mechanical harvesting. Hand suckering was required for plants treated only with FA in order to remove the interfering suckers.

Studies of residue levels of FA showed that they could not be detected 26 days after treatment. Tancogne[72] showed that C_8 and C_{10} alcohols rapidly decreased even in the absence of rainfall and high temperatures. Tso and Chu[73] report that residues of ^{14}C-labeled FA were about 1 ppm compared to a 7000-ppm natural fatty acid fraction. About 10 to 25% of the alcohol was converted to the acid fraction and from 7 to 15% was recovered in the ester fraction by the time the Maryland tobacco was harvested. Surfactant levels needed to emulsify FA were relatively high. Residue levels of ^{14}C-labeled polyoxyethylene [20] sorbitan monooleate[73,74] ranged from 0.4 to 1.4 ppm for the three tobacco classes studied, and nearly all of the residue material was hydrolyzed to free polyol and fatty acids.

In general, the contact-type sucker control agents used alone have less effect on leaf chemical and physical properties than does MH.[14,15] This may be due partly to the less effective control obtained with the contacts, even though MH may cause additional physiological effects. When comparisons were made with hand-suckered flue-cured plants, contact agents significantly lowered cured leaf total soluble nitrogen, nicotine, water-soluble acids, and total ash. For burley tobacco, use of a contact control agent resulted in lowered cured leaf nicotine, and increased alpha amino nitrogen, total volatile bases minus nicotine, and alkalinity number of water-soluble ash.

C. Sucker Control with Sequential and Combined Applications of Systemic and Contact Chemicals

Problems of controlling sucker growth and development vary depending upon the tobacco class being produced. In the U.S. producers of burley, Maryland, and some of the cigar types find that one application of MH at the recommended rate will usually control sucker growth effectively until the crop is harvested. One application of FA without hand suckering may not be acceptable because secondary and "escape" suckers may grow rather large before the plants are ready to harvest. It is necessary to remove such suckers because these air-cured classes of tobacco are harvested by cutting the entire plant near the ground and curing the whole plant with leaves attached to the stalk.[3]

With flue-cured tobacco, it is essential to have sustained control of sucker growth during the entire harvesting period, especially when harvesting is done by machine. Flue-cured tobacco ripens at the rate of approximately two to four leaves per plant per week. Because leaves are harvested about every 7 to 10 days the harvesting period extends for 6 to 8 weeks and, at times, longer.[3,75] Under U.S. production conditions neither MH nor FA alone will provide the required degree and length of control. Therefore, a program of sequential applications of contact followed by systemic chemicals on flue-cured tobacco has been developed over the past 10 to 12 years. This method also has found acceptance for use on dark tobacco.

The sequencing of the application of contact and systemic chemicals is based upon their mode of action for controlling sucker growth. The FA contact type chemical can be applied to plants relatively early, in fact before topping when only about 50% of the plants are in the button stage. Under good conditions and with the FA at the proper concentration, the sprays will contact and kill some succulent terminal buds as well as the small suckers without excessive damage to normally harvested leaves. However, some leaves that are removed during the topping operation are usually damaged by sprays at this time. On the other hand, MH should be applied only to plants at the full flower stage of maturity. If applied earlier, MH may inhibit full development and expansion of the harvestable youngest leaves because it affects cell division. However, if chemical sucker control (using FA) is not started before the full flower stage, hand suckering will be required to remove large suckers from the more mature plants. The sequential method can provide nearly complete sucker control over the entire harvesting season so no pretopping suckers or suckers which normally develop later need be removed by the producer. Collins et al.[54] outline a six-step fertilization, chemical application, and topping program designed for use with flue-cured tobacco to provide effective sucker control during the harvest season but keep MH residue levels on cured leaf at a minimum.

The six steps of their program are as follows:

1. Apply from 60 to 80 lb/acre of nitrogen plus adjustments for leaching.
2. Apply a FA contact sucker control solution (either the 85% a.i. or the approximately 78% a.i.) at a 4% concentration before topping when about 50% of the plants reaches the button stage.
3. Top plants that are ready for topping immediately after application of the contact solution.
4. Apply a second application of the FA contact (either the 85% a.i. or the approximately 78% a.i.) at a 4 to 5% concentration 3 to 5 days after the first application particularly in fields that are growing and flowering irregularly.
5. Top remaining plants that were not topped during the first topping operation.
6. Apply a product containing only MH or one which contains MH mixed with a contact (see below) about 7 days after the last contact application, preferably in the morning about 2 days after a rain or irrigation.

Programs such as the one outlined are developed from data and information obtained from numerous on-farm experiments and tests conducted under the supervision of Cooperative Extension Service personnel throughout the tobacco producing areas of North Carolina.

A product which contains a mixture of MH and FA also has been developed[76] and is being used on flue-cured tobacco in step six as outlined above. When applied at recommended rates this product supplies 11% less MH compared to products which supply MH only. It might be expected that MH residues would therefore be somewhat lower, but control of sucker growth may also be less under some conditions. The application of a solution of this chemical at greater than recommended rates may cause leaf damage because it contains the C_{10} FA. The excessive application of MH is therefore also discouraged.

The sequential method provides an approach to the effective control of sucker growth on flue-cured tobacco so that most U.S. flue-cured producers use this method. Such a treatment sequence did not change MH residue levels on either flue-cured[46] or Maryland tobacco.[27] MH is applied to flue-cured tobacco in the sequential system usually no nearer the time of harvest than it would be if MH were used alone.

The use of the sequential systems provides cured leaf with chemical and physical properties very similar to those of cured leaf from plants on which suckers have been effectively controlled by MH alone.

D. Other Sucker Control Chemicals

The *ad hoc* Regional Tobacco Growth Regulator Committee (RTGRC — see Reference 3, page 2 and Reference 10, page 82) has conducted and coordinated the evaluation of potential new growth regulating chemicals for use on U.S. tobacco since the early 1960s. Among the potential new sucker control chemicals is chlorpropham, a carbamate (Figure 2). Since this chemical may cause leaf distortion and other effects if applied to immature tobacco leaves, it most likely will find a place as part of a sequential system, especially on flue-cured tobacco.[19] In fact it may be most beneficial for flue-cured tobacco when it is applied after one or two applications of a contact and the MH application. This approach may help eliminate the likelihood of overtreatment with MH which in turn may help reduce excessive MH residues. Chloropropham for control of suckers has been applied in the same manner as a contact agent. However studies are continuing into the extent of its translocation and into its effectiveness as a systemic type chemical. No published residue data from the application of this material to tobacco are available but it now is registered for use on tobacco in the U.S.

The *N*-benzylnitroaniline coded CGA-41065 (Figure 2) is a promising new chemical for the control of suckers.[77-79] The RTGRC has also evaluated this contact-systemic chemical over a number of years and tobacco classes. It has been effective in controlling axillary bud growth over a wide range of environmental conditions. Contact-systemic type sucker control chemicals must wet suckers to be effective because they are not readily translocated. Control of sucker growth is physiological or metabolic since buds are not usually killed. CGA-41065 may also affect the development of young expanding leaves. It is therefore being evaluated for application at the mid- to full-flower stage. In a sequential program it would be applied after one or two applications of a contact and would replace the MH application. Combinations of CGA-41065 and MH have also been investigated by Steffens[80] for effectiveness of control of tobacco sucker growth. Results of greenhouse and field studies showed that combining the two inhibitors provided a broader range of sucker control than either one alone. Published residue data from the application of CGA-41065 to tobacco are not available. Leaves from plants treated with it have chemical and physical properties similar to those from the standard chemical sucker control treatments. CGA-41065 is not yet registered for use on tobacco in the U.S.

Two dinitroanilines, butralin and pendimethalin (Figure 2), have also been evaluated by the RTGRC for use as sucker control agents. Seltmann[81] showed that a dinitroaniline type chemical used as a tobacco sucker control agent can inhibit the development of a wheat cover crop after it is used at a dosage high enough to obtain an effective degree of sucker control. Upper leaves on tobacco plants treated with dinitroaniline sucker control agents tend to be greener at time of harvest compared to the controls. One report[82] showed that application of butralin as a local spray to upper leaves resulted in about 1 ppm of residue after air curing. When applied at 5,000 to 10,000 g/ha (2.2 to 4.4 lb/A) as an overall spray, residues could reach 30 ppm, however. Cured leaf from treated plants evaluated in the U.S. by the RTGRC generally compared favorably with leaf from plants treated with standard sucker control chemicals with respect to chemical and physical properties. Nevertheless, a tendency for increased levels of total alkaloids in dinitroaniline-treated leaf was observed by Seltmann.[81]

The RTGRC is evaluating other new potential sucker control agents both in advanced and preliminary tests.[83] One promising chemical that has recently been registered for use on tobacco is isodecyl alcohol, a contact type.

III. TOBACCO LEAF YELLOWING AND CURING AGENTS

Ethylene, a natural plant growth regulator, is endogenously produced by plants. It has long been known to promote fruit ripening and senescence of plant tissue.[84-86] Exogenously applied ethylene has been shown to cause leaves to turn yellow[87] as well as to abscise,[88] and older leaves were found to be more susceptible to ethylene than younger ones. The role of ethylene has been included in a recent treatise on senescence in plants.[89]

The use of ethylene gas to accelerate the yellowing of detached tobacco leaves under various curing conditions was reported in the 1930s. Rossi,[90] Pfutzer and Losch,[91] Kraynev,[92] and Asmev[93] have all reported that fumigation of tobacco leaves with ethylene gas accelerated yellowing. More recently it has been found that ethylene releasing agents applied to maturing tobacco plants in the field and ethylene generators operating in flue-cured barns during the curing process accelerate yellowing of leaves or serve as a tobacco leaf curing aid.

A. Ethephon

The chemical structure of the water-soluble ethylene releasing agent ethephon is shown in Figure 2. Free ethylene ($CH_2 = CH_2$) is released by this molecule above pH 4, and with increasing pH, there is an increasing evolution of ethylene per unit of time. The application of ethephon was first reported to "yellow" tobacco leaves on flue-cured plants in 2 to 4 days after treatment[94,95] only if the leaves had reached a certain stage of physiological maturity. Other tobacco classes were found to respond to a lesser degree. Also, growing region and seasonal effects cause differing responses.[96] Because flue-cured tobacco is most responsive to ethephon and because its use on flue-cured tobacco was considered to have the most economic potential, it has been evaluated more extensively on that class.

Initially ethephon was applied to flue-cured tobacco which was maturing so that it yellowed in the field in about 3 days. More recently, it has been found that lower dosages of ethephon can be applied to mature leaves which are then harvested the next day. This allows the yellowing response to occur in the curing barn rather than in the field. Depending on application time and dosage, and time between treatment and harvest, ethephon is used as a management tool by the producer to promote leaf yellowing and uniform ripening. It reduces curing time and the amount of green colors in leaves after curing.

Producers are encouraged to use test sprays on their tobacco to determine whether leaves are mature enough to yellow within 2 to 4 days after ethephon treatment.[54] If treated leaves are not mature they will not respond or will yellow only partially. It is also suggested that 50 gal of spray per acre be applied as a fine mist over the top of the plant. Drop nozzles can be used for sprays directed to bottom or middle leaves. It has been found that the best responses are obtained when ethephon is applied on warm, bright, sunny days. Domir and Foy[97] showed that ^{14}C-labeled ethephon penetrated mature leaf tissue and was translocated away from the application site. Ethephon was rapidly degraded by the leaf (as much as 92% after 1 day) and no detectable metabolites of ethephon were found.[98] Mature flue-cured tobacco leaves treated with ethephon contained greater amounts of reducing sugars and lower levels of starch relative to untreated leaves at time of harvest.[95] After curing, comparisons between treated and untreated leaf showed that only small differences existed in total nitrogen, total alkaloids, starch, and reducing sugars, although protein was lower in treated leaf. In another field study[99] ethephon caused some changes in chemical properties, but they were modified by amount of nitrogen fertilizer applied. Other workers[94,100] have noted changes in flue-cured leaf chemical composition which were related to leaf maturity at time of treatment, length of time between application and harvest, and curing conditions. Yield losses have been associated with ethephon-treated flue-cured tobacco and these losses have been attributed to accelerated respiration.[101] Chlorophyll is rapidly reduced as chlorophyllase activity is increased.

Burley tobacco[102] and other air-cured classes also respond to ethephon treatment, although results are not always consistent. When applied to mature leaves it can hasten ripening which allows for early harvest. It can also hasten wilting after harvest and reduce curing time which reduces the possibility of losses from sunburn, barn rot, and other curing problems. Again, sprays should thoroughly cover the leaves for best results, and harvesting should follow application within 16 to 24 hr. Yield losses have also been associated with treatment of burley tobacco with ethephon, but no yield losses occurred when tobacco was harvested the day after application.[103]

Ethephon has been applied to tobacco seedlings for the purpose of temporarily stopping growth in order to extend the transplanting season.[104] Such treatment also may inhibit early flowering. Ethephon should be applied shortly before transplanting when plants are about 5-in. tall.

B. Ethylene Generators

Ethylene gas produced by catalytic generators is being used to speed up yellowing of flue-cured tobacco leaves in curing barns. However, no reports of systematic studies on the effects of such treatments are available. Ethylene is liberated into curing barns during the first 24 hr of curing for the purpose of reducing the curing time by 10 to 15%.[102] The treatment, however, does not speed up yellowing of immature leaves. For best results the temperature during the 24-hr treatment time is to be maintained between 70 and 90°F. The shortened coloring time of treated leaves is reported to show up when the temperature in the curing barn reaches 105°F. In addition to coloring faster, the leaves also color more uniformly.

IV. PLANT GROWTH REGULATORS IN WORLD TOBACCO PRODUCTION

All reports indicate that the use of plant growth regulators for the production of tobacco has increased during the last 5 years and will likely continue to increase at least for the next several years. Outside of the U.S. the increase has been mostly in the use of sucker control agents, both systemic and contact types. In the U.S. there has been an increase in usage of the contact sucker control agents on flue-cured tobacco (multiple applications), and an increase in usage of the ethylene releasing agents and ethylene generators as curing aids for flue-cured tobacco.

In 1976 it was estimated that 3.8 million lb of MH was produced in the U.S.,[105] and about 80% — more than 3 million lb — was used for the control of sucker growth on U.S.-grown tobacco. No more recent figures are available[106] but efforts are being made to restrict the use of MH on U.S. produced tobacco to label-recommended rates in order to eliminate problems of excessive residues on cured leaf.

The use of sucker control agents in many of the lesser developed countries is increasing because production efficiency can be increased considerably with their use. Even though more labor is available for hand suckering in these countries than in more developed countries, such labor is becoming relatively more expensive. As production efficiency increases, the product becomes more competitive in the world market. As tobacco producers in developing countries become more experienced and sophisticated, they will also use more sucker control and yellowing agents as well as other agricultural chemicals.

In those tobacco producing countries where the use of MH is not restricted, its use has increased, especially in the past 3 years or so and further usage is being strongly encouraged.[107,108] This is especially true in North and South American countries like Mexico, Honduras, Guatemala, and Argentina. MH is banned for use on tobacco in Canada and Brazil. It is also being used extensively in Australia, Japan, and Taiwan. In countries where MH is used, contacts — usually the C_{10} FA rather than the C_8 + C_{10} FA mixture — are also being used, usually as part of a sequential system rather than alone. In South American countries excluding Brazil it is estimated that nearly equal amounts of contacts and MH are being used. The combined MH + FA product is also being used in many of these countries and its use is also expected to increase. Both the DEA and the K salts of MH are presently in use, but some countries use more of the one than the other. Contacts are exclusively used in Brazil, Canada,[109] South Africa and Zimbabwe because MH is considered to cause undesirable effects on leaf quality. In some countries like Brazil the C_{10} FA is almost exclusively used, whereas a larger portion of the C_8 + C_{10} mixture is used in the U.S. It is estimated that about 6.5 million lb of the FA is used for control of tobacco suckers in the U.S., which is up from the 5 million lb estimate of 1979.[1] The international market is now estimated to be approximately 4.5 to 5 million lb, mostly the C_{10} FA. These estimates were obtained from several personal contacts and should be considered as approximations.

Except for countries where tobacco production is relatively sophisticated, effective sucker control can usually be obtained with chemicals applied at rates lower than in the U.S. This is due to a lower sucker growth potential because of differences in fertilization, rainfall, soils, cultivars, and environmental conditions. This, together with the fact that some tobacco probably is not treated with any sucker control chemical, usually means that cured leaf from these countries has lower sucker control chemical residues. Because such tobacco has lower residues, especially MH, foreign manufacturers concerned about levels of MH in their final products, blended the higher U.S.-residue leaf with leaf from countries where residues were lower in order to obtain a product considered acceptable from the standpoint of MH residues. The manufacturers are finding this more difficult to do because the use of MH is increasing worldwide.

Limited quantities of the dinitroaniline, butralin, is being used for controlling tobacco sucker growth in South Africa, Australia, Taiwan, and some Latin American countries. In Australia it is formulated as a combination with FA.

At present ethephon is used as a yellowing and curing aid on flue-cured tobacco almost exclusively in the U.S. and Canada although experiments and plans are underway for expanding its use in other countries. When first used on flue-cured tobacco, it was used mostly in the more southern U.S. flue-cured growing areas. Its use has since expanded to all U.S. flue-cured areas and Canada. In Canada it is used primarily to aid in curing "tip and undertips" (the top six to seven leaves).

The use of ethylene generators in flue-cured tobacco barns is also almost exclusively restricted to the U.S. and Canada. Expansion of their use into other countries is being pursued, however. Even though the ethylene generators have been used for a relatively few years, estimates have been made that nearly 10% of the U.S. and may be 20% of the Canadian flue-cured tobacco crop are cured in barns equipped with these generators.

V. CONCLUSIONS

This review illustrates how growth regulating chemicals are used in the production of tobacco. For the most part, their uses are designed to increase both production efficiency and visible leaf quality factors. Problems concerning growth regulator residues are important. These problems along with the development of machines for harvesting the flue-cured crop are providing an impetus for the development of new methods of using existing plant growth regulators and for the development of new ones for use in tobacco production. Because the use of plant growth regulators can significantly increase production efficiency, their usage pattern has markedly changed over the past 3 to 5 years. This pattern is expected to continue to change as tobacco production in some of the lesser developed countries becomes more sophisticated. A moderate increase in the amount of plant growth regulators used for tobacco production may be expected over the next decade.

ACKNOWLEDGMENTS

The comments and constructive criticisms provided by E. L. Moore and H. Seltmann are gratefully acknowledged. The assistance provided by Mrs. A. H. Risdon and Mrs. P. M. Keithley in preparing the manuscript is much appreciated. The author also especially thanks members of commercial companies who provided invaluable information about the usage of plant growth regulators in the worldwide production of tobacco.

REFERENCES

1. **Steffens, G. L.,** Influence of growth regulators and herbicides on the chemistry of tobacco, in *Recent Advances in Tobacco Science,* Vol. 3, Tso, T. C., Ed., 33rd Tob. Chem. Res. Conf., Lexington, Ky, 1979, 133.
2. **Steffens, G. L. and Seltmann, H.,** Plant regulators for tobacco growth modification and improved safety, in *Chemical Manipulation of Crop Growth and Development,* McLaren, J. S., Ed., Butterworths, London, 1982, 193.
3. The Biologic and Economic Assessment of Maleic Hydrazide, *Tech. Bull. No. 1634, U.S. Department of Agriculture,* Washington, D.C., 1979.
4. **Skoog, F. and Thimann, K. V.,** Further experiments on the inhibition of the development of lateral buds by growth hormones, *Proc. Natl. Acad. Sci. U.S.A.,* 20, 480, 1934.
5. **Decker, R. D. and Seltmann, H.,** Axillary bud development in *Nicotiana tabacum* L. after topping, *Tob. Sci.,* 15, 144, 1971.
6. **Seltmann, H. and Kim, C. S.,** Anatomy of the leaf axil of *Nicotiana tabacum* L., *Tob. Sci.,* 8, 86, 1964.
7. **Seltmann, H.,** Application of Maleic Hydrazide to the Untopped Plant, presented at 23rd Tob. Workers' Conf., College Park, Md., 1970.
8. **Steffens, G. L. and McKee, C. G.,** Chemically topping Maryland tobacco, *Tob. Sci.,* 13, 48, 1969.
9. **Marshall, H. V., Jr. and Seltmann, H.,** Time of topping and application studies with maleic hydrazide on flue-cured tobacco, *Tob. Sci.,* 8, 74, 1964.
10. **Seltmann, H.,** Modern methods of tobacco sucker control, in *Proc. 5th Int. Tob. Sci. Congr.,* CORESTA, Hamburg, Germany, 1970, 77.
11. **Chaplin, J. F., Ford, Z. T., and Currin, R. E.,** Some effects of topping heights and suckering flue-cured tobacco, *S.C. Agric. Exp. Stn. Bull.,* No. 510, 1964.
12. **Wolf, F. A. and Gross, P. M.,** Flue-cured tobacco. A comparative study of structural responses induced by topping and suckering, *Bull. Torrey Bot. Club,* 64, 117, 1937.
13. **Woltz, W. G.,** Some effects of topping and suckering flue-cured tobacco, *N.C. Agric. Exp. Stn. Bull.,* No. 106, 1955.
14. **Steffens, G. L., Spaulding, D. W., Clark, F., Ford, Z. T., Lundy, H. W., Miles, J. D., Rogers, M. J., Seltmann, H., and Chaplin, J. F.,** Regional tests with contact and systemic tobacco sucker control agents. I. Flue-cured tobacco, *Tob. Sci.,* 13, 113, 1969.
15. **Steffens, G. L., Spaulding, D. W., Atkinson, W. O., Bortner, C. E., Link, L. A., Nichols, B. C., Ross, H. F., Seltmann, H., and Shaw, L.,** Regional tests with contact and systemic tobacco sucker control agents. II. Burley tobacco, *Tob. Sci.,* 13, 117, 1969.
16. **Spaulding, D. W., Steffens, G. L., and Hoyert, J. H.,** Regional tests with contact and systemic tobacco sucker control agents. IV. Maryland tobacco, *Tob. Sci.,* 14, 98, 1970.
17. **Smith, H. C., Link, L. A., Steffens, G. L., and Atkinson, W. O.,** Regional tests with contact and systemic tobacco sucker control agents. III. Fire-cured tobacco, *Tob. Sci.,* 15, 91, 1971.
18. **Elliott, J. M.,** The effect of stage of topping flue-cured tobacco on certain properties of the cured leaves and smoke characteristics of cigarettes, *Tob. Sci.,* 19, 7, 1969.
19. **Watson, S.,** Sucker control: a new generation of chemicals is on the way, *Flue-Cured Tob. Farmer,* 18, 7, 1981.
20. **Petersen, E. L.,** Controlling tobacco sucker growth with maleic hydrazide, *Agron. J.,* 44, 332, 1952.
21. **Vickery, L. S.,** Some effects of maleic hydrazide on flue-cured tobacco quality, *Tob. Sci.,* 3, 79, 1959.
22. **Birch, E. C. and Vickery, L. S.,** The effects of maleic hydrazide on certain chemical constituents of flue-cured tobacco, *Can. J. Plant Sci.,* 41, 170, 1961.
23. **Zukel, J. W. (Compiler),** A Literature Summary on Maleic Hydrazide 1949 to 1957, *MHIS No. 8, Naugatuck Chem. Div., U.S. Rubber Co.,* Naugatuck, Conn., 1957.

24. **Seltmann, H. and Peedin, G. F.**, Application time during the day influences chemical sucker control, *Tob. Sci.*, 16, 88, 1972.
25. **Zukel, J. W. (Compiler)**, A Literature Summary on Maleic Hydrazide 1957 to 1963, *Naugatuck Chem. Div., U.S. Rubber Co.*, Naugatuck, Conn., 1963.
26. **Frear, D. S. and Swanson, H. R.**, Behavior and fate of [^{14}C]-maleic hydrazide in tobacco plants, *J. Agric. Food Chem.*, 26, 660, 1978.
27. **Cheng, L. S. and Steffens, G. L.**, Maleic hydrazide residues in Maryland tobacco, *Tob. Sci.*, 20, 90, 1976.
28. **Bush, L. P. and Sims, J. L.**, Morphological and physiological effects of maleic hydrazide on tobacco, *Physiol. Plant.*, 32, 157, 1974.
29. **Coupland, D. and Peel, A. J.**, Maleic hydrazide as an antimetabolite of uracil, *Planta*, 103, 249, 1972.
30. **Nooden, L. D.**, The mode of action of maleic hydrazide: inhibition of growth, *Physiol. Plant.*, 22, 260, 1969.
31. **Nooden, L. D.**, Inhibition of nucleic acid synthesis by maleic hydrazide, *Plant Cell Physiol.*, 13, 609, 1972.
32. **Alfimova, R. A.**, Effect of MH on the content of nucleic acids in leaves of tobacco plants, *Vliyanie Fiziol. Aktiv. Soedin. Obmen Veshchestv Prod. Rast.*, p. 70, 1973; *Chem. Abstr.*, 81 (21), No. 131515, 1974.
33. **Hughes, C. and Spragg, S. P.**, The inhibition of mitosis by the reaction of maleic hydrazide with sulphydryl groups, *Biochem. J.*, 70, 205, 1958.
34. **Baker, J. E.**, Study of the action of maleic hydrazide on processes of tobacco and other plants, *Physiol. Plant.*, 14, 76, 1961.
35. **Epstein, S. S., Andrea, J., Jaffe, H., Joshi, S., Falk, H., and Mantel, N.**, Carcinogenicity of the herbicide maleic hydrazide, *Nature (London)*, 215, 1383, 1967.
36. **Epstein, S. S. and Mantel, N.**, Heptocarcinogenicity of the herbicide maleic hydrazide following parenteral administration to infant Swiss mice, *Int. J. Cancer*, 3, 325, 1968.
37. **Liu, Y. Y., Schmeltz, I., and Hoffmann, D.**, Chemical studies on tobacco smoke. XXVI. Quantitative analyses of hydrazine in tobacco and cigarette smoke, *Anal. Chem.*, 46, 885, 1974.
38. **Smith, W. T., Jr., Mayer, C. F., Kook, C. S., and Patterson, J. M.**, Controlled Pyrolysis of Maleic Hydrazide, presented at 29th Tob. Chem. Res. Conf., College Park, Md., 1975.
39. **Smith, W. T., Haider, N. F., Braun, L. and Patterson, J. M.**, presented at 32nd Tob. Chem. Res. Conf., Montreal, Canada, 1978.
40. **U.S. Environmental Protection Agency**, Maleic hydrazide: position document. I, *Fed. Reg.*, 42(208), 56920, 1977.
41. **Guthrie, F. E.**, Pending legislative restrictions on the use of agricultural chemicals on tobacco, *Beitr. Tabakforschung*, 7, 195, 1973.
42. **Moore, E. L.**, Maleic hydrazide controls suckers but leaves high residues, *Tob. Farmer* (Tifton, Ga.), 12(1), 1, 1975.
43. **Weber, K. H.**, The significance to the tobacco trade of the ordinance on residues from plant protectives on foodstuffs of plant origin, *Tab. J. Int.*, 6, 389, 1974.
44. **Wittekindt, W.**, Fixing of tolerances for plant-protective agents for plant treatment of tobacco in the F.R.G., *Tab. J. Int.*, 5, 323, 1977.
45. **Davis, D. L., Atkinson, W. O., and Smiley, J.**, Maleic hydrazide residues from air-cured tobacco, *Crop Sci.*, 14, 109, 1974.
46. **Hunt, T. W., Sheets, J. T., and Collins, W. K.**, MH residues on flue-cured tobacco, *Tob. Sci.*, 21, 128, 1977.
47. **Sheets, T. J. and Seltmann, H.**, unpublished data, 1979.
48. **Sheets, T. J., Leidy, R. B., Mesick, P. L., Hoyes, K. A., and Scheviak, L. A.**, Pesticide residue in tobacco, tobacco products and main-stream smoke. Also, personal communication, *N.C. State Univ. Annu. Rep. NC 02502*, Raleigh, 1978 to 1980.
49. **Hayes, K. A.**, Effects of Redying and Storage on Residues of MH and Ethoprop in Flue-Cured Tobacco, M.S. thesis, North Carolina State University, Raleigh, 1979.
50. **Davis, D. L., Atkinson, W. O., and Everette, G.**, Maleic hydrazide residues on burley and dark flue-cured tobacco, *Ky. Agric. Exp. Stn. Annu. Rep.*, 90, 37, 1977.
51. **Haeberer, A. F. and Chortyk, O. T.**, Gas-liquid chromatographic determination of maleic hydrazide in tobacco and tobacco smoke, *J. Assoc. Anal. Chem.*, 62, 171, 1979.
52. **Liu, Y. Y. and Hoffman, D.**, Quantitative determination of maleic hydrazide in cigarette smoke, *Anal. Chem.*, 45, 2270, 1963.
53. **Iwasaki, M., Miyaoka, T., Tsuda, S., Shirasu, Y., and Harada, T.**, Effects of maleic hydrazide on cigarette smoke inhalation toxicity in Syrian Golden Hamsters, *J. Pest. Sci.*, 6, 17, 1981.
54. **Collins, W. K., Peedin, G. F., and Smith, W. D.**, Agronomic production practices, in *1982 Tobacco Information*, N.C. Agric. Ext. Serv., Raleigh, 1982, 31.

55. **Seltmann, H.,** New Aspects of Chemical Sucker Control with Maleic Hydrazide, presented at 29th Tob. Workers' Conf., Lexington, Ky., 1981.
56. **Steffens, G. L.,** Tobacco sucker control: enhanced effectiveness of maleic hydrazide with gibberellic acid, *Tob. Sci.*, 18, 115, 1974.
57. **Coulson, D. A.,** Some effects of maleic hydrazide on flue-cured tobacco quality, *Tob. Sci.*, 3, 69, 1959.
58. **Moseley, J. M.,** The effects of maleic hydrazide when used as a sucker control agent upon the quality of flue-cured tobacco, *Tob. Sci.*, 3, 73, 1959.
59. **Jeffrey, R. N. and Cox, E. L.,** Effects of maleic hydrazide on the suitability of tobacco for cigarette manufacture, *U.S. Dept. Agric. Crops Res. Div. ARS-34-35*, Washington, D.C., 1962.
60. **Chaplin, J. F.,** Influence of various degrees of sucker control on flue-cured tobacco, *Tob. Sci.*, 11, 45, 1967.
61. **Peedin, G. E,,** The Effects of Various Degrees of Manual and Chemical Sucker Control on Certain Agronomic Chemical and Physical Characteristics of Flue-Cured Tobacco, M.S. thesis, North Carolina State University, Raleigh, 1969.
62. **Dawson, R. F.,** Nicotine synthesis in excised tobacco roots, *Am. J. Bot.*, 29, 813, 1942.
63. **Seltmann, H.,** Comparison of cured leaf from tobacco plants treated with various sucker controlling agents under conditions of poor and good control, *Tob. Sci.*, 22, 46, 1978.
64. **Aycock, M. K., Jr. and McKee, C. G.,** Effects of contact and systemic sucker control chemicals on Maryland tobacco cultivars, *Tob. Sci.*, 19, 104, 1975.
65. **Seltmann, H., Ross, H., and Shaw, L.,** Time of topping and methods of suckering on yield, value and alkaloid content of burley tobacco, *Tob. Sci.*, 13, 6, 1969.
66. **U.S. Department of Agriculture,** The effects of maleic hydrazide on the suitability of tobacco for cigarette manufacture, *U.S. Dept. Agric. Crops Res. Div. ARS-34-29*, Washington, D.C., 1961.
67. **Tso, T. C.,** Plant-growth inhibition by some fatty acids and their analogues, *Nature (London)*, 202, 511, 1964.
68. **Steffens, G. L., Tso, T. C., and Spaulding, D. W.,** Fatty alcohol inhibition of tobacco axillary and terminal bud growth, *J. Agric. Food Chem.*, 15, 972, 1967.
69. **Steffens, G. L. and Cathey, H. M.,** Selection of fatty acid derivatives-surfactant formulations for the control of plant meristems, *J. Agric. Food Chem.*, 17, 312, 1969.
70. **Collins, W. K., Hawks, S. N., and Kittrell, B. U.,** Effect of contact and systemic sucker control agents on yield and value of flue-cured tobacco, *Tob. Sci.*, 14, 65, 1970.
71. **Seltmann, H.,** Sucker Control Options in Flue-Cured Tobacco, presented at 29th Tob. Workers' Conf., Orlando, Fla., 1979.
72. **Tancogne, J.,** Evaluation of the residues of aliphatic alcohols used as sucker control agents on topped dark tobacco, *Tab. Ann., Sec. 2*, 11, 231, 1974.
73. **Tso, T. C. and Chu, H.,** The fate of fatty compounds and surfactants used as sucker control agents on field tobacco, *Beitr. Tabakforschung*, 9, 58, 1977.
74. **Tso, T. C., Chu, H., and DeJong, D. W.,** Residue levels of fatty compounds and surfactants as suckering agents on tobacco, *Beitr. Tabakforschung*, 8, 241, 1975.
75. **Hawks, S. N., Jr.,** *Principles of Flue-Cured Tobacco Production*, Hawks, S. N., Jr., Ed., North Carolina State University, Raleigh, 27607, 1970.
76. **Collins, W. K.,** FST-7 — a new sucker control chemical, *Prog. Farmer*, 94(3), 86, 1979.
77. **Kennedy, P. C., Seltmann, H., Atkinson, W. O., Whitty, E. B., and Wilcox, M.,** Retardation of axillary bud growth in tobacco with a trifluoromethylbenzeneamine, in *Proc. 5th Annu. Meet. Plant Growth Reg. Working Group*, Blacksburg, Va., 1978, 172.
78. **Wilcox, M., Chen, I. Y., Kennedy, P. C., Li, Y. Y., Kincaid, L. R., and Helseth, N. T.,** Control of axillary buds in tobacco by means of *N*-alkylnitrophenylhydrazines and *N*-benzylnitroanilines, in *Proc. 4th Annu. Meet. Plant Growth Regul. Working Group*, Hot Springs, Ark., 1977, 194.
79. **Wilcox, M., Whitty, E. B., Li, Y. Y., Chen, I. Y., Kincaid, L. R., Hensley, J. R., and Kennedy, P. C.,** Control of axillary buds in tobacco by CGA-41065 and *m*-phenylenediamine analogs, in *Proc. 5th Annu. Meet. Plant Growth Regul. Working Group*, Blacksburg, Va., 1978, 167.
80. **Steffens, G. L.,** Growth regulator combinations for tobacco sucker control: maleic hydrazide and a *N*-benzylnitroaniline (CGA-41065), *Tob. Sci.*, 24, 102, 1980.
81. **Seltmann, H.,** What's New in Tobacco Sucker Control, presented at 27th Tob. Workers' Conf., Atlanta, 1977.
82. **Tancogne, J., Chouteau, J., and Cozamajour, F.,** Residues left by butralin used as a sucker control inhibitor, *Tab. Ann., Sec. 2*, 14, 217, 1977.
83. **U.S. Environmental Protection Agency,** Pesticide program guidelines for registering pesticides in the United States. Control of axillary bud (suckers) growth on tobacco, *Fed. Reg.*, 40(123), 26855, 1975.
84. **Burg, S. P.,** The physiology of ethylene formation, *Ann. Rev. Plant Physiol.*, 13, 265, 1962.
85. **Pratt, H. K. and Goeschl, J. D.,** Physiological roles of ethylene in plants, *Ann. Rev. Plant Physiol.*, 20, 541, 1969.

86. **Lieberman, M.,** Biosynthesis and action of ethylene, *Ann. Rev. Plant Physiol.*, 20, 533, 1979.
87. **Zimmerman, P. W., Hitchcock, A. E., and Crocker, W.,** The effect of ethylene and illuminating gas on roses, *Contrib. Boyce Thompson Inst.*, 3, 459, 1931.
88. **Doubt, S. L.,** The response of plants to illuminating gas, *Bot. Gaz.*, 63, 209, 1917.
89. **Thimann, K. V., Ed.,** *Senescence in Plants*, CRC Press, Boca Raton, Fla., 1980.
90. **Rossi, U.,** Etilenozione del tobacco, *Boll. Tecnico. 1st Sper. Leonardo Angeloni*, 30, 222, 1933.
91. **Pfutzer, G. and Losch, H.,** Veredlung deutschen Tabaks, *Umschau*, 39, 202, 1935.
92. **Kraynev, S. I.,** The effect of ethylene on the hydrocarbon exchange and carbonhydrase during the starvation of tobacco and other vegetative objects, *Proc. Agric. Inst. Krasnodar*, 6, 101, 1937.
93. **Asmev, P. G.,** The effects of ethylene on the gas exchange and respiration ferments during the starvation period of tobacco leaves and other vegetative objects, *Proc. Agric. Inst. Krasnodar*, 6, 49, 1937.
94. **Cutler, H. G. and Gaines, T. P.,** Some preliminary observations on greenhouse-grown tobacco treated with 2-chloroethylphosphonic acid at varying pHs, *Tob. Sci.*, 15, 100, 1971.
95. **Steffens, G. L., Alphin, J. G., and Ford, Z. T.,** "Ripening" tobacco with the ethylene-releasing agent 2-chloroethylphosphonic acid, *Beitr. Tabakforschung*, 5, 262, 1970.
96. **Long, R. C., Weybrew, J. A., Wotz, W. G., and Dunn, C. H.,** Effects of 2-chloroethylphosphonic acid on the development and maturation of flue-cured tobacco, *Tob. Sci.*, 18, 70, 1974.
97. **Domir, S. C. and Foy, C. L.,** Movement and metabolic fate of [^{14}C] ethephon in flue-cured tobacco, *Pest. Biochem. Physiol.*, 9, 9, 1978.
98. **Domir, S. C. and Foy, C. L.,** Study of ethylene and CO_2 evolution from ethephon in tobacco, *Pest. Biochem. Physiol.*, 9, 1, 1978.
99. **Miles, J. D., Steffens, G. L., Gaines, T. P., and Stephenson, M. G.,** Flue-cured tobacco yellowed with an ethylene releasing agent prior to harvest, *Tob. Sci.*, 16, 71, 1972.
100. **Domir, S. C. and Foy, C. L.,** Effects of ethephon on ripening, curing, and chemical constituents of flue-cured tobacco, *Tob. Sci.*, 20, 158, 1976.
101. **Sisler, E. C. and Pian, A.,** Effects of ethylene and cyclic olefins on tobacco leaves, *Tob. Sci.*, 17, 68, 1973.
102. **Atkinson, W. O., Link, L. A., Nichols, B. C., and Peedin, G. F.,** The effects of ethephon on ripening and certain quality components of burley tobacco, *Tob. Sci.*, 24, 71, 1980.
103. **Link, L. A.,** personal communication, 1982.
104. **Link, L. A.,** Effects of ethephon applied to burley tobacco seedlings, *Tob. Sci.*, 20, 104, 1976.
105. **Fowler, D. L. and Mahan, J. N.,** *The Pesticide Review 1977*, ASCS, U.S. Department of Agriculture, Washington, D.C., 1978.
106. **Fowler, D. L.,** personal communication, 1981.
107. **Westcott, J. S.,** Desuckering tobacco: the case for use of maleic hydrazide, *Tob. Int.*, 182, 49, 1980.
108. **Fuller, G. B.,** Maleic hydrazide: its registration and safety worldwide, *Tob. Int.*, 183, 55, 1981.
109. **Rosa, N.,** Sucker control chemicals commonly used in Ontario, 1967 to 1976, *Tob. Sci.*, 24, 9, 1980.
110. **Sheets, T. J. and Seltmann, H.,** Effect of sprinkler irrigation on tobacco sucker control and residue from MH, *Tob. Sci.*, 26, 106, 1982.
111. **Chopra, N. M., Verma, M. M., and Zuniga, T. H.,** On the fate of maleic hydrazide in tobacco smokes, *J. Agric. Food Chem.*, 30, 672, 1982.

Chapter 6

CONCEPTS AND PRACTICE OF USE OF PLANT GROWTH REGULATING CHEMICALS IN VITICULTURE

John A. Considine

TABLE OF CONTENTS

I. INTRODUCTION

Grapevines provide an outstanding example of the effectiveness of plant growth regulating chemicals (PGRC) as modifiers of the agronomic characteristics of crop plants. This is a reputation which has arisen through the importance in viticulture of cultivated forms of grapevine which bear seedless fruit: fruit which often have an inherent capacity to respond to supplementary levels of PGRCs, especially auxins, gibberellins and cytokinins.

The aspects of plant development which are most open to control are those of fruit set, growth, and development. However, even in these aspects of viticulture, plant growth regulators have not been a universal panacea and have sometimes created new problems. We look to PGRCs as a means of increasing yields, of improving quality, of increasing labor efficiency, and of minimizing the deleterious effects of adverse environments. That is we have hoped and perhaps still aspire to create the perfect plant without recourse to the tedium and expense of plant breeding and selection. Despite failure to achieve all our ideals, PGRCs are important in viticultural practice and there is yet much scope for development.*

* With few exceptions, no attempt has been made to adopt a standard of naming grapevine cultivars and the names used in the original articles have been cited. Two exceptions are Black Corinth (syn. Zante Currant) and Sultanina (syn. Sultana, Thompson Seedless). Some doubt has, however, been expressed about the validity of the synonymity of Black Corinth and Zante Currant[139] though the fruit of the two cultivars is similar anatomically and physiologically.

A. Concepts of Productivity, Yield Partitioning, and Quality

Productivity is a key element in the economic viability of horticultural enterprises but limits to productivity are imposed by efficiency of light interception, of CO_2 fixation, and translocation, and by demands of other vine parts which compete for the products of fixation.[299,310] The role of PGRCs in improving productivity is confined to instances where genetic, environmental, disease, or management associated factors limit production of mature fruit.

Instances of genetic limitations to productivity of cultivars with otherwise desirable attributes include limited flower initiation,[40,208] and limited fruit development or numbers of fruit set. Environmental conditions may restrict many of the processes of reproduction from inflorescence initiation,[40,211] to development of the reproductive structures leading to development of the syndrome of *coulure*.[338,284] Conversely, situations in which fruit growth is excessive for the numbers of fruit set also lead to problems of physical damage of berries and to subsequent invasion by *Botrytis*.[160,287]

Management decisions such as use of an unsuitable rootstock, trellising or training system, or provision of excessive levels of inorganic nutrients can lead to problems such as excessive vegetative growth and low floral initiation.[206,207] Such problems of excessive vigor can sometimes be ameliorated by use of an appropriate PGRC to restrain excessive vigor, limit or promote numbers of inflorescences initiated or fruit set and their development or otherwise modulate the problem.

There is strong evidence for the concept of an optimum yield[193,299] a yield which is sustainable[376,401] and which utilizes the full potential of the available energy and nutritional resources of the vine. This concept is often expressed in terms of the harvest index, the proportion of dry matter accumulation which is harvested vs. that which is apportioned to vegetative growth. It is a mistake though to consider that the harvest index is a constant, a mistake which may arise through concern with the productivity of mature vines. Grapevines, like other perennial crop plants show a trend in harvest index which increase in time from zero to an asymptote, in vines, at about 5 to 6 years of age. Once the plant has achieved its particular asymptotic value of dry matter partitioning, only minor variations will occur according to seasonal conditions. PGRCs may aid in the fine tuning of this balance to minimize seasonal variations in yield or they may be used to alter the distribution to maximize the proportion of energy devoted to harvestable fruit, rather than to excessive vegetative growth which may be lost at pruning.

Scope also exists for a more radical approach such as has been used for apples[196,421] in the production of "meadow orchards". In this approach a plant growth retardant is used to induce precocious fruiting of a plant, at a much smaller plant size, with increased scope for mechanization and greatly increased bearing potential per hectare. Thus while the principles and applications described here concern principally the fine-tuning of productivity about a fortuitous plant mass, these principles also enable, in theory at least, for manipulation of size of that mass and the time that it takes to achieve.

Yet productivity in terms of dry matter production per acre or per hectare is not a sufficient measure of the productivity of a modern enterprise; labor productivity also contributes in an increasingly important manner to the economic efficiency of a horticultural enterprise. Here too PGRCs are being applied, principally as aids to mechanized harvesting, either as an aid to fruit removal or through management of uniformity of ripening and timing of harvest date.

Quality, as a concept, varies according to the use the fruit is to be put and to regional customer preferences. Wine making, the end use for the bulk of grape production, requires that fruit mature, in the sense that the fruit attains the required content of sugar, balance of sugar and acid, and, especially if a colored grape, a high degree of anthocyanin and tannin content. Sugar, while its accumulation may be highly dependent on environmental conditions, is also dependent on the relationship between fruit volume and leaf surface area. Fruit numbers, size, shapes, and coloring are each factors which have been shown to respond to particular PGRCs as will be described in the sections that follow.

The table grape industry, perhaps more than any other component of viticulture, has requirements which at present can only be met by use of PGRCs. Consumers demand and pay a premium for large, sweet, seedless, and sound fruit.[290] It is in these areas that research on the application of PGRCs in viticulture has been concentrated, an effort which in most instances has met with conspicuous success as have treatments to advance ripening to secure the early, prime market prices. Fruit color can also be important and though largely controlled by environment factors, can be improved by chemical treatments.

Requirements of the seedless dried fruit industries are not commonly met by application of PGRCs[288] yet opportunities exist, especially in instances where for reasons of adverse climatic conditions, disease or management, the distribution of dry matter between fruit and plant is not optimal, or there is a desire to minimize risk of disease or of certain physiological disorders. Treatments outlined below, to increase fruitfulness, set, cluster looseness or fruit size may each have an application in particular circumstances.

II. FLOWER DEVELOPMENT

A. Inflorescence Initiation

Inflorescence number and size are not widely regarded as factors which impose limitations to productivity and indeed probably as much research has been carried out on the use of PGRCs to reduce fruit number as there have to increase it. Nevertheless there are a number of situations where control of inflorescence number per vine and size of inflorescences would be particularly useful. Three notable examples are: (1) vines grown in environments which are marginal for inflorescence initiation and development;[211] (2) particular cultivars such as Sultanina in which inflorescence initiation is restricted genetically; and (3) seedling plants (to enable early evaluation of genotype[320]).

1. General Physiology

Inflorescence initiation in grapes has been reviewed by Srinivasan and Mullins[322] who emphasized morphological and physiological aspects and by Buttrose[40] who examined principally environmental effects.

Capacity for inflorescence initiation is a quantitative genetic character and degree of inflorescence initiation is measured in terms of both number of inflorescence primordia per latent or dormant bud,[39,208] and proportion and position of latent buds bearing inflorescence primordia. Different species also vary, differences being associated with a varying fraction of nodes bearing tendrils on shoots.

Initiation of inflorescences in latent buds during summer may be depressed by temperature less than about 25°C, by low light intensity, water stress, and improper mineral nutrition,[38,37,40] though species differ in their tolerance of low temperatures and low light intensity (see review by Srinivasan and Mullins).[322] Also, grape vine seedlings competent to form inflorescences at an early stage of growth, rarely do so under natural conditions.[315,320]

Excessive inflorescence initiation has not been viewed as a problem which warrants the use of PGRCs,[193,194] though means exist to do this. However, because of the risk of excessive reduction in inflorescence initiation such a problem is probably best treated in the first instance through modification of agronomic practice, as may the case where inflorescence initiation is limiting.[211,299]

Three categories of PGRC have been demonstrated to promote inflorescence initiation in particular circumstances, the plant growth retardants, the cytokinins, and particular nucleotide and related bases; and one other PGRC, GA, has been shown to reduce and delay inflorescence initiation.

Table 1
SUMMARY OF TREATMENTS WITH CCC WHICH HAVE ENHANCED INFLORESCENCE AND FLOWER PRODUCTION IN *VITIS*

Target bud[a]	Application time/state	Concentration of CCC (mg a.i./ℓ)	Cultivars/species	Ref.	Comments
Primary/secondary	Shoot 15—40 cm	500(mult.)[b]	Muscat of Alexandria	328	20°C/long days
Primary/tertiary	Pre-bud burst	4500	Cardinal	189	
Secondary	2 Weeks preanthesis	300	Muscat of Alexandria	73	
Tertiary	Anthesis	300	Carbernet Sauvignon	211	Cool environment/long days
Primary/tertiary	10—15 Leaves	30—1000 (mult.)	Muscat of Alexandria	318	Low temperatures (ca. 20°C)

[a] Primary bud — the primary axis in the season of application of treatment: Secondary bud; the secondary axis or axillary shoot in season of growth (prompt or summer shoot): Tertiary bud; the latent or dormant bud which becomes the primary axis in the season following that of application.
[b] Mult — multiple applications.

2. Growth Retardants

The retardant CCC appears to have particular promise as a means of stimulating inflorescence initiation (Table 1). The effectiveness of CCC has been confirmed in vitro,[315] under controlled environment conditions,[189,318,328] and in the field.[211] One other related growth retardant, ACPC, has been shown to be perhaps even more effective than CCC.[320] It is conceivable that other related growth retardants may be even more effective and perhaps more specific than CCC.[164] Alar has been shown to be ineffective in this process in *V. vinifera,*[320,360] though it has been reported to be effective in *V. lubrusca.*

CCC may act at two separate stages of inflorescence formation: (1) at inflorescence initiation (anlage formation), during formation of either the prompt axillary bud or the latent axillary bud; or (2) it can affect differentiation of the primary axis tendril initials during growth of the axis. The timing and concentrations of the spray application and environmental conditions required for an effect on each of the three axes differ (see Table 1). Coombe[73] who was first to report that CCC could promote inflorescence initiation, obtained the response as a side effect in experiments designed to promote fruit set in Muscat of Alexandria. In this instance, the response was noted principally on lateral growth as a ''second crop'' and in a few instances on the primary shoot. No effect was noted on subsequent crops. Initiation on primary shoots can be enhanced either by applying higher concentrations[189] or by applying multiple applications at short time intervals.[318,328] These latter treatments are also those required to reliably promote inflorescence initiation in latent buds.

Responsiveness to CCC is not solely a matter of timing or total amount applied, but is also affected by environmental conditions and probably also by the genetic characteristics of particular cultivars. For example, the responsiveness of Muscat of Alexandria is greatest at temperatures which normally preclude inflorescence initiation[16] and under long day conditions.[318,328] The importance of cultivars in determining response to treatment with CCC has not been specifically reported, but by analogy with the effects of cytokinins in relation to cultivars and species[319] such an effect may be expected. Attention should also be drawn to the deleterious effects of repeated application of high concentrations of CCC to vine and leaf morphology.[321,360]

CCC and ACPC have been applied to induce precocious flowering in grapevine seedlings[320] but are only effective in this instance if applied in association with a synthetic cytokinin, PBA or BA.

The mode of action of CCC in inducing inflorescences has been investigated in part by Srinivasan and Mullins[322] who proposed from evidence collected from various sources that CCC played a dual role by promoting synthesis of cytokinins while at the same time reducing the level of endogenous gibberellins to a suitable level.[160,282,308,316,318]

3. Cytokinins

Cytokinins have been shown to be powerful promotors of the processes of differentiation which convert uncommitted analage to the inflorescence pathway of development. In vitro, a complex mixture of naturally occurring cytokinins plus a synthetic cytokinin have been shown to be essential[315] but in the in vitro situation only one of the two common synthetic cytokinins appears to be required, either BA or PBA.[318,319] However, multiple applications of highly concentrated solutions of these substances are required to produce consistent and worthwhile effects and it is unlikely that either of these substances will replace CCC as a practical compound for promoting inflorescence initiation in the field, though cytokinins may provide a valuable adjunct to CCC in plant breeding programs.[322]

Sensitivity to cytokinins is by no means uniform throughout species and cultivars of *Vitis* and of those cultivars and species tested, variation in sensitivity was at least partially explained on the basis of plant sex: male vines in general being more responsive than female or hermaphroditic plants.[319]

4. Other PGRCs

A number of substantial reports have been published which show that application of specific purine and pyrimidine bases may enhance inflorescence initiation.[18,152,169,170] In each instance, the concentration applied (50 $mg \cdot \ell^{-1}$) was well below the amount needed to "explain" the response on the basis of a nutritional effect. Kessler and Lavee[170] for example reported a two-fold increase in the number of floral primordia in cultivar Dabouki, a cultivar which usually is devoid of inflorescences in the basal buds of a shoot. Of the compounds tested, uracil gave the most consistent response between the several reported experiments; response to xanthine and adenine were unpredictable however; Kessler and Lavee[170] report a depression of inflorescence initiation with these two bases while Balasubrahmanyam and Khanduja[18] reported a positive effect on Sultanina.

A point of difference between the two studies was a major difference in timing for whereas Kessler and Lavee[170] applied their treatments during the early floral initiation period, Balasubrahmanyam and Khanduja[18] applied their treatments soon after bud burst and well before the normal onset of inflorescence initiation. It is worth noting though, that their timing coincided with the period of sensitivity of some cultivars to GA induced inhibition of inflorescence initiation.[6]

Caffeine also showed potential as a stimulant of inflorescence initiation.[18,170] The mode of action of these compounds is obscure, though some of the bases possess a degree of cytokinin-like activity.

5. Gibberellins

While there is some evidence that gibberellins are an essential part of the system of endogenous growth regulators which together control first anlagen formation and then its differentiation into an inflorescence,[322] anlage formation has not been shown to be a factor which limits inflorescence initiation even in seedlings.[321] There is, however, considerable evidence that in practice applied GA causes barrenness by promoting the tendril pathway of anlagen development and it may also cause young inflorescences to behave in a tendril-like manner.[4,160,315,318]

Sensitivity of cultivars of *Vitis* to GA-induced barrenness varies widely (Table 2). There appears to be no simple rule which can be used to predict sensitivity though in the reports

Table 2
COMPARISON OF SUSCEPTIBILITY OF SOME CULTIVARS OF *V. VINIFERA* TO GA-INDUCED INHIBITION OF INFLORESCENCE DEVELOPMENT

Tolerant[a]	Sensitive
Pinot Blanc[160]	Chasselas[160]
Pinot Gris[160]	Gewurztraminer[160]
Auxerrois[160]	Sylvaner[160]
Riesling[160]	Carignane[370]
Sultana[370]	Red Malaga[370]
Black Corinth[370]	Ribier[370]
Zinfandel[370]	Tokay[370]
	Muscat of Alexandria[218]
	Waltham Cross[324]

[a] Tolerant — 50 mg·ℓ$^{-1}$ gives a reduction of less than 25%; sensitive, 10 to 50 mg·ℓ$^{-1}$ gives a reduction of greater than 25%.

examined only seeded cultivars were demonstrated to be sensitive while either seedless or seeded cultivars showed various degrees of tolerance. Most seedless cultivars appear to be tolerant in view of the widespread use of GA to promote fruit growth in such cultivars (see below). In examining the sensitivity of species other than *V. vinifera*, Alleweldt[4] concluded that male vines were more sensitive in general than either female or hemaphrodite vines. Some cultivars such as Gewurztrauminer have been shown to be particularly intolerant of applied GA with concentrations as low as 10 mg · ℓ$^{-1}$ causing a marked reduction in infloresence initiation.[160]

Timing of GA application is an important factor and sensitivity is greatest during early shoot growth.[5,6,160,370] GA acts to delay the onset of inflorescence initiation on a shoot as well as reducing the size and number of inflorescence initials which occur in a given latent bud.[14,160] This delay in the pattern of inflorescence initiation is illustrated in Figure 1 which shows the increasing delay of onset of initiation and degree of barrenness induced in Chasselas by increasing concentrations of GA.

Weaver and McCune[370] found that while sensitivity of Red Malaga to applied GA was greatest when shoots were about 40 cm long, sensitivity of basal buds persisted until a few weeks after anthesis. In their experiments, use of a spur pruning method meant that buds other than basal buds were not assessed and it is highly likely that sensitivity of buds at higher points of insertion persisted for a much greater period. It has, however, also been noted that inflorescences once formed are relatively tolerant of GA.[413]

It is conceivable that sensitivity of cultivars to GA-caused disorders is due to a relative inability to metabolize the applied GA, though there have been no reports of this nature, rather attempts have been made to correlate effectiveness of GA with content of endogenous gibberellins. These attempts have been unsuccessful.[117]

B. Inflorescence Development

Inflorescence development in terms of number of flowers carried per inflorescence is open to manipulation in degree of ramification, flower sex, megaspore development (abortion), pericarp development and style elongation, and elongation, thickening and degree of lignification of the rachis and pedicel. However, the process by which many of these factors

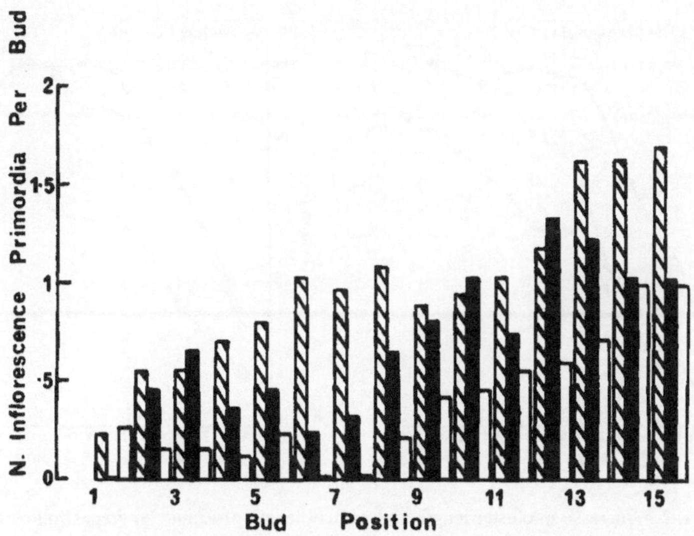

FIGURE 1. Depression of inflorescence initiation in buds of shoots of Gewurztraminer treated with zero (▧), 10 (■), or 50 (□) mg·ℓ⁻¹ GA. Diagram modified from Julliard and Balthazard.[160]

are influenced is in many instances poorly understood, and often PGRC effects which may influence one aspect positively may have detrimental side effects on one or more other aspect of inflorescence development.

1. Inflorescence Ramification

Inflorescence size as measured by number of flowers per inflorescence is, in grapevines, a quantitative character, varying between cultivars, between seasons and according to bud insertion position along a cane and to temporal sequence of initiation on a particular bud axis.[20,38,39,208]

The factors which control inflorescence development and ramification are principally genetic and environmental[40,318] though the mode of action through which the genetic and environmental factors operate may be safely assumed to be those discussed in the previous section which influence the initiation of inflorescence branching. Certainly the effects of GA and CCC appear to be carried through in this way.[160,211]

The inflorescence primordium as developed in the latent bud is not immutable and its development may be enhanced during early bud burst or, especially in cuttings, its growth may be retarded to the extent that it abscises.[226,227] Development of inflorescences in terminal buds, either natural or contrived by pruning, is often greatly enhanced beyond that predicted by initial primordium weight[20] which suggests that potential for enhancement of development exists during early bud-burst.

Inflorescences may be stimulated to continue to develop, either by suppressing shoot growth or by stimulating inflorescence growth. Suppression of shoot growth by applying thiosemicarbazides[272] or ABA[228] is effective, as is physical removal of leaves and shoot tip.[227] The thiosemicarbazide effect has not been pursued, though one of them, 2,4-dichlorobenzaldehyde thiosemicarbazide has been shown to have weak cytokinin-like activity.[446]

2. Development of the Inflorescence Framework

The framework of the grapevine inflorescence, its rachis and pedicels, respond to manipulation with specific PGRCs. The two chemicals which have been shown to have the greatest effect are GA and CCC, though BA has also been shown to affect inflorescence

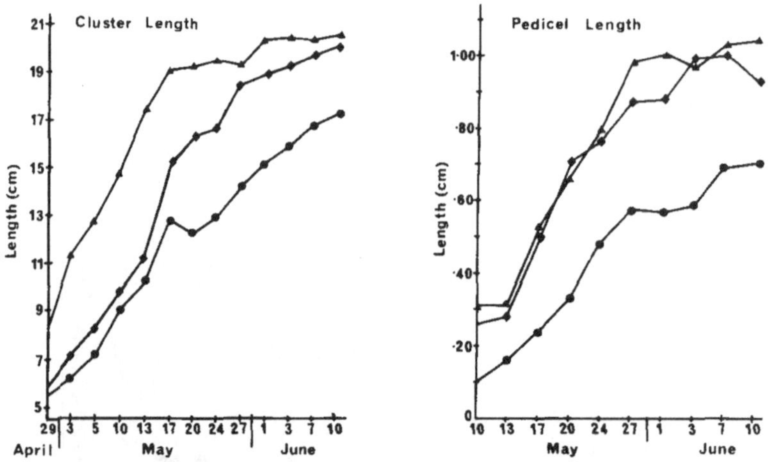

FIGURE 2. Increase in cluster length and pedicel length of Zinfandel grapes following treatment with 10 mg·ℓ⁻¹ GA. Treatment applied 27 days before bloom (▲), 15 days before bloom (♦), or not treated (●). Diagram modified from Weaver.[361]

growth.[342,343] The significance of rachis elongation in viticultural practice lies in the compactness of the mature clusters of many cultivars and hence of the proneness of the fruit of these cultivars to mechanical damage by growth forces and to fungal diseases such as *Botrytis*, both by virtue of poor aeration and the damage.

Enhancement of inflorescence growth was one of the first positive effects noted from the application of GA[3,283,354] though enthusiasm for this treatment quickly diminished on recognition of the deleterious side effects that the treatment had on some cultivars.[4,160,355,370] Nevertheless treatment with GA remains a potentially useful procedure for those cultivars which are relatively tolerant of applied GA (see Table 2) or where economics permit individual dipping of clusters to minimize the injurious side effects.

One cultivar which has been thoroughly studied is Zinfandel.[361,367] Responsiveness of the rachis of this cultivar to 100 mg·ℓ⁻¹ GA decreases with time from soon after bud-burst through to anthesis and no significant increase occurs in the 2 week period preceding anthesis (although there was an apparently anomalous response to an application at early flowering). In the more recent publication[361] it was shown that the primary response to a more realistic concentration of GA was an increase in the rate of rachis extension, and there was only a minor and not statistically significant increase in overall length (Figure 2). On the other hand, pedicel length was enhanced, especially by treatments applied earlier than 2 weeks before full bloom. This report is supported by an earlier report[365] in which the response of a number of cultivars was examined including Zinfandel, and in which the response of pedicels of Zinfandel were shown to exceed that of the rachis and that the responsiveness was retained to lesser but important extent through to anthesis.

While the results described above are indicative of the response of grapevines generally, important differences exist between cultivars in degree of response and responsiveness at particular times. In Weaver and McCune's study,[365] Black Corinth and Sultaninina were shown to be less responsive than Zinfandel or Tokay. Similarly, Julliard and Balthazard[160] found that only two of the seven cultivars they treated at the 4-leaf state of bud-burst showed a significant response to 50 mg·ℓ⁻¹ GA (Sylvaner and Gewurztraminer). However, other reports have shown that these and some other cultivars do show a significant response to GA applied either 2 weeks before anthesis or at anthesis.[54,55] In this particular instance Sultanina and Cape Currant (probably a seedless Red Muscadell)[54], were the most responsive while statistically significant responses were obtained from Doradillo, Muscat of Alexandria, and Black Corinth. In these experiments pedicel length was found to be more responsive than rachis length.

Specific recommendations have been made for a number of cultivars, both seeded and seedless: Italia,[22] Perlette,[47] Sultanina,[154] Delaware and Campbell Early,[223] Pusa Seedless,[337] Muscat Bailey A,[343] and Zinfandel.[361] However reduction in set of berries was in many of these examples a more important aspect of the creation of less tightly packed fruit clusters and this will be elaborated in the sections which follow.

Auxins also affect cluster framework development, promoting in particular increase in pedicel and peduncle diameter and lignification[386] though applications at a very early stage of inflorescence development were observed to have little or no effect.[227] One auxin, BTOA, has an effect analogous to that of high concentrations of GA in that it causes the rachis to respond thigmotropically like a tendril and to coil and lignify.[75]

Cytokinins are now finding commercial use to promote parthenocarpic fruit development of table grape cultivars.[342,343] Their use in this instance limits the stimulation of rachis growth which normally occurs when GA is applied alone. This effect is contrary to that obtained by Mullins[227] in his studies of the factors controlling sustained growth of young inflorescences on cuttings.

A detrimental side effect of applied growth retardants, both CCC and Alar, is that, almost inevitably, they restrict rachis extension further compounding cluster compactness[55,73,338]. The degree of effect of the growth retardants varies with time of application and with cultivar and generally the effects are relatively minor.[54,55]

3. Pistil Development

In general little research has been devoted to enhancing pistil growth in the period before bloom and in theory considerable scope exists because the pre-bloom growth phase provides the "capital" of cell number and cell volume on which future growth is based. Evidence for the significance of this period of growth comes from the studies of Harris et al.[124] who demonstrated that differences in size of berries between plants grown in a glasshouse and those grown in the field were determined before bloom. It is probable that this effect was due to the raised temperatures existing in the glasshouse because temperature before anthesis is inversely related to final pericarp size.[217] Also fruit of cultivars grown in hot, desert environments produce fruit which is often considerably smaller than fruit of the same cultivar grown in a milder climate.[168] Pistil development has also been shown to be positively correlated with shoot vigor.

A variety of PGRCs have been reported to affect pericarp development in the period leading up to anthesis. Coombe reported enhanced development of Black Corinth ovaries treated 10 days before bloom with 4-CPA, GA, or with a combination of both chemicals.[75] However he did not report whether the effect persisted to fruit maturity. Enhanced growth of the ovaries and of other flower parts of some seeded cultivars has also been reported. Gibberellin applied to Delaware and Campbell Early to induce parthenocarpic fruit development also caused enhanced growth of the pistil, especially of axial growth, and of the style, anther filaments, and operculum.[223,231]

Cytokinins have been shown to be necessary for near complete flower development of Concord flowers grown in vitro,[268] but when applied early in the period of inflorescence growth caused the production of fused pistils.[316] Application of cytokinins to enhance set of fleshy, parthenocarpic berries leads to larger but radially broadened pistils,[219,397] which is comparable to the effect on growth obtained by addition of the auxin 4-CPA.[410]

Shape of seeded and seedless berries is subject to the influence of GA if applied either 2 weeks before anthesis or at anthesis but in seeded berries the effect is due to a redistribution of growth rather than to a promotion of growth in a particular dimension.[55] Shape of berries of seedless cultivars or seedless berries on seeded cultivars is especially responsive to application of GA applied either at anthesis or up to 2 weeks before anthesis,[55] a time scale which corresponds to the period of intensive rib-meristem activity within the pericarp.[59] The degree of response varies from about 5% for Doradillo to about 20% for Zante, Sultanina, and Muscat, to nearly 50% for Cape Currant. Responsiveness thus varies over quite a wide range and is not solely related to degree of seed development (Muscat and Doradillo, seeded; Sultanina stenospermocarpic; and Zante and Cape Currant, parthenocarpic).

4. Sex Conversion

Degree of sexual development of flowers of *Vitis* varies considerably and though there are but three principal floral forms in functional terms, male, female, and hermaphroditic, there exist cultivars with various degrees of maleness or femaleness.[138,139,239,241,247,327,331] Stability of most of these forms is firm but occasionally plants which normally bear functional male flowers, form functional hermaphroditic flowers.[240]

In exploring the possibility of finding a reliable means of converting male plants of *V. vinifera* ssp. *sylvestris* to a functionally female plant, Negi and Olmo[240] compared the effect of a number of common PGRCs and environmental and nutritional treatments. The only treatment which worked effectively on a number of *Vitis* genera was the cytokinin, PBA. Treatment was most effective when applied at the megaspore mother cell stage of development. Of the cytokinins examined only Zeatin was as effective as PBA in inducing set of seeded fruit, though Kinetin gave an equivalent number of apparently complete flowers, set of fruit was much lower than for the other two cytokinins, though still well above the natural untreated level of fruit set.[240] GA also promoted pistil development but to a much lesser degree than any of the cytokins and did not result in seeded fruit production while auxins and the auxin transport inhibitor, TIBA enhanced maleness. Srinivasan and Mullins[319] also found that application of cytokinins to stimulate inflorescence formation caused the conversion of male hybrid vines to hermaphrodite vines.

A number of cultivars of *V. vinifera* ssp. *sativa* produce pollen with a high degree of infertility[110,252] and dioeciousism poses problems to plant breeders wishing to use ssp. *sylvestris* or female plants of other related *Vitis* species in plant improvement programs. However there are no reports of successful conversion of female plants to hermaphrodite plants compared with that for male plants as discussed above. In their work on inflorescence development, Srinivasan and Mullins[319] reported enhanced inflorescence development in a female vine, Katakourgan, which subsequently set a significant number of berries. However they did not report whether the possibility of cross fertilization from other vines was excluded.

5. Ovule Development/Abortion

As vital as complete ovule formation is to normal fruit set and development, the formation of seeds is often seen to pose a problem for table grape and dried grape purposes. Control of ovule development thus has a potential role, both in overcoming climate and disease-induced poor set (millerandage and coulure), and as an aid to reducing set in situations where mature fruit clusters are too tight and as a means of inducing parthenocarpy.

Few direct studies have been made of imperfect ovule development in relation to climate and disease[224] though occasionally increased set and seed development obtained by use of plant growth retardants[73,233] have been attributed to an enhancement of the proportion of normal ovules.

A number of studies have been carried out on the relationship between PGRC-induced parthenocarpy and growth and development of the ovules[75,147,231,408,410] but this stimulation of growth usually leads to faulty ovules which have defective embryo sacs[47,231] which may be compared with those described by Pearson[258,259] in the naturally parthenocarpic Black Corinth.[258,259]

6. *Pollen Development/Sterility*

Pollen development is defective in vines bearing females flowers and to various degrees in many apparently perfect flowered cultivars. Despite this situation few attempts seem to have been made to overcome the block in development of normal pollen in these plants, despite the potential usefulness of such a treatment in plant breeding. Most reports concerning the effect of PGRCs on pollen development describe deletarious effects.

Gibberellin may be used to prevent formation of germinable pollen[82,146] in many but not all cultivars (e.g., Neomuscat pollen is tolerant of applied GA[145]) though by any criteria other than germinability, the pollen is normal following such treatments.[231] The reduction in seed formation following GA-treatment of inflorescences before bloom is however not causally related to the effect of GA on pollen germination, but rather to effects on the ovules.[147,231] The effect that GA has on ovule receptivity precludes the use of GA as a male sterilant in breeding programs. However some other PGRCs have been shown to be effective as male sterilants without seriously depressing ovule fertility: MH, TIBA, Na 1,2-dichloroisobutyrate.[149] GA and IBA at low concentrations will enhance pollen germination and viability, but higher concentration (i.e., greater than 5 to 10 $mg \cdot \ell^{-1}$) seriously depress pollen function especially in those cultivars with a high natural rate of germination.[21]

Few records have been made of the effect of plant growth retardants on pollen germinability but those that have been made demonstrate little or no significant enhancement, even when applied to cultivar Muscat of Alexandria which has only a moderate proportion of viable pollen grains.[15,233,237]

7. *Bloom*

Timing and degree of normal function of the process of anthesis may be modified by application of certain PGRCs, especially by those known to promote development and to have catabolic effects on metabolism. Unfortunately, the treatments examined so far act to enhance the rate of inflorescence development, bringing the date of anthesis forward, but have associated deleterious effects on the normal process of pollination and syngamy. Yet another, less direct approach to manipulation of date of anthesis, is to modify the timing of commencement of shoot growth (see section on bud-burst).

a. Time

The effect of GA application on advancement of bloom have been well documented.[292,231,361] In Zinfandel the injurious side-effects of GA sprays on subsequent fruit development and flower initiation were minimized by use of a low concentration of GA ($10 mg \cdot \ell^{-1}$) and bloom occurred 3 to 4 days earlier in response to an application of GA between 10 and 20 days before bloom, earlier or later applications being less effective.[272]

b. Calypra Abscission

Open pollination in *Vitis* is facilitated by abscission of the calyptra prior to anther dehissence and attainment of stigma receptivity. However self-pollination may possibly be ensured by treatment with PBA[316] which prevents normal abscission of the calyptra and provides an opportunity for cleistogamous pollination. Other PGRCs which have been shown to inhibit calyptra abscission, namely NAA, 4-CPA and GA,[406] also interfere with normal seed formation which precludes their use for the purpose of inducing self-pollination. Interference with calyptra abscission appears to be correlated with effectiveness of PGRC stimulation of parthenocarpic development in grapes.[406,408]

III. FRUIT SET AND GROWTH

A. General Physiology

Fruit set embraces three component elements: (1) non-abscission; (2) establishment of normal pericarp growth; and (3) of normal pericarp softening and maturation. Each component shows a degree of interdependence and each is subject to PGRCs in a manner which interacts with the vines natural physiology. The physiology of the subject has been reviewed by a number of authors e.g. Coombe[76].

Fruit set and development in untreated vines is known to be affected by both functional and nutritional/competitive conditions.[74,75] Functional limitations may arise through impaired ovule and/or pollen development,[34,247,249,251,252,259] isolation,[34] or through environmental stress.[122,284] Pollination and syngamy are considered to be near absolute requirements for set and development of fruits of *Vitis* and apomictic seed development to be rare or nonexistant.[251,253] Even in instances when PGRCs are applied to surplant normal seed development as a requirement for pericarp growth, the effectiveness of the chemical is usually enhanced in the presence of normal pollination.[251,231,271,369]

The significance of competition and supply of metabolites as determinants of fruit set has probably been recognized since historical times and was established in modern literature by Müller-Thurgau.[224] The theme has since been elaborated by a number of scientists e.g., Coombe,[69,70,74] Mullins,[225] Shaulis and Oberle,[298] and Skene,[307] though the significance of competition for set of seeded berries of Kyoho has been disputed by Naito and Hayashi[233] who demonstrated that reduced competition increased set of seedless rather than seeded berries.

Environmental conditions which may lead to poor fruit set includes extremes of temperature, especially of high temperatures above about 30°C[116,172,338] though low temperature (below about 13°C) may also have specific effects on function such as reducing pollen viability.[173,174] Water deficit stress at time of fruit set, shortly after bloom, can have an especially devastating effect on fruit set, and it may be that the poor set obtained in conditions of high temperature is an effect of water stress.

Only a few studies have specifically studied the potential use of PGRCs to offset the ruinness effects of a capricious climate.[284,338] Further research on this topic is warranted.

B. Methods of Estimating Set

Elucidation of the effect of PGRCs on fruit set and development is often obscured by application of loose definitions. The term set is applied here to mean the development of a flower pistil to form a fruit of normal size and composition. To distinguish between the process of set and the numbers or proportion of flowers set, the term is specifically qualified as number or proportion of ovaries on flowers set (abbreviation: n. set).

The definition of set may be qualified, to distinguish between set of parthenocarpic or seeded berries but has also been further qualified by some authors who examined the effectiveness of PGRCs in preventing the abscission of ovaries and of establishing degrees of ovary growth which ranged from nil, to some growth but without subsequent softening (fleshy parthenocarpic berries), and up to normal growth.[28,213,217,218,358] The term set is also used to describe treatment-enhanced development of nonabscised ovaries.[96,62,354,386] In instances where fleshy, soft fruit development is initiated, then such instances come within the terms of the definition. Instances which simply involve enhanced fruit growth are not considered to fall within the scope of the term "set" and are not so considered here.

Many authors, when studying fruit set, omit measuring flower number or even pruning clusters to equal flower numbers. Problems arise in such studies if treatment effects are confounded with effects on inflorescence size (number of flowers per inflorescence) because percent of flowers set is closely related to number of flowers per cluster.[69] Set is also strongly influenced by leaf area[69] and by shoot vigor.[307] Precision and accuracy of interpretation demand that either steps be taken to ensure uniformity of experimental material across the treatments or that base measurements be made of the determining covariants to enable elimination of covariant effects in analysis.

Table 3
SUMMARY OF APPLICATION OF CINTURING TO ENHANCE THE NUMBER AND GROWTH OF BERRIES SET IN *VITIS*

Cultivar	Time	n.Set	Size	Yield	Ref.	Comments
Seedless						
Black Corinth	90%bl	0	+	+	13	Comparison with PGRC's
	+3d	+	+	(+)[a]	49	
	±50%bl	0	+	+	99	Time vs. response
	75%bl	+	+	+	100	Cf. 4-CPA, topping etc.
	75%bl		+		113	Anatomical effects
	±bl	+	+	+	150	Time/resp., n.set, size
	70%	+	+	+	348	Cf. 4-CPA
Sultanina	+7d	0	+	+	12	Resp. declined with time
	+10d		+		30	± Thinning
	±bl	+	+	+	150	Time/resp., n.set, size
	+10d				215	Method ± GA
	+10d	0	+	+	348	Strength of berry attachment
	+10d		+	+	377	Timing, ± GA
	+10d	0	+	+	366	± 4-CPA
	+10d	0	+	+	388	± 4-CPA, thinning
Monukka	±bl	+	+	+	150	Time/resp., n.set, size
Seeded						
Black Prince	+10d		(+)	(+)	86	Cane vs. trunk cincture
	+7d	0	+	+	87	Promotion of ripening
Cardinal	+24,39d	0	+	+	297	+24d incr. size, maturity
Chardonnay	75%bl	+(sdls)	0	+	188	Width of cincture
Clairette	bl	+	+	+	25	♀
Hunisa	±bl	+(sdls)	+	+	151	Time/resp., sdd, sdls

Note: n.set, number of fruit set per inflorescence; ±nd, days before or after bloom, bl; +, increase; −, decrease; 0, no effect; blank, not reported; sdd, seeded; sdls, seedless

[a] Brackets indicate uncertainty of response or response inferred.

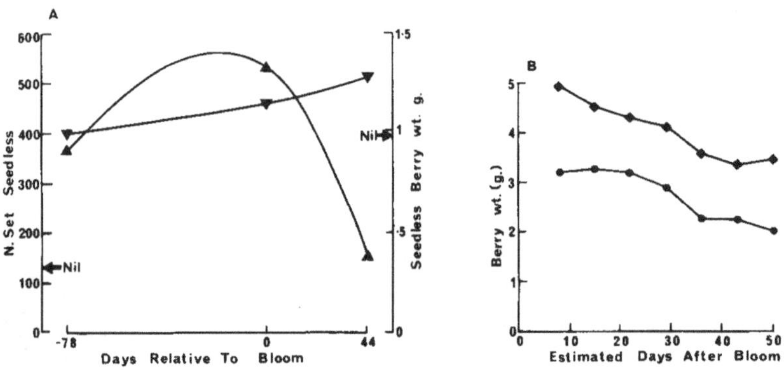

FIGURE 3. Response to cincturing of Hunisa and Sultanina grapes. A. Effect of time of cincturing on the number of seedless fruit set and their subsequent growth in Hunisa. ▲, numbers of fruit set; ▼, berry fresh weight. Data from Jacob.[151] B. Growth of Sultanina fruit in response to time of treatment by cincturing, with (♦) or without (●) GA treatment (15 mg·ℓ⁻¹) at bloom plus 40 mg·ℓ⁻¹ at bloom + 15 days). Data from Weaver and Pool.[377]

Treatments to modify fruit set also encounter a peculiar problem which requires use of additional control treatments beyond use of either a nonsprayed or dipped control or a sprayed without active ingredient. Problems of interpretation arise because application of even a water spray can cause a significant reduction in fruit set[287] and it is also possible that wetting and spreading agents may also cause a further reduction in set because of their toxic nature.[391] Ideally then, experiments to examine the effects of growth substances on fruit set should include up to three controls, a nonsprayed, a water only, and a water plus wetting agent. An additional fault, frequently found in literature on fruit thinning, is use of insufficient replication to establish statistically significant effects at degrees of thinning which have practical significance.

Measurements of numbers of fruit set per cluster are not ideal estimates in experiments where the objective is to reduce the numbers of berries set. Commonly, indices are constructed which may be visual, or based on numbers of berries set per length of cluster branch. In these experiments, the objective is to reduce cluster density to enable unrestricted growth of berries and aeration of the cluster, and thus the method adopted by Samish and Lavee is probably close to ideal because it integrates the effects of berry number and berry volume.[287] The Samish and Lavee method is to measure the number of berries per cluster(n) and mean berry volume (presumably not total berry volume as stated in the original paper) and then to estimate the volume of a cone with dimensions equivalent to that of the fruit cluster(V). Thus index of density (I_d) is found by:

$$I_d = \frac{n \cdot \bar{v}}{V}$$

$$V = \frac{\pi \cdot w^2 \ell}{12}$$

where w is the maximum width of the fruit cluster and ℓ is the length of the fruit bearing portion of the cluster. Total berry volume may also be found by immersion if the individual components of n and $\bar{v}$ are not required. This equation gives values from 0 to 1 (or 0 to 100 if multiplied by 100) and Samish and Lavee obtained values for Shaslas and Queen which ranged from about 20 to 80, with 50 being near ideal.

Table 4
SUMMARY LIST OF AUXIN-LIKE CHEMICALS WHICH HAVE BEEN TESTED FOR THEIR EFFECT ON FRUIT SET IN GRAPEVINES

Compound	Ref.
Active and Useful	
4-Chlorophenoxyacetic acid (4-CPA, PCPA)	62, 386
2,3,4-Trichlorophenoxyacetic acid (2,3,4-T)	11
Naphthaleneacetic acid (NAA)	66, 386
Benzothiazol-2-oxyacetic acid (BTOA)	66, 351
Moderate or Slight Activity, not Useful	
2,4-Dichlorophenoxyacetic acid (2,4-D)	62, 386
β-Naphthoxypropionic acid (BNOPA)	66, 386
α-Naphthoxypropionic acid (ANOPA)	387
Ethyl-β-naphthoxyacetic acid (EBNOA)	62, 66
2,4,6-Trichlorophenoxyacetic acid (2,4,6-T)	11, 62
2,3,5-Trichlorophenoxyacetic acid (2,3,5-T)	11
2,3,5-Triiodobenzoic acid	387
Phenoxyacetic acid (PAA)	66
2-Methyl-4,6-dichlorophenoxyacetic acid (MCPA)	62, 66
Methyl-2-methyl-4-chlorophenoxyacetic acid	66
2-Methylphenoxyacetic acid	62, 66
Indole-3yl-acetic acid (IAA)	62, 386,79
Indole-3yl-butyric acid (IBA)	62, 386
N-2-chlorophenylphthalamic acid	387
N-1-chlorophenylphthalamic acid	387
N-phenylthalimide	387
Oestrone	250
Hexoestrol	62
Keto-hydroxy-estrone	66
Inactive or Toxic	
β-Naphthoxyacetic acid (BNOA)	62, 386
2,4,5-Trichlorophenoxyacetic acid (2,4,5-T)	85, 386
2,4,5-Trichlorophenoxypropionic acid (2,4,5-TP)	287, 387
2,3,6-Trichlorophenoxyacetic acid (2,3,6-T)	11
2-Chlorophenoxyacetic acid (2-CPA)	386
2-Chlorophenoxypropionic acid (2-CPPA)	386
4-Chlorophenoxypropionic acid (4-CPPA)	387
3-Chlorophenoxyacetic acid (3-CPA)	283
2-Methyl-6-chlorophenoxyacetic acid	66
Ethyl-4-chloro-2-methylphenoxyacetic acid	66
Indole-3yl-propionic acid	387
4-Chloro-o-toloxyacetic acid	387
Naphthaleneacetamide	387
O-isopropyl-N-phenylcarbamate	387
α-(2,4-Dichlorophenoxy)-propionic acid	387
4-Chloro-(2-methyl)-phenoxyacetic acid	66
2-Methyl-4-chlorophenoxyacetic acid	66

C. Cincturing

Cincturing (girdling, ringing) is considered here because it has been the standard method of improving fruit set and development of grapes, the standard against which the effectiveness of PGRCs have been judged. The history of this practice has been summarized by Winkler[402] and by Branas et al.[33] Table 3 lists some of the recent information on the subject.

In general, increases in numbers of berries set per cluster occur if cincturing is carried out either just before or during bloom[99,150] and size of seedless berries is enhanced most by treatment at shatter, about 10 to 14 days after bloom.[150,151] If seeded cultivars are cinctured during this period then only parthenocarpic berry set is increased (Figure 3A).[151,188] The practice was most successful on Black Corinth, but if carried out during this period on other cultivars can lead to excessive set and cluster compactness.[373] For this reason later treatment was generally recommended and then in normally well setting clusters, only if combined with cluster trimming and berry thinning.[30,215,377,388]

The size increase obtained by cincturing is additive to that obtained by application of 4-CPA[366,388] or GA[215,377] in the production of enlarged Sultanina fruit for fresh consumption (Figure 3B).

D. Auxins

1. Comparative Activity

Auxins were tested in the late 1940's and early 1950's as a means of enhancing the development of berries and of stimulating the set of parthenocarpic fruit.[62,386,387] At least 40 different synthetic auxin analogues were tested (Table 4) but only one, 4-CPA, came into common use while three others have been recommended for particular and usually different uses. Only one auxin other than 4-CPA was found to be useful for enhancing the growth of parthenocarpic berries (2,3,4-T) but its lower toxicity was not a sufficient benefit to outweight its additional cost and the prior establishment of 4-CPA.[11,14]

BTOA though active as a stimulant of numbers of berries set, was less active than 4-CPA as a means of increasing fruit growth.[66,351] It had particular advantages, which were not followed up, in that it induced a more uniform set and growth and fruit treated with it were considered to be less prone to damage, especially to rainfall-induced fruit splitting.[66] It has since been recommended as a means of delaying fruit maturity (see later).

NAA has been used occasionally in attempts to promote fruit growth but usually without success (Table 5). The principal use of NAA is as a fruit thinning agent, though it has been largely surplanted by GA (see later). 2,4-D and 2,4,5-T both possess slight activity and have been recommended, but both display a very narrow range of positive activity and even slight excess can lead to severe vine damage.[13,14,62,85,386]

2. Toxicity

Toxicity is a characteristic of nearly all the synthetic auxins examined.[363,386,387,392,393] This statement applies even to those which are used commercially such as 4-CPA.[63,159,386] Foliage sensitivity to 4-CPA is about twice as great at early bloom as at fruit set[66] and the maximum concentration which can safely be used is regarded to be about 50 $mg \cdot \ell^{-1}$.[66,346,387]

3. Choice of Cultivars

Use of auxins to promote fruit set and development has been almost wholly confined to the principal seedless cultivars, Black Corinth, and Sultanina (see Table 5). Only one report of a positive growth response of seeded fruit has been made and that has not been followed up.[392] Responses demonstrated in other seeded cultivars include either no response and inhibition of seed development,[392] or a promotion of set of parthenocarpic berries which may or may not soften and mature (see Table 5).

4. Practice

The nature of the response/concentration curve obtained through application of 4-CPA is typical of that of auxins in general, an initial phase of enhancement of response followed by a progressive reduction in response with increasing concentration (Figures 4A and B). The maximum increase in weight of fruit of Black Corinth was obtained with 15 to 20$mg \cdot \ell^{-1}$ 4-CPA. A parallel response curve is obtained for numbers of fruit set per cluster if the 4-CPA is applied at or soon after full bloom.

A wide range of application times have been quoted in the literature, varying from 18

Table 5
SUMMARY OF THE USE OF SYNTHETIC AUXINS TO INCREASE THE NUMBER OF FRUIT SET AND FRUIT DEVELOPMENT

Cultivar	Time	Conc. (mg·ℓ⁻¹)	n.Set	Size	Yield	Ref.	Comments
4-CPA							
Seedless							
Zante	70%bl	50	+	+	+	62	Survey of activity of analogues
	±bl	20	+	+	+	63, 64, 65	Survey of activity of analogues, time/conc. assessment
	70%bl	20				70	Effect of leaf removal
	100%bl	20	+	+	+	96	± GA 1 mg·ℓ⁻¹
	fb or +14d	20		+	+	98	± GA, CCC
	100%bl	20	−	+	−	99	± GA
	50%bl	20	−	+	−	100	± Cincturing, topping
	50%bl	20	−	+	+	101	± Pruning severity, cincturing
	−3d	40		+		286	Anatomical changes
	fb	20			+	159	Vine vigour, time of applic.
	+4d	20	+	+	+	348	± Girdling, time/conc. on seed development
	fb	50		+		386	Survey of activity, maturity
	fb, +4d	50		+	+	387	Survey of activity, time/conc. on seed development
	70%bl	20		+		113	Anatomy of seed, pedicel
	bl	10		+		303	40 mg·ℓ⁻¹ at bl toxic, ± defoliation
	bl	15	+	+	+	396	Effect of leaf removal
	70%bl to +24	25		+		346	Time/conc. on seed development
White Corinth	bl	50		+		387	No seed development, cf. Zante
Sultanina	70%bl	20	+			70	Effect of leaf removal
	−18d	20	+	+	+	154	Prevented flower drop
		5		+		163	Retards maturation; ± GA, cincturing
	+10d	40		+		286	Anatomy, shape
	+21d	20		+		314	± GA, cluster trimming; shape
	+10d	15		+	+	348	Time, interaction with cincturing, pedicel attachment
		15		+		364	Crop level, site of application
	±bl	10	(−)			383	Effect as a thinning agent

Table 5 (continued)
SUMMARY OF THE USE OF SYNTHETIC AUXINS TO INCREASE THE NUMBER OF FRUIT SET AND FRUIT DEVELOPMENT

Cultivar	Time	Conc. ($mg \cdot \ell^{-1}$)	n.Set	Size	Yield	Ref.	Comments
	+10d	20	+	+	+	386	Survey of activity, effect on pedicels
	+10d	20		+		387	Survey of activity, toxicity
	+10d	20	0	+		388	± Thinning, cincturing
Black Monukka	+10d	50		+	+	387	
Pusa Seedless	+2d	20		+	(+)	81	± GA
Seeded							
Khalili	−bl	20	+	+(sdls)	+	106	
Aneb-e-Shahi	fb	10		(+)	(+)	243	No effect on % seedless
Ohanez	+bl	15		+		44	Increases shininess, toughness, and berry attachment
Muscat of Alexandria	bl	50	+(sdls)			387	Survey of analogue activity
	bl	20	+(sdls)			396	Effect of defoliation
Hunisia	bl	50	−	−	−	387	Survey of activity
Grenache	bl	20	0	−		396	Effect of defoliation
Pinot Chardonnay	bl	20	+(sdls)			396	Effect of defoliation
Ribier	+10d	25	+				
BTOA							
Zante	bl to +18d	20	+	+	+	351	Time/conc.; cf. girdling toxicity
	+10d	30	+	(+)	+	351	Uniformity, resistance to damage
NAA							
Muscat	bl, +bl		+			323	Weather, glasshouse crop
Brighton Grape	bl	0.5—100	0	0	0	411	
Queen of Vineyard	+23d	20		0		183	Relation to seed no per berry
2,3,4-T							
Zante	fb	40			+	11	Delay maturity; toxicity
Sultana	+10d	50		+	+	14	

Note: n.set, number of fruit set per inflorescence; ±nd, days before or after bloom, bl; fb, full bloom; +, increase; −, decrease; 0, no effect; blank, not reported; sdd, seeded; sdls, seedless

days before bloom[154] to 24 days after bloom.[346] Some controversy exists concerning the optimum time for application and this matter has not been thoroughly resolved. Jawanda et al.[154] reported that 4-CPA applied either 18 or 10 days before bloom increased the proportion of flowers that set and subsequently fruit growth. Jones[159] claimed an equal effect on yield of Black Corinth whether 4-CPA was applied at 20% bloom or at any time thereafter up to about 2 to 3 weeks after bloom. The relative response after bloom is confirmed by the data published by Weaver[346] (Figure 4C). Coombe[66] however found that response was greatly reduced by application of the chemical at about 20% bloom. The decreased set of Black Corinth reported by El-Zeftawi and Weste[100,101] could support Coombe's observations as

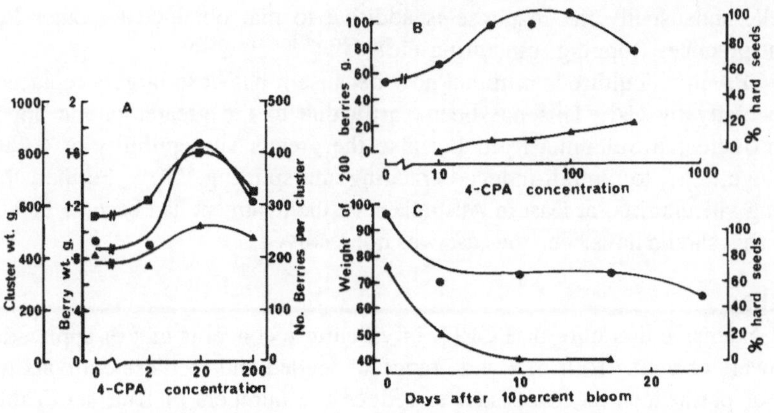

FIGURE 4. Response of Sultanina and Black Corinth grapes to application of 4-CPA. A. Increase in number of berries set, growth, and cluster weight of Sultanina fruit treated with 4-CPA following fruit set, about 10 to 15 days after bloom. Data from Weaver and Williams.[386] Number of fruit per cluster (▲), mean fruit weight (■), mean cluster weight (●). B. Growth of Black Corinth fruit and hard, lignified seed-like structures in response to increasing concentrations of 4-CPA. ●, mean fruit weight; ▲, percentage of fruit which contained the seed-like structures. Data from Weaver.[346] C. Effect of time of application of 4-CPA on growth of Black Corinth berries and the development of hard seed-like structures. Data from Weaver.[346] Symbols as in B.

could those of Weaver[383] who applied 4-CPA at 20% bloom to Sultanina to thin the fruit clusters.

The optimum time of application for increased berry size in Sultanina is generally considered to be at or just after the time of shatter of the non-set ovaries (see Table 5). A somewhat later time of bloom plus 3 weeks was recommended by Sproule and Stannard[314] to avoid the problem of excessive set of hard, nonfleshy (shot) berries.

Application of 4-CPA can stimulate the development of the ovule integuments causing development of a hard, sclerified seed-like structure[113] in Black Corinth but not in the related White Corinth nor in Sultanina or Black Monukka.[387] Development of these structures is greatest if the auxin is applied at 70% bloom and declines to a minimum level at about 7 to 10 days after bloom (see Figure 4C). Development of the sterile seeds is also concentration dependent, increasing linearly with the logarithm of the concentration of 4-CPA (see Figure 4B).

Analysis of tissue sections has shown that 4-CPA increases both cell division and cell expansion, especially in the region of the inner pericarp.[286] The thickening of the pedicel[348,386] is due to enhanced activity of the vascular cambium which produces additional secondary phloem and xylem.[113] The additional thickening has been related to degree of berry attachment[348] which is greatly enhanced in Sultanina fruit. The type of growth induced by 4-CPA is qualitatively different from that caused by application of GA in that growth of the transverse axis is enhanced relative to the longitudinal axis giving spherical fruit.[286]

The site of action of 4-CPA is not entirely clear because it exerts almost as much activity if it is applied to the leaves as if it is applied to both leaves and fruit: application to the fruit only gives a lesser response.[364] No attempt has been made to correlate its effects on fruit set and growth with that on vegetative growth as has been done with the plant growth retardants.[74]

Degree of response obtained by application of 4-CPA depends on crop load: vines which are fully or over-loaded with crop potential respond poorly if at all to application of 4-CPA.[364] However, vines which for reason of genetic constitution, detrimental environmental conditions, or management practices such as pruning level, cluster thinning and pruning, respond well, and usually the response is additive to that obtained by other PGRC's or management practices (topping, cincturing etc).[96,98,99,101,314,348,386]

Use of 4-CPA in viticulture is minimal now because it has been largely replaced by other PGRCs especially by GA. This has been partly due to the greater public appeal of the elongated, GA-treated Sultanina fruit, but also the greater susceptibility of 4-CPA-treated fruit to damage, e.g., to rainfall-induced cracking and splitting.[57,96,163] Further, there was a feeling among viticulturists, at least in Australia, that the treatment had become unreliable[100,175] though why that should have been the case was not resolved.

E. Gibberellins

The most extensive literature on PGRCs in viticulture concerns that of application of GA to grape flowers or young fruit in a wide range of seeded and seedless cultivars to enhance the growth of parthenocarpic fruits and to reduce the numbers of fruit set (Tables 6, 11, and 12). Successes however are not universal and the application of GA is not without its problems (e.g., inhibition of inflorescense initiation).

1. Effect on Numbers of Fruit Set

Authentic examples of GA-induced increase in numbers of fruit set are rare and generally reference to GA-induced increase in fruit set in the literature refers to GA-enhanced fruit growth. The instances where GA has been reported to increase fruit set are those in which the treatment was applied either before[55,106,154,337] or after bloom, at fruit set (shatter).[55,117] The pre-bloom treatments presumably act by increasing the set of parthenocarpic berries on both normally seedless and seeded cultivars (see section on stimulative parthenocarpy). However, there have been some instances of increased set of parthenocarpic berries by GA application at bloom.[199,337,365] Application of GA, though generally antagonizing seed development, has been reported to cause a substantial increase in the numbers of seeded Concord berries (approximately 30%), and also to cause a slight enlargement of those berries.[111]

The most sensitive stage of flower development in terms of susceptibility to GA-induced thinning is early flowering from about 10% bloom to 60% bloom.[26,48,49,155,197,198,305,385] Response of Black Corinth flowers (Figure 5) illustrates this response and shows that the response is maximal in this cultivar at the first sign of flowering. GA-induced thinning is generally regarded as being due to an effect on ovule development/receptivity[224] but in this cultivar the response must either be due to the secondary effect on pollen viability[224,369] or to some other effect, because the ovules have nonfunctional embryo sacs.[259]

The observation that even 0.3 $mg \cdot \ell^{-1}$ is sufficient to cause significant thinning,[56] a concentration which is below that generally required to cause a significant reduction in pollen viability,[369] suggests that some other mechanism may be operating.

Apart from the study on Concord already mentioned, the increases in numbers of fruit set on inflorescences of seeded cultivars were due to enhanced retention of seedless berries (see Table 11). Such berries rarely develop to a useful size and even on seedless cultivars such as Perlette may not soften.[47] In such instances a second application of GA is usually applied to stimulate further development and softening of the berry (see below, section on stimulative pathenocarpy).

2. Comparative Studies

Examination of results obtained during the early period of research on promotion of berry growth by GA led to the idea that responsiveness was related to degree of seed development:

Table 6
SUMMARY OF PUBLISHED DATA ON RESPONSE OF SEEDED AND SEEDLESS CULTIVARS OF GRAPE TO APPLICATION OF GA

Cultivar	Time of application	Conc. (mg·ℓ⁻¹)	n.Set	Size	Yield	Ref.	Comments
Seedless							
Black Corinth	+3d	100	−	+	+	26	Time/conc. series
	+3d	10	0	+	(+)	49	
	−bl, bl, +bl	20	(−)	+	+	55	± CCC, time/Conc.
	+5d	100		+		75	$GA_3 = GA_{4/7}$
	bl, +14d	5	(+)	+	+	98	± CCC, cf. 4-CPA, time of application
	100%bl	1	(+)	(+)	(+)	96	+CCC or +4-CPA (effect not isolated)
	+3d	40		+		286	Anatomical effects
		500		+	+	354	
	70%bl	var		+	+	368	Time/conc./site of application
	90%bl	20	0	× 1.7	× 1.6	13	cf. 4-CPA, cincturing
	80%bl	3	−	+	+	56	± CCC
	bl	15	−	+	+	396	± Defoliation
	bl	10		+		356	$GA_2 < GA_1 = GA_3 = GA_4$
	bl	20	(−)	(+)		369	Pollination not necessary for response
Sultana							
Physiological and general studies							
	−bl, bl, +bl	20	+, −, +	0, +, +	0, 0, +	55	±CCC, time, conc.
	+5d	100		+		75	$GA_3 > GA_{4/7}$
	100%bl	6	−	+		97	±CCC
	bl+	400	0	+		117	cf. Effect on shoot and endogenous GA
	bl+	600		+		128	
	+7d	10		+		286	Anatomical effects
	+61	40		+		303	Offset effect of 50% defoliation
	−bl	100	−	+		325	Multiple applications
	+10d	50		+	+	354	
	−18d	50	+	+	+	154	Control of flower drop
Table grape studies							
	bl, +10d	20/40	−	+		48	Timeing, dual applications
	var	40	−	+	(+)	49	Time, conc., n.applications
	10 to 95%bl	25/30	−	+		215	Time, conc., cincturing
	+10d	20		+	+	314	± Trimming, 4-CPA, n.applications, time
	+10d	1000	−	+	+	368	Crop load, site of application, cincturing
	var	1000		+		377	Cincturing, time (biphasic response)
	bl, +10d	80	−	+		385	Thinning

Table 6 (continued)
SUMMARY OF PUBLISHED DATA ON RESPONSE OF SEEDED AND SEEDLESS CULTIVARS OF GRAPE TO APPLICATION OF GA

Cultivar	Time of application	Conc. ($mg \cdot \ell^{-1}$)	n.Set	Size	Yield	Ref.	Comments
	bl to +54d	15/40		+	+	306	n.Applications, firmness, shattering
	bl	40	−	+		155	Thinning response
	bl, +10d	20	−	+		197	Thinning response, time
	10 to 90%bl	100	−	+		198	Thinning/conc., 20 $mg\ell^{-1}$ GA gave shot berries
	bl, +10d	15		+		384	Time of thinning
Dried or Raisin Fruit Production							
	+7d	20	0	+	+	14	cf 2,3,4-T, DPU, uracil
	50%bl	10	0	+	+	165	Variance of fruit weight
	50%bl	10	0	+	0	166	
Cape Current	−bl, bl, bl+	20	− 0, 0	+	+	55	Time, conc., ±CCC
Perlette	bl	40	−	+	(−)	47	Time and thinning
	fb	50	(−)	(+)		85	
	bl, +10d	15, 80	(−)	+		168	Site, conc., n.applications, thinning effect
	bl, +10d	125	−	+		336	Conc., thinning
	+10d	1000		+		377	Poor response, cf. Sultanina
	bl+	50		0		205	Response, cf. Sultanina
	fb	125		+		84	$GA_3 \simeq GA_{4/7}$
	bl	80	(−)	+		369	Thinning but poor response
Pusa Seedless	+3d	100		+	+	81	±4-CPA, uracil, kinetin
		50	±	+	+	178	<50 Increased n.set
	−bl, bl	125	×3.2	×2.2	×6.9	337	Large positive response
Himrod	fb	50	0	0	0	24	Depressed seed rudiment growth
	+10	40	×1.15	×1.01	×1.23	192	
	+20d	100		+	+	292	
Frankental	bl+	500		+		128	Less responsive than Sultanina
Shundokhani	+14d	50		×3.0	+	285	Large seasonal variation in response
31-123F	bl+	40	×1.4	× 1.6		117	cf Effect on shoot growth and endogenous GA
38-13F	bl+	40	×1.3	×1.6		117	
42-1F	bl+	40	×1.2	×1.3		117	
43-13F	bl+	40	×1.8	× 1.0		117	
43-16F	bl+	40	×1.7	×1.0		117	
Black Kishmish	bl, +10d	500	0	+	+	199	Conc., site, rain, 50 $mg\ell^{-1}$ optimum
	bl, +10d	100		+		200	Response in relation to seed development

Table 6 (continued)
SUMMARY OF PUBLISHED DATA ON RESPONSE OF SEEDED AND SEEDLESS CULTIVARS OF GRAPE TO APPLICATION OF GA

Cultivar	Time of application	Conc. (mg·ℓ⁻¹)	n.Set	Size	Yield	Ref.	Comments
Black Monukka	+ 10d	50		+		368	
Round Kishmish	bl, + 10d	100		+		199	
Concord Seedless	bl, + 6d	200	+			271	
Seeded							
Show enhanced growth of fruit							
Aneb-e-Shahi	+ 21d	100	0	× 1.3	(+)	80	Increased width of berry
Bhokri	+ 21d	100	0	× 1.6	+	80	+ 21d, + 7d
Concord	+ 5d	200		+	+	111	± SADH
Neo Muscat	+ 20d	100		+	+	292	
Queen of Vineyard	+ 23d	20		+		183	If less than 2 seeds per berry
Seeded Thompson	+ bl	40	× 1.1	× 1.6	(+)	117	cf. Shoot growth and endogenous GA
Seibel 9549 (de Chaunac)	+ 10d	50	0	+	+	194	± Thinning, cluster position, hardiness
	+ 10d	50	−	+	+	193	± CCC, thinning
Tokay	bl	5	+ (sdls)	+		369	Pollination not required for response
Unresponsive (< 10% incr.) or growth depressed by GA application							
Alamwick	+ 21d	100		0	0	80	
Bharat Early	+ 21d	100	0	0	0	80	
Black Hamburg	+ 21d	100		0	0	80	
Campbell Early	+ 20d	100		+	+	292	Best at seed hardening (browning)
Delaware	+ 37d	100		× 1.5	+	140	
	+ 25d			+	+	141	Best at seed hardening (used 5 mg·ℓ⁻¹ at − bl to thin)
Gros Colman	+ 21d	150		+	+	80	
Muscat of Alexandria	− bl, bl, + bl	20	+ (sdls)	+ (sdls)	0	55	− bl reduced nsdd and yield
	+ 20d	100		+	+	292	
	bl	5	−	0	−	396	± 4-CPA, defoliation
Black Muscat	+ 21d	100		0	0	80	
Doradillo	− bl, bl, + bl	20	+ (sdls)	0	±	55	Time, conc. ± CCC

Note: n.set, number of fruit set per inflorescence; ± nd, days before or after bloom, bl; fb, full bloom; +, increase; −, decrease; 0, no effect; blank, not reported; sdd, seeded; sdls, seedless.

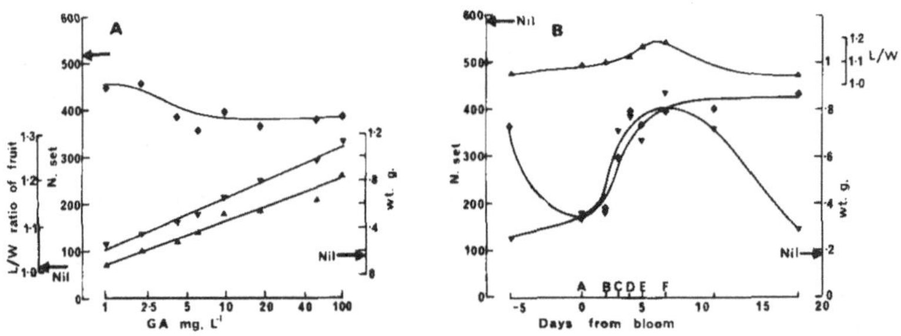

FIGURE 5. Response of Black Corinth fruit to applied GA. A. Change in growth, number of fruit set per cluster and fruit shape in response to applied GA (mean of 3 application times; 75% bloom, bloom + 3 days and bloom + 7 days). ▲, mean fruit weight; ▼, mean fruit shape, ♦, mean number of fruit set per inflorescence. Data from Bertrand and Weaver.[26] B. Change in growth, number of fruit set per cluster, and fruit shape in response to GA applied at range of stages of inflorescence and fruit development. Symbols as in A. Developmental stages listed on the x-axis are as follows; A, initiation of capfall; B, 25% capfall; C, 50% capfall; D, 75% capfall; E, 95% capfall; F, 3 days after full bloom. Data from Bertrand and Weaver.[26]

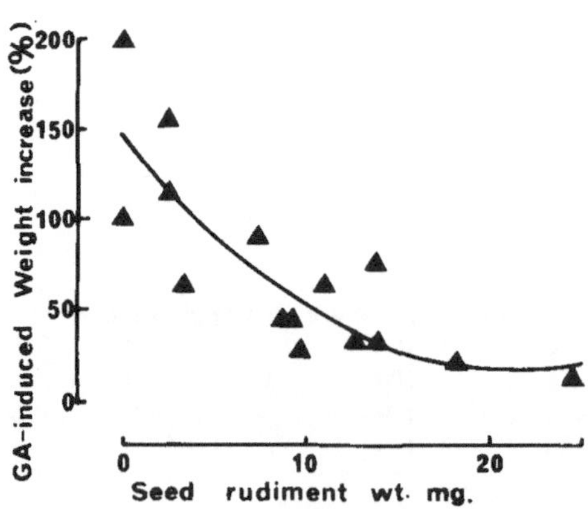

FIGURE 6. Relation between increase of growth of 15 seedless grape cultivars treated with 200 mg·ℓ⁻¹ GA at bloom and weight of seed rudiments in mature untreated fruit. Data from Smirnov and Perepelitsyna.[311] The curve represents a second order polynomial line of best fit, $y = 146.1 - 12.3x + 0.29x^2$, $R = 0.84$.

parthenocarpic fruit being most responsive, followed by those fruit which were stenospermocarpic and which responded only after the embryo had aborted.[55,75,311] Those seeded cultivars which responded, responded in a manner which was related to the degree and number of seeds per berry and the degree of their development.[183] More recent research has demonstrated considerable variation in degree of response to applied GA in both seeded and seedless cultivars.[78,80,117,140,141,144]

Attempts to predict responsiveness to applied GA on the basis of degree of seed development,[78,80,311] natural tendency to set a proportion of seedless berries,[200] responsiveness of shoot growth to applied GA,[117] or content of endogenous gibberellin-like substances[117] have been only partially successful. Smirnov and Perepelitsyna[311] established a relationship between seed rudiment weight and relative response to treatment (Figure 6) and Manankov[200] who examined the response of 102 cultivars found that responsiveness was substantially accounted for by the tendency of cultivars to set seedless fruit naturally. A regression between

Table 7
CORRELATION MATRIX OF RESPONSE OF 7 GRAPE CULTIVARS TO APPLIED GA IN TERMS OF FRUIT CHARACTERISTICS

		I	II	III	IV	V	V/VI
% Weight increase	(I)	1.00					
% Seedless berries	(II)	**0.84**[a]	1.00				
Mean seed no. per berry	(III)	**−0.77**	**−0.90**	1.00			
Mean seed weight per berry	(IV)	−0.00	−0.54	0.24	1.00		
Mean berry weight	(V)	−0.00	−0.23	−0.47	0.57	1.00	
Seed index	(V ÷ IV)	0.66	**0.89**	**−0.89**	−0.36	0.54	1.00

Note: Calculated from the data presented in the original paper.[44] Cultivars used were, Bharat Early, Black Hamburg, Black Muscat, Aneb-e-Shahi, Bhokri, Gros Colman (Pusa), and Alamwick.

[a] Values in bold print are statistically significant at $P \leq 0.05$.

relative increase in size of fruit in response to applications of GA 100mg·ℓ^{-1} applied at bloom and again about 10 days later and the percentage of seedless berries reported for 20 cultivars presented in the paper gave the relationship:

$$y = 94.3 + 1.07x,\ r^2 = 0.89$$

where y is percent increase in weight of fruit and x is percent seedless berries. However, it is most likely that the response obtained in this study reflects the enhanced growth of seedless fruit and not that of seeded fruit (see below and section on stimulative parthocarpy).

Support for these results were obtained in another study (Table 7)[78] in which a greater range of factors were examined. However, in this study the values for percent of seedless berries were those from treated clusters and thus the results are confounded with the response of the seedless berries to GA. The most significant part of this study was that a statistically significant and negative correlation was demonstrated between seed number per berry (but not seed plus seed rudiment weight!) which confirms the results obtained by Lavee for Queen of Vineyard.[183] Response was not related to berry weight.

Absence of seed is not a dependable indicator of responsiveness to applied GA (see Figure 6). Hagiwara et al.[117] cite examples of cultivars in which the range of response was about 6%, 30% or 60% depending on the cultivar. Also, Perlette,[369,377] Frankental,[128] and Himrod[24] are reported to be generally less responsive than Sultanina to GA applied at bloom or at fruit set, about 10 days later.

Presence of seed does not necessarily create a lack of growth response to applied GA though responsiveness has not been well documented. A number of reports have been made which demonstrate that some cultivars respond to GA applied between 3 and 5 weeks after bloom at a stage of development which is associated with seed hardening.[80,140,141,143,144,292] This stage probably coincides with termination of the first period of post-bloom growth. Cultivars which have been shown to respond by increased growth include, Campbell Early, Delaware, Muscat of Alexandria, Muscat Bailey A, Aneb-e-Shahi, Bhokri, and Gros Colman. Unresponsive cultivars include, Alamwick, Bharat Early, Black Muscat, and Black Hamburg.

In summary, it appears that there are genetic controls which determine responsiveness of fruit to applied GA and that these factors are independent of those which determine responsiveness of shoots to applied GA.[117] The presence and degree of seed development appears to regulate, not necessarily the responsiveness of fruit growth, but the timing of the

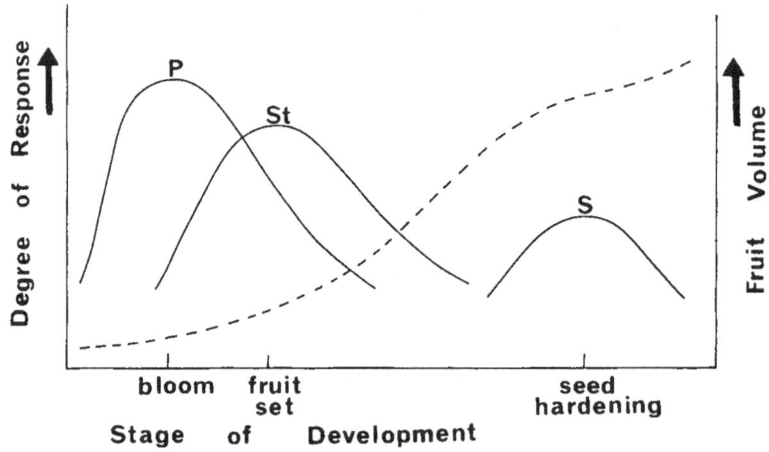

FIGURE 7. Idealized GA response curves for grape cultivars varying in degree of embryo and seed development. P, parthenocarpic; St, stenospermocarpic; S, seeded. It is unlikely that such a set of discrete curves exist in practice, but rather there is probably a continuous function with the optimum coinciding with cessation of seed development with the opportunity for relative increase declining with time.

response. The parthenocarpic cultivars which respond, do so most intensely in response to GA applied within a few days of full bloom (see Figure 5),[26,41] stenospermocarpic cultivars and seedless fruit on clusters of seeded cultivars appear generally to be most responsive at about the time of embryo abortion, at about 10 days after bloom, while those seeded cultivars which respond, do so at the end of the first phase of growth, at the end of the growth of the testa and nucellus.[68,246] Substantial growth responses have been reported in each instance (see Table 6). An idealized summary of this information is presented in Figure 7.

3. Characteristics of the Response

The relationship between concentration of applied GA and response is best described by a semi-log relationship of the type $y = a + b \log x$ where y represents for example berry weight or berry shape and x is the concentration of the applied GA. Examples of this relationship are presented in Figure 5 and Table 8. A summary of some selected data from the literature suggests that GA applied close to the optimum stage of development causes about 0.5 g increase in weight for both Black Corinth and Sultanina for each 10-fold increase in concentration tested (range about 1 to 1000 mg·ℓ^{-1}). GA applied at a much later stage to Sultanina gives less than half the response obtained at the earlier time. Double application of GA to Nimrang[199] gave a response which was approximately equal to that expected from a single application to Perlette. The response of Nimrang berries was presumably due to seedless berries though this information was not reported.

Gibberellin also exerts a specific morphogenetic effect on berry shape causing in seedless berries relatively more axial than radial growth,[286] and causing a redistribution of growth in seeded cultivars,[55,243] even in those which show no overall growth response.[55] As mentioned previously, in an earlier section, this response is presumably related to an enhancement of rib meristem activity in the maturing ovary.[59] The effect is generally greatest when GA is applied either just prior to or during the early phases of flowering.[55,26] An example of the relationship between timing and response is presented in Figure 5. This response, like that of berry weight also follows a log-linear response curve (Figure 5). This however is not a full explanation of the effect of GA because an anatomical study carried out on fruit treated at bloom plus 7 days showed a bias towards polar growth and a suggestion of increased cell division in the pericarp.[286] During this period of post-bloom development the inner pericarp

Table 8
REGRESSION OF THE RELATIONSHIP BETWEEN BERRY WEIGHT (Y, G) AND CONCENTRATION OF APPLIED GA (X, MG·ℓ⁻¹) FOR SOME SELECTED CULTIVARS

Cultivar	Equation	r^2	Ref.	Comments
Black Corinth	$y = 0.22 + 0.44 \log x$	0.99	26	Mean of three application times 75% bl + 3d and + 7d
Sultanina	$y = 2.35 + 0.46 \log x$	0.94	377	Applied about bloom + 3 to 5d
	$y = 3.13 + 0.18 \log x$	0.91	377	Applied about bloom + 23 to 25d
Perlette	$y = 1.88 + 0.25 \log x$	0.82	377	Applied about bloom + 10d
Nimrang	$y = 3.34 + 0.29 \log x$	0.92	199	Dual application, bloom and bloom + 10d

Note: bl, bloom; d, days

cells divide periclinically as a radial meristem[446] and GA presumably prompts greater activity in proximal and distal regions of the inner pericarp. GA applied at later stages of fruit development doesn't cause fruit to elongate and may even cause a regression in shape towards a more spherical form.[80] The nature of these later responses have not been studied but a trend towards a more spherical berry could be predicted on purely physical-mechanical grounds for treatments which presumably promote only cell expansion;[55] cell division being virtually complete within 14 days of bloom.[61]

Pollination is not essential for the GA promotion of fruit growth in those cultivars which are responsive and indeed may even be considered to be detrimental, in that effectiveness of GA is usually less during the presence of seeds. However, while application of GA may to a degree replace pollination in some cultivars[224,369] this is by no means true of all cultivars, a fact which limits its usefulness as a replacement for pollination of some female cultivars (see Table 6).

Gibberellin, unlike the auxins, is relatively ineffective in promoting fruit growth unless it is applied directly to the fruit. Application of GA to leaves only produces relatively little effect on the fruit of Black Corinth and Sultanina and if applied to flowers in particular portions of an inflorescence causes an effect in the treated portion only.[368]

The relationship between crop load and response to GA has not been studied in sufficient depth or precision to enable unequivocal statements concerning the relationship. The data for one study reported by Weaver and McCune suggest that the effect of GA on yield of Sultanina is independent of crop load over the range studied (16 to 48 fruit clusters per vine), but that crop load increase causes a reduction in individual berry weight and extends the period required for the fruit to attain maturity (Table 9).

Several of the known gibberellins have been tested and in general $GA_3 = GA_4 = GA_7 > GA_2$,[75,356] but one cultivar, Black Corinth is more responsive to GA_3 than to a mixture of $GA_{4/7}$[75] (Table 6).

4. Practice

Rainfall is a factor in response to GA application, and rainfall occurring within 8 to 12 hr of application may reduce the effectiveness of the spray and perhaps necessitate reapplication.[199] Also, most authors recommend inclusion of a neutral wetting agent to a concentration of 100 to 200 mg·ℓ⁻¹.

Table 9
DEPENDENCE OF DEGREE OF RESPONSE OF FRUIT OF SULTANINA TO APPLIED GA ON CROP LOAD

Yield (y, kg per vine) in relation to number of clusters per vine (x)		
Nil GA	$y = 5.5 + 0.56\ x$	$r^2 = 1.00$
20 mg·ℓ⁻¹ GA	$y = 11.3 + 0.51\ x$	$r^2 = 0.88$
Berry weight (y, g) in relation to yield per vine (x, kg)		
Nil GA	$y = 2.5 - 0.025\ x$	$r^2 = 0.83$
20 mg·ℓ⁻¹ GA	$y = 3.5 - 0.031\ x$	$r^2 = 0.88$
° Brix at harvest (y) in relation to yield per vine (x, kg)		
Nil GA	$y = 24.5 - 0.27\ x$	$r^2 = 0.85$
20 mg·ℓ⁻¹ GA:	$y = 25.2 - 0.30\ x$	$r^2 = 0.89$
Combined (Nil + GA)	$y = 24.7 - 0.28\ x$	$r^2 = 0.88$

Data re-analyzed from Weaver and McCune.[368]

The ability of GA to promote growth of fruit of responsive cultivars at a wide range of crop loads is also reflected in its ability to enhance growth additively to the effects of other PGRCs. Compounds and treatments tested for additive effects include 4-CPA,[81,96,98,314] CCC[55,56,97,98,193] cincturing,[215,368,377] and, SADH.[111] Such combinations are often advocated to maximize increase in growth of fruit at moderate levels of applied GA to minimize costs and to avoid unwanted side-effects[215,377] or to offset the effects of the plant growth retardants on fruit size.[193,111]

Combinations of treatments are often taken to their extreme in the production of table grapes. It is common practice to apply two sprays of GA, one at early to mid-bloom, to thin the cluster to provide sufficient space for growth of the berries, and to produce attractive elongated berries and to reduce the over-all crop load; and a second at the time of fruit setting about 10 to 14 days later to maximize berry growth. Also, cincturing is often practised in association with the second spray. Other treatments commonly combined with gibberellin application and cincturing include removal of the basal portions of the fruit clusters, and removal of any clusters judged to be excessive (see references listed in Table 6). Each of these treatments are necessary to produce the ideal fruit cluster for table use: one which is of moderate size, with large elongate berries, which is loose enough to enable removal of individual berries and to allow unimpeded growth of individual berries, and a cluster whose rate of maturation is not limited by total crop load.

Use of GA for purposes other than table grapes is limited by the availability of other, cultural methods of maximizing yield and cost.[14] The economics of dried fruit or raisin fruit production are closely tied to productivity per hectare and usually, productivity is best maximized by modification of pruning, training or other cultural practices. Creation of an elite market for enlarged fruit may justify use of GA for part of the crop.[165,166,167]

F. Cytokinins

Cytokinins are the only class of PGRC which have been shown to be capable of promoting substantial and in some instances almost complete set of all ovaries present at bloom.[242,284,397,408,409] However, despite demonstration of the effectiveness of some synthetic cytokinins, their potential appears to be perceived as being of curiosity value only and not worthy of detailed investigation.

Problems facing users of cytokinins include, relatively high cost, poor solubility in aqueous solutions of moderate pH (about 80 mg·ℓ⁻¹ for BA, and 200 mg·ℓ⁻¹ for PBA), poor uptake from external surfaces, and low mobility within the plant. Development of PBA has reduced the degree of some of these problems, notably, solubility, and mobility[397] but cost and degree of uptake may still pose problems. Concentrations of both BA and PBA which greatly exceed the nominal solubility of the compounds have reputedly been applied.[397,408,409,410] In these instances the active ingredient was supplied as a concentrate in isopropanol and the dilutions applied contained from 1 to 10% v/v isopropanol. Additionally, Zuluaga and his co-workers[409,410] kept the solutions at temperatures up to 45°C to maintain solution of the active ingredient.

1. Characteristics of the Response

Responses recorded for cytokinins include enhanced percentage of flowers set of both seedless berries on naturally seedless and normally seeded cultivars[397,408,409,410] and of seeded berries in a seeded cultivar, Cardinal.[187] In Cardinal, 300 mg·ℓ^{-1} PBA applied at 50% bloom doubled the number of seeded berries set, but this increased set was associated with reduced berry size and with the setting of a very substantial number of commercially unacceptable parthocarpic berries.[187] However other instances have been cited[219] of increased set of seeded berries as a result of cytokinins applied 2 to 3 weeks before bloom. In these instances, the increase in set was associated with an enhancement of the proportion of perfect flowers in a cultivar, Muscat Bailey A, which normally produces flowers which range from male to perfect hermaphrodite.[138,139,331] Use of cytokinins to produce parthenocarpic berries on seeded cultivars is discussed in the section on stimulative parthenocarpy.

Increase in the number and weight of berries set in the parthenocarpic cultivar, Black Corinth, is related to the logarithm of the concentration of the compound applied, either BA or PBA (Figure 8). The graphs are explicit examples of some of the problems associated with application of cytokinins: the results are variable and considerable amounts must be applied to obtain perceptable effects. These characteristics are presumably associated with poor penetration and unpredictability of amount actually applied when the concentration exceeds the solubility of the active ingredient in the solvent. The results suggest that PBA is more effective mole for mole than BA because the slope of the response curve for PBA is 2 to 3 times as great as that of BA. A shift in effectiveness due for example to improved penetration should be expressed in a change in intercept, rather than slope, with smaller concentrations being more effective.

The response of Sultanina to PBA applied at fruit set, about 10 days after bloom, differs from that observed for Black Corinth inflorescences treated at bloom. In Sultanina, PBA at concentrations up to 500 mg·ℓ^{-1} increased the proportion of berries set up to just more than twofold, but only the highest concentration tested (1000 mg·ℓ^{-1}) caused a significant increase in the weight of treated berries.[397]

2. Practice

The place of cytokinins in viticulture is indeed uncertain and utilization for enhancement of numbers of flowers set either directly by application at bloom or indirectly, through improved flower development, by application before bloom will probably depend on improved technology of application or development of improved analogues.

G. Plant Growth Retardants

1. Types and Use

Plant growth retardants are the only class of PGRC which has found general acceptance as a means of increasing the number of normal berries. Four types of retardant have been shown to be active in *Vitis* quaternary ammonium (CCC),[51,72] phosphonium (Phosphon D),[72] aliphatic azide (SADH),[27] and cyclic azide (MH).[69,184] Of these chemicals only two have been widely tested, CCC and SADH (Table 10).

2. Effects on Number of Berries Set

An overwhelming number of the reports examined found that CCC application to cultivars

Table 10
EFFECT OF TWO PLANT GROWTH RETARDANTS, CCC AND SADH, ON FRUIT SET AND YIELD OF A NUMBER OF CULTIVARS, SPECIES, AND THEIR HYBRIDS, TOGETHER WITH A SUMMARY OF TREATMENT CONDITIONS

		CCC					SADH						
Species and cultivars	Presence of seeds	Maximum conc. applied (mg·ℓ⁻¹)	Time of application (± nd)	Fruit number	Fruit size	Yield	Maximum conc. applied (mg·ℓ⁻¹)	Time of application	Fruit number	Fruit size	Yield	Reference	Comments
V. vinifera													
Cape Current	(+)	2000	− 14	+	0	+						55	± GA
Sultanina	−	2000	− 14	+	−	0						55	± GA
		100	b/n − 21, − 14	+								72	
		1000	− 10	+	−		1000	− 10	0	0	0	73	
		300	bl	+	−							97	± GA
		1000	− 21 and bl	+	−	(+)	4000	− 21 and bl	+	−	0	360	
Black Corinth	−	2000	− 14	(+)	0	0						55	± GA
		100	b/n − 21, − 14	+								72	
		100	+ bl	+	+	+						96	+ GA
		1000	− 10	+	(+)	+						73	
Perlette	−	2000	− bl	±	+	+						85	
Yaghooti	−						7500	bl	+		+	330	
Cabernet Sauvignon	+	100	− 20	+								74	
		1000	− 10	+	−	(+)	1000	− 10	0	0		73	
		300	bl	0(+)	0	0(+)						211	Response in next season
		200	− 7	0	0	0						284	Light/temperature stress
		200	− bl	+	+	+						307	Applied to roots in culture solution
		600	− 7	+	0	(+)						359	
		2400	− 21/− 14	+	−	+						262	Effect of cluster position
							8000	− 12/− 5/bl	+	−	0	263	

Cardinal	+	1000	ca. −5	+	−	(+)	1000	ca. −5	0	0	0	242	
Carignane	+	200	−7	0	0	0						284	Light/temperature stress
Chardonnay	+	1000	−7	+	−							359	
Ciliegliolo	+	1000	−15	+	−	+						195	
Doradillo	+	300	−14	+	−		300	−14	0	0	0	71	
		100	b/n −21, −14	0								72	
		1000	−10	+	−	±	1000	−10	0	0		73	
		100	−20	+								74	
		2000	−14	+	−	0						55	± GA
French Columbard	200	−7	0	0	0							284	Light/temperature stress
Grenache	+	100	20	+								74	
		1000	−10	+	−	±	1000	−10	0	0	0	73	
		2000	−15, −1	+	−	±						345	
Khalili	+	1200	−bl	+								106	
Müller-Thurgau	+	0.4 kg·ha⁻¹	−bl	+		(+)						51	
Muscat of Alexandria	+	2000	−14	+	−	0, −						55	Yield depressed by −bl application
		300	−14	+	−	(+)	300	−14	0	0	0	71	
		100	b/n −21, −14	+	−	+						72	
		1000	−40 to bl	+	−	+	1000	−10	0	0	0	73	Seasonal variation in yield response
		300	−20	+	−	+						74	
		300	−20 to bl	0	0	0						97	
		200	−18	+								232	
		200	−26 to +2	+								235	
		200	−20	+								237	
Muscat Ottonel	+	4000	−bl	+	−							136	Deletarious effects of high conc.
Palomino	+	300	−14	+	−	+	300	−14	0	0	0	71	
		100	−20	+	−							74	
		1000	−10	+	−	+	1000	−10	0	0	0	73	Seasonal variation in response
Pinot Noir	+	200	−7	0	0	0						284	Light/temp.stress
		1000	−7	+	−	8						359	

Table 10 (continued)
EFFECT OF TWO PLANT GROWTH RETARDANTS, CCC AND SADH, ON FRUIT SET AND YIELD OF A NUMBER OF CULTIVARS, SPECIES, AND THEIR HYBRIDS, TOGETHER WITH A SUMMARY OF TREATMENT CONDITIONS

		CCC					SADH						
Species and cultivars	Presence of seeds	Maximum conc. applied ($mg \cdot \ell^{-1}$)	Time of application (± nd)	Fruit number	Fruit size	Yield	Maximum conc. applied ($mg \cdot \ell^{-1}$)	Time of application	Fruit number	Fruit size	Yield	Reference	Comments
Riesling	+	0.2 kg·ha^{-1}	−bl	+		(+)						51	
Shiraz	+	300	−14	+	−	0	300	−14	0	0	0	71	
		1000	−10	+	−	+	1000	−10	0	0	0	73	
		100	−20	+								74	
Sylvaner	+	0.2 kg·ha^{-1}	−bl	+		(+)						51	
Traminer	+	0.2 kg·ha^{-1}	−bl	+	−	(+)						51	
Trebbiano Bianco	+	1000	−bl	+	(−)	+						274	Repeated application gave premature lignification
White Riesling	+	200	−7	0	0	0						284	Light/temperature stress
Interspecific Hybrids													
Himrod	−	2000	−9, −3	+	0	+	2000	−9, 3	(+)	(−)	(+)	24	
NY 21572	−	1500	−9, −3	+	0	+	2000	−9, −3	0	(−)	0	24	
NY 21576		1500	w9, −3	+	0	+						24	
Kyoho	+						2500	−21	+			232	
							2500	−12,bl	+	−		233	Sdd/sdls in response to time of
							2500	−25 to −4	+			235	
							2500	−18	+			237	
		1000	−26	(+)	(+)	(+)	5000	−26	+	(±)	+	238	
de Chaunac (Seibel 9549)	+500	+10	+	−	0							193	Cluster thinning ± GA

V. labrusca							
Concord	1000	−bl and bl	+	+	+	41	Summary of 7 year results
	1000	ca. −3	+	m1.6−	0	115	
	2500	−7.bl, +21	+			137	
	1000	−11	+	(−)	(+)	338	Relation to temperature
	1000	−10,bl	+	−	+	339	Relation to degree of set

Note: n.set, number of fruit set per inflorescence; ± nd, days before or after bloom, bl; +, increase; −, decrease; 0, no effect; blank, not reported; sdd, seeded; sdls, seedless.

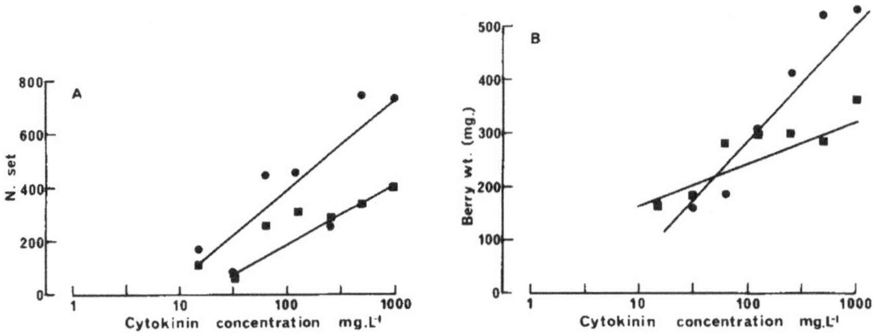

FIGURE 8. Increase in number and growth of fruit set on inflorescences of Black Corinth grapes in response to applied cytokinins. ■, BA; ●, PBA. Regressions for numbers of fruit set obtained were; $y = -94 + 168 \log (BA)$, $r^2 = 0.86$; and $y = -266 + 325 \log (PBA)$, $r^2 = 0.70$. The relationships for fruit weight (y measured in mg) obtained were: $y = 63 + 99 \log (BA)$, $r^2 = 0.83$; $y = -330 + 330 \log (PBA)$, $r^2 = 0.93$. Data from Weaver et al.[397]

of *V. vinifera* or to interspecific hybrids which include *V. vinifera* as a parent, produced an increase in the number of berries set. The only exception being a case where potted vines treated 1 week before bloom were placed under conditions of severe light and low temperature stress.[284] The effectiveness of CCC in increasing the number of berries set is independent of the presence or absence of seeds in the fruit of particular cultivars (see Table 10) and generally does not effect the ratio of seeded to seedless berries set.[55,274,235]

SADH generally does not affect the ratio of seeded to seedless berries set, though an instance of specific enhancement of set of seedless berries has been reported.[233] However this effect was confined to instances where the SADH was applied close to bloom. Earlier application produced a uniform increase in set of numbers of seeded and seedless berries.

3. Time of Treatment

Time of application is crucial if the maximum increase in numbers of fruit set is to be obtained. Coombe's studies[73] showed a relatively sharp optimum for increase in numbers of fruit set in terms of day of application before bloom: this optimum day of application varied from 8 to 28 days before bloom according to season and location. Whether the optimum day of application coincides with a particular developmental stage has not been reported. Naito[235] also found a degree of variation of time to optimum response according to location and season but the variation was only between 12 and 18 days before bloom and a satisfactorily high degree of response was obtained between about 26 and 13 days before bloom (Figure 9). In this latter instance, treatment between 7 days before bloom and bloom gave no positive response. Despite these general conclusions concerning time of application an increase in number of fruit set has been reported for applications as late as a few days after full bloom.[96]

4. Effects on Fruit Growth

Both CCC and SADH cause an unwelcome decrease in berry volume (see Table 10). A yield increase occurs only in instances where the increase in set outweighs the decrease in berry volume. It is therefore most surprising that so little research has been published on the interaction between time of application and concentration of CCC applied on number of berries set per cluster and reduction in berry volume. It has been claimed that the reduction in volume is a matter of overall yield[211] but other studies have shown that berry size (volume or weight) is decreased by application of CCC, whether applied before or after bloom[55] and whether cluster thinning was practiced.[193]

Two studies which have examined the relative response of number of berries and berry

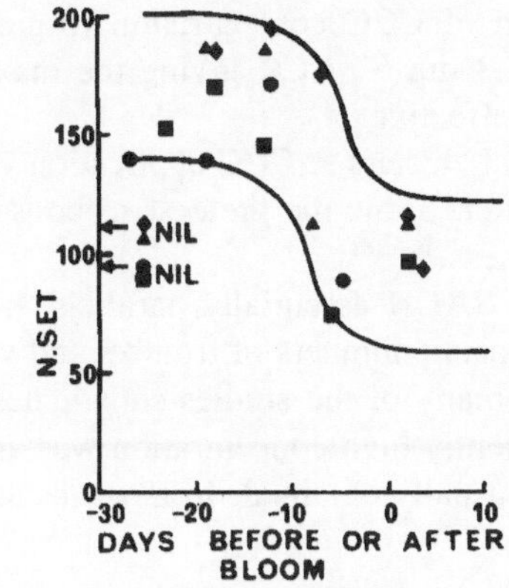

FIGURE 9. Effect of time of application of CCC to developing flowers of Muscat of Alexandria grapevines on the numbers of fruit set following bloom. The concentration applied was 200 mg·ℓ$^{-1}$ and the data represents the results obtained in 4 experiments. ●, 1977; ♦, 1978A; ▲, 1878B ■, 1979. The curves represent a "by eye" fit of the upper and lower limits of the results and indicate a sharp stepwise response. Data from Naito.[235]

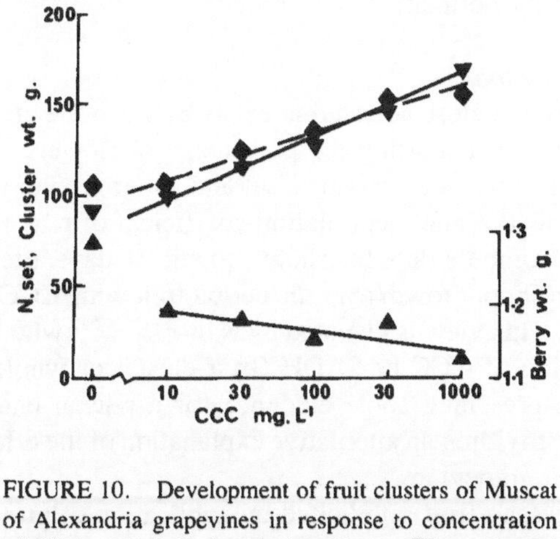

FIGURE 10. Development of fruit clusters of Muscat of Alexandria grapevines in response to concentration of CCC applied 10 days before bloom. ▼, number of fruit set per cluster; ♦, cluster weight; ▲, fruit weight. The curves represent the following relationships, in order, number of fruit set; fruit cluster set; fruit cluster weight (g); individual fruit weight (g): $y = 63.5 + 34.8 \log (CCC)$, $r^2 = 0.99$; $y = 83.2 + 26.1 \log (CCC)$, $r^2 = 0.95$; $y = 1.22 - 0.0274 \log (CCC)$, $r^2 = 0.68$. Data from Coombe.[73]

weight to a concentration series of CCC drew similar conclusions about the response of berry weight but different conclusions in relation to berry set.[73,262] Both authors showed that response of berry weight was apparently asymptotic and reached a minimum at a relatively low concentration of retardant, between about 3 and 30 mg·ℓ$^{-1}$. However, the data presented in Figure 10 is well fitted by a linear-log regression line. Extension to zero effect, suggests that a CCC concentration as low as 0.01 mg·ℓ$^{-1}$ may reduce berry weight. The example of response of numbers of berries set given[73] also shows a clear linear increase in numbers of berries set with the logarithm of CCC concentration (Figure 10). The Peterson study[262] showed a step response, with 3 mg·ℓ^{-1} CCC giving the maximum increase in set. Higher concentrations were no more effective.

Responsiveness of individual clusters to CCC application varies with position of insertion along the shoot, primary clusters show the greatest response and higher order clusters are progressively less responsiveness.[193,262]

Response of grapevines to SADH essentially parallels that of CCC with the exception that it is less effective in increasing numbers of fruit set and yield in *V. vinifera* than is CCC (see Table 10). However, in many of the studies shown, less than optimal concentrations were applied. Where a sufficiently high concentration was applied, an increase in numbers of berries set per cluster was usually obtained. In most instances, the weight of *V. vinifera* berries treated was reduced.

5. Cultivar Differences

Concord, *V. lubrusca*, shows a greater sensitivity to SADH than to CCC[32] but also shows a difference in the nature of the response, in that, depending on concentration applied, shoot growth may be either increased or decreased. The nature of this response, concentration relationship may be reflected in the general increase in berry size reported by Cahoon et al.[41] A similar positive response berry weight may occur in the interspecific hybrid, Kyoho, treated with either CCC or SADH (see Table 10). Most reports examined, however, showed a decrease which in one instance at least was sufficient to completely offset the effect of the increased berry numbers on yield.[115] In this study however, the chemical was applied rather later than is probably optimal.

6. Characteristics of the Response

It is generally agreed that a close correlation exists between the effect of CCC in reducing the rate of shoot growth and increasing the percentage of flowers set.[74,307] The regression equation obtained by Skene between arcsin % set and shoot growth rate (cm · week^{-1}) gave a regression coefficient of -0.5 and a correlation coefficient of $r = -0.9998$ ($y = 38.1 - 0.5 \times$).[307] Analysis of Coombe's data for shoot growth 21 days after bloom also yielded a close correlation between shoot growth over the period following CCC application and degree of fruit set ($r = -0.93$). This view is disputed by Naito[233,235,237] who found little relationship between site of application of CCC or SADH (fruit cluster or whole shoot + cluster) and percentage fruit set. He presented some evidence for a rise in endogenous cytokinins in response to treatment as providing an alternative explanation of the effect of growth retardants on percentage fruit set in grapevines.

The response of percentage fruit set elicited by application of growth retardants seems to vary inversely with the natural percentage set.[338,339] However most of the evidence is circumstantial and the matter requires more specific information before a firm conclusion can be reached. Certainly some of the largest responses have occurred in circumstances which have yielded a particularly low degree of natural fruit set.[195,345]

It seems therefore that growth retardants may be used to enhance percentage of flowers set in circumstances which are likely to lead to poor fruit set such as excessively vigorous shoot growth as well as possibly high temperature-induced poor set.[338] Unfortunately, the need for an application of these chemicals must be anticipated well before the period of bloom and set and this lessens their usefulness. They may however be applied as a form of insurance if circumstances are likely to justify it.

It may be inferred from the data of Coombe[73] and of Loreti[195] that benefit in terms of numbers of fruit set per cluster may be gained by applying concentrations much higher than those normally applied. However there is good evidence that repeated application of very high levels of plant growth retardants leads to vine damage and to a decline in productivity.[136]

7. Carry-Over Effects

Carry-over effects of growth retardants have not often been assessed despite their reputation for enhancing inflorescence initiation. Menary[211] demonstrated an important carry-over effect which was expressed in most components of yield: he found an enhancement of inflorescence initiation, size of inflorescence, proportion of flowers set and size of fruit. The effect was noted in a climatic region which was marginal for inflorescence initiation. Cahoon et al.[41] found no carry-over effect with SADH applied to Concord though Weaver[360] sound some carry-over effect with both CCC and SADH applied to Sultana.

8. Combined Effects of Other PGRCs

An alternative solution to the problem of reduced berry weight in response to application of growth retardants is to counter the response with an application of GA.[55,56,96,97,193] This method has been found to be effective and is part of a general recommendation for a fruit setting spray to replace 4 CPA on Black Corinth.[96] To be effective, the spray of GA should be applied at the optimum time for enhancing growth of the particular cultivar, e.g., at about full bloom for parthenocarpic cultivars, at fruit shatter for stenospermocarpic cultivars[55] and at seed hardening for seeded fruit.[140,141,144,292]

H. Miscellaneous Chemicals

Many chemicals have at one time or another been applied to fruiting grapevines and occasionally they have been reported to affect growth or numbers of fruit set. Most of the studies listed here have been restricted for one reason or another and so absence of an effect should not be construed as indicating that the chemical will be ineffective under all conditions. Thus CEPA (2-chloroethyl phosphonic acid) and ABA have been reported not to affect growth of fruit of Muscat of Alexandria[378] but these observations were made as an aside in an experiment designed to study the thinning effect of particular chemicals. One of the chemicals tested in that experiment, IT3456, a morphactin, caused a reduction in berry weight and this factor would need to be taken into consideration in choosing a thinning spray.[378]

Compounds which have been shown to exert a positive effect on numbers of fruit set or the development of set berries includes urea,[67] uracil,[81] boron,[249] and succinic acid.[212] Of these, only boron has been shown to exert a consistent effect and even this is probably limited to regions which are deficient or marginally deficient in boron. Boron is considered to be required for normal pollen germination and growth.[402]

It has been claimed that ethylene, by virtue of its effect in restricting shoot elongation will maximize the numbers of fruit set.[202,303] Thorough studies of this application for CEPA have not been published but one study which specifically approached this proposal found little to support the idea.[379]

IV. PARTHENOCARPY

Seedless fruit often command both a higher market price[290] and a wider market base than seeded fruit. The idea then of creating an ideal product from an otherwise acceptable but seeded fruit has led to considerable attention being devoted to solving this problem by the application of PGRCs with the objective of preventing normal seed development and then replacing the need for seeds by an applied PGRC (Table 11). The full potential of this technique has not been achieved and to the author's knowledge only one cultivar, Delaware, is treated with PGRCs on a large scale to produce "seedless table grapes" from an otherwise seeded cultivar.

Table 11
LIST OF CULTIVARS IN WHICH INDUCTION OF PARTHENOCARPY HAS BEEN ATTEMPTED AND A SUMMARY OF THE TREATMENTS USED

		Treatment to induce parthenocarpy			Treatment to promote growth				
Cultivar[a]	% Natural parthenocarpy	Time of treatment	GA conc. ($mg \cdot \ell^{-1}$)	Response (% parthenocarpic)	Time of treatment	GA conc. ($mg \cdot \ell^{-1}$)	Growth response in weight	Ref.	Comments
Cultivars reported to show an economically useful degree of response									
Aneb-e-Shahi		−10d	50	30	+21d	100	×1.25	78	
		−8d	50	30	+3d	50	+	82	
		fb	100	+			×1.4	243	± 4-CPA, Berry elongation
		−2d	100	62			×1.6	344	40 $mg \cdot \ell^{-1}$ optimum
Bhokri		−10d	100	98	+21d	100	×1.35	78	
		−10d	50	98	+7d	50	+	82	
		−bl	100	+				312	
Bicane 700		−15d	80	+	+7d	80	+	410	± Auxins, cytokinins
Chaush (♀)		fb	500	?	+10d	500	+	199	
		bl	20	?				335	Replacement of pollination
	10.4	fb	100	53	+10d	100	×0.98	200	
Delaware	2	−17d	100	98	+16d	100	+	52	Time, method of application, early maturity
		−bl	(100)	+	+10d	(100)	+	140	Optimum time for growth coincides with end of cell division, rain
		−bl	100	+	+15d	100	+	143	Time of treatment
		−12d	100	+	+15d	100	+	144	Time of treatment
		−21d	100	+				145	−21 to −16 optimum for parthenocarpy

	0.3	−10d	100	100				146	± Auxins, time, rainfall, pollen and ovary viability
	0.3	−10d	100	100	+15d	100	+	147	Time, endogenous IAA, GA, growth GA_3, GA_7
		−bl	100	100			+	217	Temperature, applied at tetrad stage
		−15d	100	+	+10d	100	+	218	± cAMP, small effect
	14	−14d	100	100				220	Set of fleshy vs shot
		−15d	100	100				222	Time vs fleshy:shot
		−13d	100	+				229	Pollen and ovule development
		−13d	100	+				230	Pollen and ovule development
		−13d	100	+	+10d	100		231	Most complete study of physiology and practice
		−11d	100	+	+10d	100	+	234	± 1.5% Urea, 200 $mg \cdot \ell^{-1}$ BA
		−14d	100	+	+10d	100	+	236	± 1.5% Urea, 200 $mg \cdot \ell^{-1}$ BA
		−bl	100		+bl	100		289	Maturity effects
		−14d	100	+	+10d	100	+	290	Maturity, economics
DK51	27	−14d	100	91				220	
Delicia de Caucaso		−15d	80	+	+7d	80	+	410	± Auxins, cytokinins
Flame Tokay		−3d	30	+	+7d	100	+	407	± Auxins, cytokinins
		−3d	30	+	+bl	100	+	408	± Auxins, cytokinins
		−3d	80	+	+7d	80	+	409	± Auxins, cytokinins
		−15d	80	+	+7d	80	+	410	± Auxins, cytokinins
Fredonia	7	−14d	100	96				220	Fleshy vs shot
Mase 7	44	−14d	100	100				220	Fleshy vs. shot
Monarca		−15d	80	+	+7d	80	+	410	± Auxins, cytokinins
Madeleine Angevine (♀)			200	+		200	+	412	Conc., n.applications
		−15d	80	+	+7d	80	+	410	± Auxins, cytokinins
	26	fb	100	86	+10d	100	×1.13	200	

Table 11 (continued)
LIST OF CULTIVARS IN WHICH INDUCTION OF PARTHENOCARPY HAS BEEN ATTEMPTED AND A SUMMARY OF THE TREATMENTS USED

		Treatment to induce parthenocarpy			Treatment to promote growth				
Cultivar[a]	% Natural parthenocarpy	Time of treatment	GA conc. (mg·ℓ⁻¹)	Response (% parthenocarpic)	Time of treatment	GA conc. (mg·ℓ⁻¹)	Growth response in weight	Ref.	Comments
Ohanez 2-2		−15d	80	+	+7d	80	+	410	± Auxins, cytokinins
Ohanez 2-4		−15d	80	+	+7d	80	+	410	± Auxins, cytokinins
Pis de Chevre B		−15d	80	+	+7d	80	+	410	± Auxins, cytokinins
Pizzutello B		−15d	80	+	+7d	80	+	410	± Auxins, cytokinins
Pink Muscat		−bl	Auxins	+	+7d	20	+	407	4-CPA, NAA
Rose Muscat		−15d	80	96	+7d			408	± Auxins, cytokinins, time
		−3d	30	+	+7d	100	+	407	(95) ± 4-CPA
		−3d	80	+	+7d	80	+	409	± Auxins, cytokinins
Trebbiano		10% bl	100	+			+	203	
Cultivars reported to show an incomplete or low degree of response or uncertain response									
Alamwick		−10d	100	33	+21d	100	0	78	
		−5d	150	33				82	
Auvergne	1	bl	100	3	+8d	100	×1.01	200	
Auxerrois		−10d	50	0				160	
Bakhtiori	12	bl	100	22	+8d	100	×1.14	200	
Bharat Early		−10d	75	2	+21d	100	×1.08	78	
		−15d	75	2				82	
Bicane (♀)	13	bl	100	48	+8d	100	×1.13	200	
Black Hamburg		−10d	75	3	+21d	100	×1.03	78	
		−15d	75	3				82	
Black Muscat		−10d	75	1	+21d	100	×1.06	78	
Campbell Early		−17d	100	88	+bl	100		145	Time of application

		−bl	100	0				217	Induced shot non-fleshy fruit, temperature, application at tetrad stage.
		−15d	100	+	+10d	100	+	218	± cAMP, sl. GA sparing activity
		−15d	100	+	+10d	100	+	219	+ 100 mg·ℓ⁻¹ BA, time
	3	−14d	100	100				220	Set of fleshy vs shot
		−15d	100	95				222	Time (−30 to 0 d)
		−15d	100	+	+10d	100	+	231	Increased L/W of berry, time, conc.
		−15d	100	99	+bl	100	0	292	Poor growth of seedless
		−bl	(100)	29	+bl	(100)	0	291	Poor response
Cardinal		−15d	80	(+)	+7d	80	(+)	410	± Auxins, cytokinins, poor response
Cereza		−15d	80	(+)	+7d	80	(+)	410	± Auxins, cytokinins, poor response
Concord		bl	250	+	+3d	250	+	181	
	7	−14d	100	68				220	Set of fleshy vs shot
Criolla Grande		−3d	80	(+)	+7d	80	(+)	408	± Auxins, cytokinins, poor response
de Chaunac		fb	100	0	+bl	100	0	313	
Dol'cheto	0.1	bl	100	4	+8d	100	×0.76	200	
Fintendo		−15d	80	(+)	+7d	80	(+)	410	± Cytokinins, auxins, poor response
Fredonia		bl	100	+				271	
Golden Queen		−7d	200	−	+21d	100	0	79	Reduced n.set, no parthenocarpy ± auxins and cytokinins
Gewurztraminer		−bl	50	+	+bl	50	+	160	Effect on flower initiation

Table 11 (continued)
LIST OF CULTIVARS IN WHICH INDUCTION OF PARTHENOCARPY HAS BEEN ATTEMPTED AND A SUMMARY OF THE TREATMENTS USED

		Treatment to induce parthenocarpy			Treatment to promote growth				
Cultivar[a]	% Natural parthenocarpy	Time of treatment	GA conc. ($mg \cdot \ell^{-1}$)	Response (% parthenocarpic)	Time of treatment	GA conc. ($mg \cdot \ell^{-1}$)	Growth response in weight	Ref.	Comments
Gros Colman		−10d	100	80	+21d	100	×1.3−	78	
		−6d	75	80				82	
		−15d	80	+	+7d	80	(+)	410	± Auxins, cytokinins, poor response
Italia		−15d	80	(+)	+7d	80	(+)	410	± Auxins, cytokinins, poor response
Karaburnu	16	bl	100	30	+8d	100	×1.35	200	
Katta-Kurgan (♀)	12	bl	100	41	+8d	100	×1.26	200	
Khalili		−bl	100	+			+	106	
Kyoho		−6d	100	+	+bl	100	+	140	
	15	−14d	100	98				200	Set of fleshy vs shot
Madresfield Court		−bl	100		+bl	100		291	Reduced cracking
Merlot	0.01	bl	100	1	+8d	100	×0.60	200	
Moldavskii (♀)	0.6	bl	100	3	+8d	100	×1.02	200	
Molinera		−15d	80	(+)	+7d	80	(+)	410	± Auxins, cytokinins, poor response
Muscat Hamburg	9	−14d	100	98				220	6.3Set of fleshy vs shot berries
		−10d	50					231	Increased L/W of berries
Muscat Bailey A		−bl	(100)	90	+bl	(100)	+	141	Seasonal var., n.applic. seed-like structure

	1	−14d	100	93	+bl	100	+	143	Time, advance of maturity (4 wk)
		−24d	100	75	+bl			144	Time of application
	26	−14d	100	97				220	Set of fleshy vs shot
		−bl	(100)	77	+bl	(100)	+	291	Maturity advance (4 wk)
		−15d	100	76	+bl	100	+	292	Maturity advance
	1	−16d	100	76	+bl	100	+	332	Maturity advance (5 wk)
		−15 and −8d	100/200	97	+10d	100	+	341, 342, 343	Abnormal seed-like growth
		−15d	100	60				404	
		−bl	200	98	+bl	200	+	340	Relation to shoot vigour
Muscat of Alexandria		−10d	50					231	Increased L/W of berry
		−bl	(100)		+bl	(100)		291	Poor response
		−3d	80	(+)	+7d	80	+	408	± Auxins, cytokinins, poor response
		−15d	80	(+)	+7d	80	(+)	410	± Auxins, cytokinins, poor response
Neomuscat	5	−8d	100	71	+bl	100		143	Time of application
		−18d	100	75	+bl	100		144	Time of application
		−12d	100	84	+bl	100		145	Time, effect on pollen/ovule
	34	−14d	100	55				220	Set of fleshy vs shot
		−bl	(100)	(+)	+bl	(100)	(+)	292	Low respiration, rachis hardened, maturity
Niagara	38	−14d	100	62				220	Set of fleshy vs shot
Nimrang (♀)		bl	100	?	+10d	100	+	199	Conc., site of application, rainfall
	12	bl	100	39	+8d	100	×1.03	200	
Pionnier		−14d	100	99				220	Set of fleshy vs shot
Pinot Blanc	4	bl	100	+	+8d	100	0	28	

Table 11 (continued)
LIST OF CULTIVARS IN WHICH INDUCTION OF PARTHENOCARPY HAS BEEN ATTEMPTED AND A SUMMARY OF THE TREATMENTS USED

		Treatment to induce parthenocarpy			Treatment to promote growth				
Cultivar[a]	% Natural parthenocarpy	Time of treatment	GA conc. (mg·ℓ$^{-1}$)	Response (% parthenocarpic)	Time of treatment	GA conc. (mg·ℓ$^{-1}$)	Growth response in weight	Ref.	Comments
Pukhyukovskii (♀)	0.02	bl	100	2	+8d	100	×0.84	200	
Red Millenium		−bl	(100)	+	+b−	(100)	+	140	
Regina de Malvesia		−3d	80	(+)	+7d	80	(+)	408	± Auxins, cytokinins, poor response
Royal Vineyard	0.5	bl	100	1	+8d	100	×0.9	200	
Ruby Cabernet		bl	18	0			0	67	+Cincturing, urea
Savagnin Rose	11	bl	100	(−)	+8d	100	+	28	
Semillon		−3d	80	(+)	+7d	80	(+)	408	± Auxins, cytokinins, poor response
Tashley (♀)	15	bl	100	67	+8d	100	×1.09	200	
Tavriz	6	bl	100	15	+8d	100	×1.06	200	
Tokay		bl	500	+	+42	25	0	365	
		−27d	2,4-D	+				392	8 mg·ℓ$^{-1}$ 2,4-D active over a wide range of time of application
Tolstokoryi (♀)	0.2	bl	100	2	+8d	100	×0.88	200	
Varyushkin	0.2	bl	100	2	+8d	100	×0.96	200	
White Reisling	15	bl	100	+	+8d	100	0	28	

Note: n.set, number of fruit set per inflorescence; ± nd, days before or after bloom, bl; fb, full bloom; +, increase; −, decrease; 0, no effect; blank, not reported; sdd, seeded; sdls, seedless; d, days.

[a] See Schwabe and Mills[293] for some cultivars not listed here.

A. Physiology

1. Steps in the Process

The steps which must be followed to successfully induce the set of a cluster of completely parthenocarpic fruit are prevention of normal seed growth, prevention of abscission of the ovary in the absence of seed growth, stimulation of growth of the ovary wall and stimulation of the softening and maturation of the retained seedless berry. These processes are affected by the choice of PGRC, the timing of their application with respect to flower and fruit development, the interaction of these factors with particular environmental factors such as temperature, and especially with the genetic make-up of the particular cultivar.

Each of the three classes of growth promoting regulators, auxins, gibberellins, and cytokinins have been tested for their ability to induce development of large, fleshy seedless berries, but only GA has been exhaustively tested (see Table 11).

2. Formation of Seedless Berries

Gibberellin applied during a period from about 30 days before bloom to at or just after bloom has been shown by many workers to reduce the set of seeded berries. Optimum timing of application has been shown for the four cultivars which have been studied thoroughly, Delaware, Muscat Bailey A, Campbell Early, and Neo Muscat, to lie between about 20 to 10 days before bloom (Table 11). Examples of the results obtained for two cultivars are presented in Figure 11 which demonstrates the effect of time of application on percentage of ovaries set which are seedless, and the proportion of those set which are fleshy (i.e., which soften) vs. those which are hard and nonfleshy (shot). This period also coincides with the period of greatest advance of bloom (about 4 to 5 days) and overlaps with the period of enhancement of fleshy pericarp development.[222]

3. Cultivar Characteristics

The contrast between Delaware and Campbell Early in proportion of fleshy seedless berries which develop is representative of the results obtained with a wide range of cultivars and grouped in Table 11 as those showing a useful degree of response and those showing only partial or zero response. A number of papers have been published which attempt to either explain these differences in responsiveness or to predict response.[78,200,220] Manankov[200] examined the response of 106 cultivars of grape and claimed that responsiveness was closely related to predisposition to set a proportion of seedless berries (for the 20 cultivars presented, $r = 0.92$ for the relation between percentage of seedless untreated and seedless GA treated). However, a similar analysis on the data presented by Motomura and Hori[220] shows no correlation between these two parameters or between any other of the parameters measured ($r = 0.06$, percentage of seedless untreated vs. seedless GA treated). The percentage of seedless berries set under natural conditions is unlikely to be a reliable measure of the genotype of a particular cultivar because of the dependence of this value on such issues as disease status and weather conditions.

Motomura and Hori[220] propose the existence of three classes of grape: (1) those cf Delaware in which GA application prevents set of any seeded berries and causes the set of a high proportion of seedless fleshy berries; (2) those cf Campbell Early, in which GA, though effective in preventing set of seeded berries, fails to stimulate growth and softening of the set ovaries; and (3) those like Neo Muscat which set a low proportion only of seedless berries, berries which fail to develop and soften after the GA treatment. In effect they propose that there are two independent characters; one which determines the proportion of seedless berries set in response to GA application and a second which determines the sensitivity of those berries to GA induced growth and softening. These conclusions are in accordance with the conclusions drawn by Hagiwara et al.[117] in their studies on the responsiveness of the berries of some seedless cultivars to applied GA.

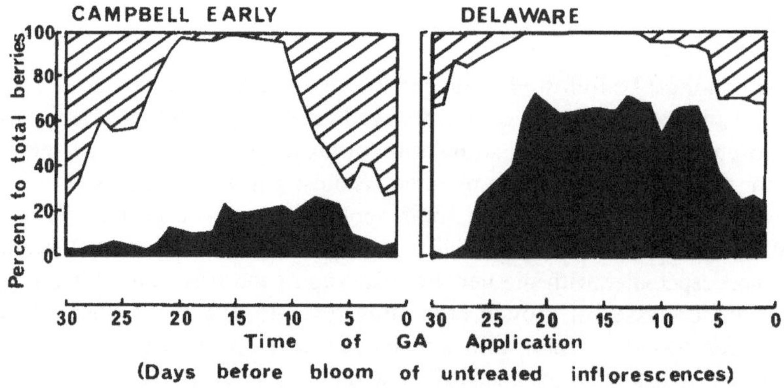

FIGURE 11. Effect of time of application of 100 mg·ℓ$^{-1}$ GA on the proportion of berries set which are seeded (upper cross hatched areas), seedless, but non-fleshy (middle open area) or seedless and fleshy (lower filled area). The main difference between the two cultivars Campbell Early and Delaware is the proportion of seedless berries set which are fleshy. (Modified from Motomura, Y. and Ito, H., *Tohoku J. Agric. Res.*, 24, 151, 1973. With permission.)

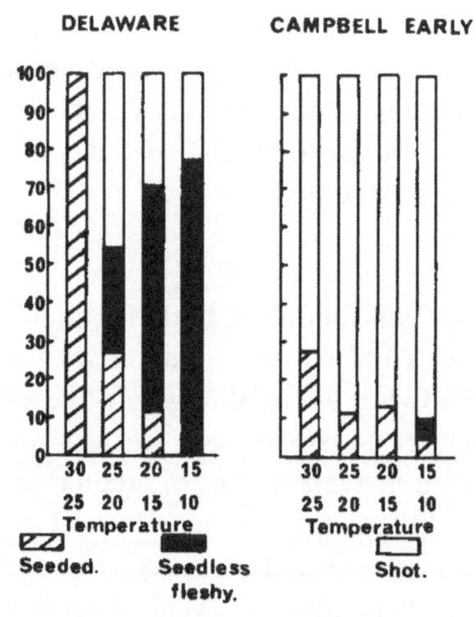

FIGURE 12. Relationship between the development of fleshy seedless berries on inflorescences of Delaware and Campbell Early grapevines in response to 100 mg·ℓ$^{-1}$ GA applied at the tetrad stage of pollen development and the temperature applying during the remaining period of flower growth. (Figure redrawn from Motomura and Hori.[217] With permission.)

4. Genotype Environment Interactions

Two factors have been shown to be related to the degree of response obtained in development of fleshy berries resulting from a pre-anthesis application; one environmental, temperature, and a second, physiological, rate of metabolism of the applied GA. For the three cultivars studied, the rate of metabolism of GA was in the order, Delaware (I) < Campbell Early (II) < Niagara (III).[216,220] Temperature also has a marked effect which is superimposed on the genotype effects described above (Figure 12). In Delaware, a day/night regime of 30/25°C give zero fleshy seedless berries in response to GA application whereas in a regime of 15/10°C, 78% of the seedless berries set were fleshy.[217] In Campbell Early only the lowest temperature gave a set of any fleshy berries.[217] These temperature effects are correlated with rate of metabolism of applied GA which increases with raised temperatures[216] and with degree of ovary growth which increases with lowered temperatures. [217]

The inference of these studies is that rate of metabolism is a significant factor in responsiveness of cultivars to applied GA and this idea has been given some support by the improvement of the response given by Campbell Early[220] and Muscat Bailey A when the prebloom application of GA was supplemented by a further application at bloom.[341,342] Unfortunately, the additional application to Muscat Bailey A was associated with development of a hard seed-like structure which is unacceptable commercially.

5. Characteristics of the GA Response

Gibberellic acid applied prebloom acts to promote the growth of all parts of the ovary and to enhance the overall rate of growth.[228,231] It also tends to enhance the length of the ovary and of the subsequent fruit.[231] GA applied during the optimum period, between about 20 and 10 days before bloom causes a failure of reproductive development. The pollen, though of normal size and shape shows zero or near zero germination in ideal conditions and is ineffective in pollination.[229,230,231] Growth of the ovule and especially of the integuments is promoted, but the embryo sac is defective and lacks an egg cell and polar nucleii.[231] Its appearance is similar to that of the parthenocarpic White Corinth described by Pearson.[258,259] Time of application appears to be important in this process because pollination of Muscat Bailey A ovaries treated at 10 days or nearer to bloom will form a proportion of seeds if pollinated with fertile, untreated pollen.[404]

GA, though one of the most effective and widely used PGRC for the purpose of inducing parthenocarpy, is not universally effective and has some disturbing side effects, such as reduced inflorescence initiation and distortion and hardening of the rachis of some cultivars.[292] Addition of BA may reduce the severity of some of these problems.[342]

6. Use of Other PGRCs

Zuluaga and his colleagues have examined a range of other plant growth regulating chemicals and found that the effectiveness of GA was enhanced by addition of 30 mg/ℓ^{-1} 4-CPA and that 4-CPA on its own was perhaps even more effective as an inducer of seedlessness and parthenocarpic fruit development.[406,407,408,409,410] A conspicuous effect of 4-CPA was that it induced a high proportion of berries without seed remnants (about 2:1) whereas GA caused the production of soft, stenospermocarpic-like seed structures (about 1:2). Benzyl adenine, when applied in conjunction with either or both of the other two PGRCs had an additive effect on set, but the ratio of parthenocarpic to stenospermocarpic berries was about that of GA alone, or if the BA was applied with 4-CPA only then the degree of seed development matched that of 4-CPA.[408] The optimum time of application of BA or of PBA was found to be at about early bloom, at which time the number of ovaries set on Rose Muscat for example was nearly equal to the total number of flowers present and about 10 times greater than that set by either 4-CPA or GA alone.[408] This degree of set is clearly much greater than that required commercially and a concentration of BA considerably less than the 800 mg/ℓ^{-1} applied would probably suffice. Alternatively, it may be that PBA, which shows greater activity than BA[397] could be used,[410]

Response of a range of cultivars to inclusion with GA of 4-CPA or BA is variable and only about half of the cultivars tested gave a worthwhile response.[410] Whether use of PGRCs other than GA on cultivars which have been shown to be poorly responsive to GA will effect a useful response is as yet unascertained, though one cultivar which has been tested by application of both GA alone[231,291] and with the inclusion of other PGRCs[408,410] has been shown to be intractable under both schemes. Inclusion of BA and urea or cAMP with GA applied to Delaware did not lead to a marked enhancement of response.[218,234,236]

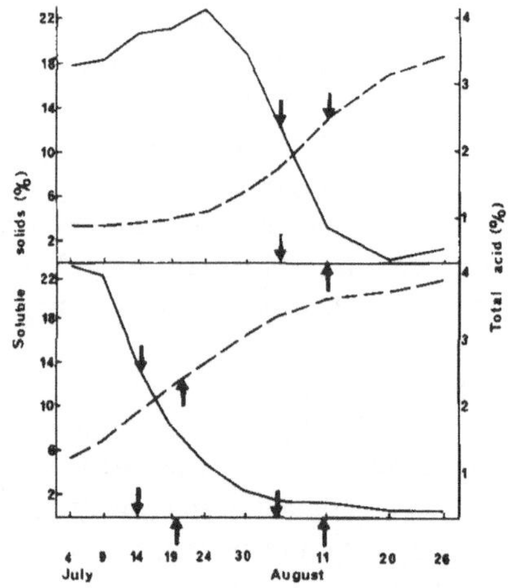

FIGURE 13. Comparison of the timing of the sequence of sugar accumulation and acid depletion in GA-treated seedless and untreated seeded Delaware fruits. A, data for seeded fruit and B, corresponding data for GA-induced seedless fruits. The downward pointing arrows indicate the $t_{1/2}$ for acid depletion and the upward pointing arrows, the $t_{1/2}$ for sugar accumulation. Data from Muranishi.[231]

7. *Promotion of Berry Growth*

The treatments used to induce parthenocarpy usually depress yield and produce unacceptably small berries. Treatments such as those normally applied to naturally seedless cultivars such as the stenospermocarpic Sultanina are usually applied to stimulate further growth (see Table 11). The most common treatment used is 100 $mg\cdot\ell^{-1}$ GA applied at about 10 days after bloom. This has been shown to be the most effective time of application,[231] at a time which coincides with the end of cell division and the commencement of the first period of post-bloom cell expansion.[140] Other treatments are also effective, for example cincturing and 4-CPA.[408]

8. *Maturity Effects*

An important bonus which is associated with the production of parthenocarpic fruit with PGRCs is a greatly enhanced earliness of maturity (Figure 13). This response is largely due to the absence of the normal post-bloom lag period associated with seed hardening and embryo growth and maturity can be advanced by as much as 28 to 35 days.[143,231,291,292,332] The response is not due simply to the absence of seeds because maturity of even seeded berries is also advanced.[231] Coloring of treated berries is reported to occur in advance of the normal decline in acidity which can mislead,[52,231] however, the data in Figure 9 shows that this trend is not maintained and the treated, seedless fruit have a higher sugar:acid ratio at maturity than do untreated seeded fruit. This is also supported by the observations of Looney[192] and Peynaud and Ribereau-Gayon.[265]

B. Practice

The concentrations of GA required to obtain a commercially acceptable result are high enough to cause damage to even the less susceptible vine cultivars, especially when applied during early shoot growth, before anthesis. This observation then provides one of the bases for the practice in Japan of dipping individual inflorescences in a solution of GA. A more pressing reason though is the requirement for a complete and uniform wetting of the inflorescence or fruit cluster and this is difficult to achieve with an overall spray or with a hand spray even with addition of 100 to 200 mg·ℓ$^{-1}$ of a wetting agent.[52,231] There is also an economic reason for choice of an individual dip rather than a spray because Muranishi[231] found that dipping used about half the amount of material required for spraying and that dipping 12000 flowers required only 64 hr compared with 108 hr if the treatment was by hand spraying. It is also important to include a wetting agent at between 100 to 200 mg·ℓ$^{-1}$ to ensure adequate wetting.[52,231]

V. FLOWER AND FRUIT THINNING

A. Place in Viticulture

Flower and fruit thinning form an important aspect of the management of many cultivars and yet the topic is inadequately documented for all except a few particular cultivars such as Sultanina, Zinfandel, and Carignane. Even in these cultivars much of the documentary literature which is readily available examines the problem on a small experimental scale and has utilized techniques such as cluster dipping, techniques which are not feasible on a field scale.

Reasons which have been advanced for thinning practices include: prevention of disease and physical damage of fruit in over-crowded and poorly aerated fruit clusters, prevention of vine overload, improvement of juice quality, improvement of cluster presentation and fruit growth, and reduction of competition between the primary and the second crop.

B. Choice of Chemical and Use

1. Required Characteristics

Factors which should be considered when selecting a chemical thinning agent include:

1. Absence of phytotoxicity
2. Uniformity of thinning over the individual cluster
3. Reliability
4. Timing, need for precision and whether before, during or after bloom
5. Formation of abnormal berries, e.g., shot berries
6. Safety and degree of control (slope of the dose/response curve)

A summary of information gleaned from published literature is presented in Table 12. The views presented in Table 12 represent a personal judgment drawn from the literature, rather than from practical experience, and thus should be used with discretion. Table 13 lists factual information for the two better established thinning agents, GA and NAA.

2. GA

Gibberellic acid is the most widely used of the thinning agents though its usefulness is limited by the susceptibility of many cultivars to GA-induced barrenness,[160] by induction of the retention of deformed and shot berries on many cultivars (Table 13), by variation in degree of sensitivity of cultivars (Figure 14), and a limited period of sensitivity which in many cultivars is coincident with the beginning of flowering of the inflorescence (e.g., Black Corinth, Figure 14; and Sultanina[29]). Nevertheless, it is used successfully on many cultivars, either by virtue of their tolerance of GA or by applying a minimum concentration to reduce injurious side-effects to a tolerable level (Table 13).

Reported dose/response curves for GA indicate a linear-log relationship between either yield or numbers of berries set and concentration of GA applied (Figure 14). The concentration range between 0.1 and 10 mg·ℓ$^{-1}$ appears to be the significant range, though the general assumption of a linear response has meant that this range has been inadequately

Table 12
SOME PROPERTIES OF THINNING AGENTS AND CLASSES OF THINNING AGENTS USED TO THIN FLOWER OR YOUNG FRUIT CLUSTERS OF GRAPEVINES

Attribute	GA	NAA	CEPA	Pollenicides	Morphactins	Water
Phytotoxicity	C. Pass[a]	C. Pass	Pass	Pass	Pass	Nil
Uniformity	Good	Pass	Fail	Pass	Pass	?
Reliability	C. Pass	Pass	Fail	Fail	?	?
Timing	Pass	Good Range	Poor	Pass	Good	?
Abnormal Berries	C. Pass	Pass	Pass	Good	?	Pass?
Safety	Pass	Pass	Fail	Poor	Fail(?)	Pass

[a] C. Pass represents conditional pass.

tested. Concentrations higher than this range are only significant if enhanced growth of seedless berries is desired,[48,49] though if this is the case, then the application should be delayed to later in the flowering period to reduce the response to a desirable level, to maximize growth, and to minimize the degree of formation of shot berries (Table 13 see also section on berry set).

3. NAA

Examination of Table 12 leads to a suggestion that NAA may provide a satisfactory alternative to GA, especially for those cultivars which are particularly sensitive to GA-induced injury. The only real problem which appears in the literature is some phytoxicity[358] but this needn't be a problem, provided relatively low concentrations are applied and application before bloom is avoided.[358] A particular advantage possessed by NAA is that it is equally or nearly equally as effective when applied at shatter or fruit set about 10 to 14 days after bloom.[214,287]

The dose-response curve for NAA, like that for GA is linear-log and the significance of this in choosing a concentration and designing experiments is illustrated in Figure 15. Most of the concentrations tested lie in the overthinning range.

4. CEPA/Pollencides

The ethylene producing chemical, CEPA, has also been tested as a thinning agent, and though active from bloom through to set, tends to be difficult to control and to produce nonuniform thinning.[378,383] Pollenicides such as dinitro-*o*-cresol (DNOC) and dinitro-*sec*-butylphenol have been used with marginal success, due to unpredictability of the response and the steepness of the dose-response curve.[287,358]

5. Morphactins

While a number of herbicides and other PGRCs have been examined,[350,383] the only other class of compound which appears to hold promise is the morphactins or chlorofluorenols.[374,375,378,383,379] Response of inflorescences or fruit clusters treated with four of this class of chemical regulator is presented in Table 14 as the estimated concentration required to give a 50% reduction in numbers of fruit set. Much of the information was derived from activity survey reports and more extensive research is required to establish effectiveness and degree of unwanted, or injurious side-effects.

As a class, the morphactins tend to exert an all or nothing response, with a concentration of 1 or even 0.1 mg·ℓ^{-1} being sufficient to cause abscission of all flowers. Usually they are most effective when applied at or just after fruit set. This is a desirable time of application because it allows degree of natural set to be estimated before the thinning agent is applied.

Table 13
SUMMARY OF RECOMMENDED PROCEDURES FOR THINNING GRAPEVINE FRUIT CLUSTERS WITH EITHER GA OR NAA

Cultivar	Recommended time	Recommended conc. ($mg \cdot \ell^{-1}$)	Relative response	Ref.	Comments
Gibberellic acid					
Sultanina	20 to 75% bl	20	×0.55	48,49	Adequate thinning with 5, but 20 gave greater berry weight, some shot berries set
	fbl	6	×0.75	97	
	20 to 80% bl	10	×0.60	414	Noted difficulty in achieving correct timing in practice
	10 to 70%	10	×0.75	197,198	Shot berries set if conc. > 20 $mg \cdot \ell^{-1}$
	bl	15	×0.54	213	Analysis of components of set affected
	60% bl	15	×0.62	215	Conc. applied and timing
	40% bl	10	×0.6	385	
Perlette	50% bl	10	×0.6	47	Later application gave fewer shot berries
	40% bl	50	×0.85	336	
Aneb-e-Shahi	−2d	40	×0.47	344	
Auxerrois	−11d	50	0	160	Inhibited inflorescence initiation
Carignane	50% bl	10	×0.41	49	7-Fold increase in ovary retention
	128d	5	×0.75	390	
Chasselas	−11d	50	0	160	Inhibited inflorescence initiation
Delaware	−bl	5	—	145	Use to prevent cracking due to overcrowding
Gewurztraminer	−11d	20	×0.61	160	Inhibited inflorescence initiation
Golden Queen	−7d	50	×0.72	79	
Grenache	bl	5	×0.87	396	Increased set of shot berries
Gros Semillon	−15d	5	—	406	
Muscat of Alexandria	bl	5	×0.31	396	
Ohanez	fbl	20	×0.75	97	
Pinot Blanc	−11d	50	0	160	
Pinot Chardonnay	80% bl	5	—	396	Decreased n.sdd berries
Pinot Gris	−11d	50	0	160	
	−15d	5	—	406	

Table 13 (continued)
SUMMARY OF RECOMMENDED PROCEDURES FOR THINNING GRAPEVINE FRUIT CLUSTERS WITH EITHER GA OR NAA

Cultivar	Recommended time	Recommended conc. (mg·ℓ$^{-1}$)	Relative response	Ref.	Comments
Reisling	−11d	20	×0.65	160	
Ribier	fbl	5	×0.64	365	
Seibel 5455	25% bl	10	×0.85	132	Effect offset at higher conc. by increased growth
Sylvaner	−11d	50	0	160	Inhibited inflorescence initiation
Tinta Madeira	−28d	5	0	390	
Tokay	fbl	5	+	365	Increased set and no. shot berries
	−bl	2.5	×0.75	382	Wetting agent (Triton® B-1956) also effective
Zinfandel	−bl	5	×0.62	213	3-Fold increase in no. rudimentary ovaries set
	50% bl	10	?	361	Earlier application reduced berry weight
	−31d	10	×0.66	365	
	100% bl	5	×0.80	382	Conc. time, effect of Triton B-1956
	bl	5	×0.89	390	
Naphthalene acetic acid					
Carignane	70% bl	5	×0.67	214	
Queen of Vineyard	+10d	5	×0.67	287	Water effective at bloom
Shaslas Dore	+10d	5	×0.67	287	Water effective at bloom, increased berry size by 20%
Zinfandel	bl or +10d	10	×0.46	358	Prebloom application toxic to foliage

Note: n.set, number of fruit set per inflorescence; ± nd, days before or after bloom, bl; fbl, full bloom; +, increase; −, decrease; 0, no effect; blank, not reported; sdd, seeded; sdls, seedless; d, days.

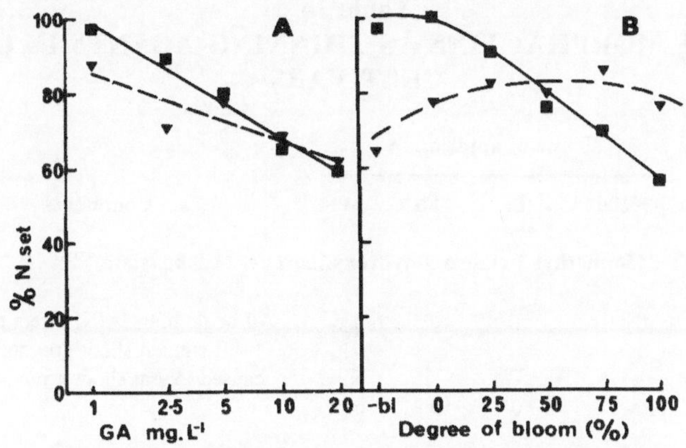

FIGURE 14. Thinning of flowers of Tokay and Zinfandel with GA. A. Effect of concentration of GA on thinning. GA applied at 50% bloom to Tokay and 60% bloom to Zinfandel. The curves represent the following fitted equations where y represents percentage of flowers set as fruit, respectively for Tokay (— —) and Zinfandel (——): $y = 85.4 - 17.5 \log (GA)$, $r^2 = 0.79$; $y = 99.5 - 31.2 \log (GA)$, $r^2 = 0.98$. B. Dependence of thinning response on time of application for Tokay and Zinfandel. Data averaged over all concentrations as in A. Data for A and B from Weaver and Pool.[382]

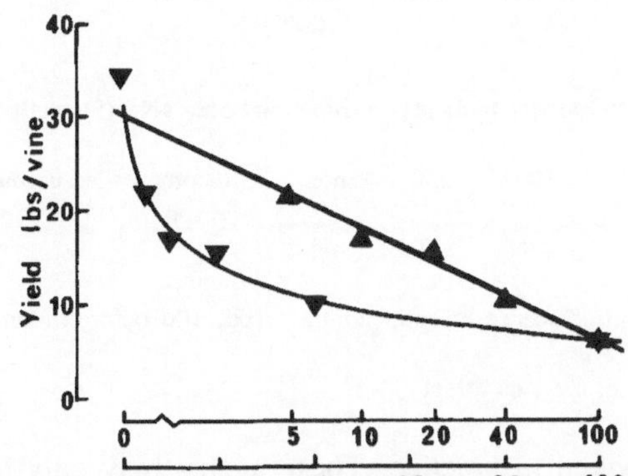

FIGURE 15. Depression in yield of Zinfandel grapevines associated with application of NAA at 20% bloom. Data presented both on a semilog scale (▲) and as retransformed arithmetic values (▼) to demonstrate the importance of recognizing the logarithmic nature of the response when choosing appropriate concentrations to achieve a desirable proportion of thinning. The equation fitted was $y = 29.8 - 11.9 \log (NAA)$, $r^2 = 0.98$. Data from Weaver.[358]

The morphactin analogues which possess a high degree of physiological activity may also exert a number of side-effects such as retarding growth of the primary shoot and stimulating growth of axillary buds.[378]

6. Water

The effectiveness of water or water plus a wetting agent such as Triton B-1956[287,382] has

Table 14
ACTIVITY OF MORPHACTINS AS THINNING AGENTS IN GRAPEVINE CULTIVARS

Cultivars	Time of application: −10d[a]	Bl	FS	PS	Comments
IT 3456(methyl-2-chloro-9-hydroxyfluorene-9-carboxylate)[374,375,378]					
Carignane	A/N[b]	3	10	n.e.[b]	All or nothing (A/N) effect at bl or −bl; stunted shoot growth and caused lateral shoot growth
Muscat of Alexandria	A/N	0.05	10	n.e.	
Sultanina	1	A/N	10	n.e.	
Perlette			0.05		
IT 4433(2-chloro-9-hydroxyfluorene carbonic acid-(9)-p-chlorophenoxy ethyl ester)[383]					
Carignane	200	100	4	n.e.	
Sultanina	100	100	30	20	
IT 5732(2-chlorofluorene carbonic acid-(9)-methyl ester)[383]					
Carignane	A/N	A/N	A/N	n.e.	Too active
Sultanina	A/N	5	3	1000	
IT 5733(2,7-dichloro-9-hydroxyfluorene carbonic acid-(9)-methyl ester)[383]					
Carignane	200	100	250	n.e.	Recommended because of low dose-response curve slope[383]
Sultanina	A/N	A/N	50	n.e.	

[a] Stage of development of inflorescence at application; −10d, 10d before bloom; Bl, bloom; FS, fruit set (+10d); PS, post set (+35d).

[b] A/N, all or nothing response; n.e., no effect.

been substantially overlooked in practice. These agents may only be effective during the time of anthesis and pollen germination, thus creating an acute problem of timing the spray. Nevertheless, in view of their innocuousness, the treatments seem to be worth pursuing.

VI. FRUIT MATURATION, HARVESTING, AND STORAGE

A. Outline of Industry Needs

Increasing sophistication of management, scale of production, and the ever-increasing search for a competitive edge in quality or market nitch, have added to demands for improved methods of manipulation of fruit composition and time of harvest. These demands come now, not only from the table grape industry which has traditionally sought to advance maturity and enhance consumer appeal, but also from the wine grape and the dried fruit industries.

In the wine grape industry, the demand comes from two concurrent developments, the trend to extensive farm and wine-making enterprises comprising a few highly desirable cultivars and a heightened consumer awareness of quality leading to insistence of large-scale wine-makers on the technical quality of the musts used for bulk wines i.e., standards of sugar and acid content and balance, color intensity (if a colored grape), and pH. Treatments to delay or advance maturity may help the vigneron obtain an ideal must quality over the extended period required to harvest a large planting of a single cultivar, though development of high-capacity mechanical harvesters has to a large extent reduced the necessity of harvest management for this purpose.

In marginal climates, such as those of northern Europe, treatments to advance maturity could help to stabilize variations in quality due to seasonal influences, and help to limit the need for additives to musts. Conversely, many warm climate regions, produce wines of only moderate quality because of an imbalance in the sugar/acid balance and of the raised pH due to an enhanced content of potassium tartate and reduced malate concentration. Here, treatments to limit acid metabolism are of potential use.

The concept of scale-associated problems applies to other ''processing'' industries which have in addition problems related to their individual requirements. The limited period for solar drying of grapes is a problem for many dried fruit industries, especially for those in marginal climates such as the Australian industry. Advancement of maturity could be of considerable benefit for 1 day at the beginning of the harvest season is equivalent to 3 or even 4 days at the end of the season.

Scale of production, sophistication, and a desire to extend the available marketing period have led to attempts to store harvested fruit in its natural state for increasingly extended periods of time, and to mechanized harvesting of fruit for processing. In storage, the objectives are to maintain fruit quality and attachment to the pedicel while in mechanical harvesting it is desirable to reduce the degree of berry attachment. Reduced berry attachment will lead to an increased harvest efficiency, and to reduced fruit damage. Fruit damaged during harvesting is undesirable for wine-making purposes and unacceptable for dried fruit production. Thus fresh fruit industries could benefit from enhanced fruit attachment and reduced senescent deterioration, in storage, while most other forms of grape use would benefit from abscission enhancing procedures and perhaps from maturity (senescence) advancing techniques. Abscission promotion may also be of value in the table grape industry should there be a trend towards a single berry trade. Development of this trade would probably be wholly dependent on fruit abscission promoting treatments.

B. Maturation

1. The Ripening Process

The ripening process in fruit of the grape is a syndrome of many co-ordinated and tightly linked processes: growth, accumulation of reducing sugars, depletion of organic acids, color change, wall degradation and softening, and increased membrane porosity.[60,75,76,265,416] Ripening terminates by softening and shriveling of the fruit associated with loss of water and vascular blockage. In the main, PGRC effects on ripening are due to an alteration of the timing of the whole process, rather than with extent, rate, or coordination. There are some exceptions to this general observation which are associated principally with coloring. Coloring appears to be one of the most variable of the ripening processes, responding to temperature, water deficit stress,[122] light intensity,[265] and number of seeds.[362] However, each of these factors influences the extent of coloring but not the propensity to color which remains tightly linked to each of the other processes.

Seeds appear to be the most important internal factor regulating the timing and perhaps the extent and rate of the maturation processes.[281,362] Seeds appear to affect maturation directly by influencing the timing of the onset of the maturation process[281] and indirectly by affecting the fruit volume.[362,253] The principal effect of seeds is to extend the period of the lag phase (the period of near zero growth separating the two periods of fruit growth), for example in Sylvaner, by 5 to 6 days for each seed beyond one seed.[281] Not all cultivars are equally affected by the presence of seeds and the effect of seeds on the timing of maturation in GF 31-17-115 was substantially less than that which occurred in Sylvaner.[281] In Red Malaga, the difference between no seeds and 1 seed was greater than that between 1 and 4 seeds.[362]

The presence of seeds also exerts a consistent effect on fruit composition, leading usually to lower soluble solids content and raised levels of acidity, specifically of malic acid.[281] Altered composition may be due to the confounding effect of increased fruit volume which could slow the rate of sugar accumulation and acid metabolism but the effect of seeds on malic acid metabolism may rule out such a nonspecific effect. The effect of seeds on fruit volume change is however significant in itself and means that while sugar concentration may be reduced, the total content of sugar per berry may be increased by the presence of seeds.

Rate of metabolism of acids or of accumulation of sugars is not usually modified by changes in seed number from 1 to 4[281,362] but the absence of seed may lead to a greatly enhanced initial rate of sugar accumulation.[362] This disproportionate effect of the difference between no seeds and number of seeds on sugar metabolism is related to a similarly disproportionate effect of the presence and number of seeds or absence of seeds on berry volume.[362]

2. Indirect Methods of Manipulating Ripening

Traditional methods of modifying maturity have included limiting yield by thinning, and reducing competition between ripening fruit and the root system by cincturing. The effect of total yield per vine on rate and extent of maturation are well known and treatments which cause an excessive yield depress sugar accumulation on a concentration basis and vice versa[287] (see Table 9 and sections on fruit set and development). Cincturing, while a complex phenomenon, demonstrably causes raised soluble carbohydrate concentrations in the parts of the plant above the site of the cincture, depletes that of parts of the vine below the cincture and causes an associated increase in sugar accumulation in the ripening fruit.[366]

The relationship between seed development and maturation provides another route to the indirect manipulation of fruit maturation. The use of gibberellin and other PGRCs to prevent seed development and stimulate parthenocarpic fruit growth is associated with an advance of maturity of Delaware grapes which may be as great as 5 to 7 weeks. This effect is almost solely related to reduction in extent of the lag phase of fruit growth and it occurs despite the use of post-bloom application of GA to promote near normal growth of the fruit (see section on parthenocarpy and Figure 13).

Cultivars which have been reported to respond to cincturing at or about veraison include Black Prince,[87] Cardinal,[297,373] Muscat Hamburg,[417] Red Malaga,[256,347,353,366,373] and various *V. lubrusca* cultivars and *V. lubrusca* × *vinifera hybrids*.[248] Maturity effects reported include advance in maturity, increase in final concentration of soluble solids and concentration of anthocyanins (if a colored cultivar), and increased yield.

3. Direct Methods of Manipulating Ripening

The direct methods of manipulating the ripening process are those which from the available evidence appear to impinge directly on the process of initiation of ripening, which once started proceeds unless growth is physically constrained.[75] These methods involve application of particular PGRCs which may either delay (auxins) or advance the timing of the onset of the maturation process (ABA, CEPA or ethylene).

a. Auxins

Not all auxin analogues tested exert similar effects on maturity. The best known is BTOA[118,352] which has been shown to cause a long delay in the advent of ripening of both seeded and seedless grape cultivars. Other auxin analogues which have a similar effect are 2,4,5-TP,[254,333] NAA,[112] and 4-TNA.[304] NAA, however, apparently exerts an additional effect on the physiology of the ripening fruit and acts to delay termination of the maturation phase of growth, possibly preventing the onset of vascular blockage and the resultant shrively

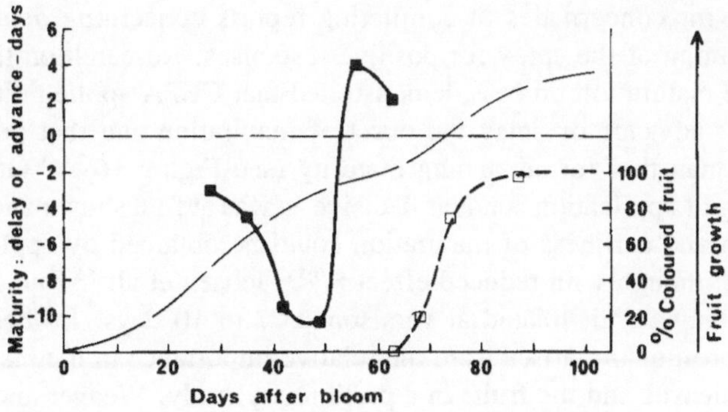

FIGURE 16. Effect of time of treatment Shiraz fruit with CEPA on degree of advance or retardation of the time to attain 40% of coloured berries (■). Superimposed are data for the time-course of veraison as measured in term of % coloured fruit (□) and from that an estimate the course of fruit growth (--, data not provided). Data from Hale et al.[120]

and softening.[112] This report may have important implications for table grape production and would seem to be worthy of further study and confirmation.

One other auxin analogue which may prove to be an exception to the rule that all auxins delay fruit maturation is βNOPA (beta-naphthoxypropionic acid).[386] This compound was rejected in early studies on the effect of auxins on fruit growth and development because it was not as effective as 4-CPA in promoting an increase in the number of berries set. However, it apparently caused those berries which were set to mature well in advance of any other treatment. At present this observation remains a curiosity but it may be worth some further study.

In many ways the effects of BTOA mimic those of seeds in that it delays the onset of the ripening process but not the rate of sugar accumulation following veraison and enhances accumulation of malate.[118] BTOA, and presumably other auxins also, is most effective in delaying maturity if applied shortly before the onset of veraison[415] but a useful delay may be obtained if the chemical is applied as early as fruit set.[364] The pattern of response with time is similar to that obtained with CEPA (Figure 16), but without the final advance in ripening.[415]

The concentrations applied have ranged from 10 to 100 mg·ℓ$^{-1}$ but concentrations greater than about 20 mg·ℓ$^{-1}$ have been reported to cause foliage damage, especially if applied early in the first post-bloom phase of fruit growth.[352] The site of effect of BTOA is an unresolved issue in that auxins applied to leaves are often as effective as if they were applied directly to the fruit.[352] While this possibility has not been specifically resolved, evidence obtained from local application to particular berries within a fruit cluster suggests that each fruit must be treated and that the site of action lies within the fruit.[352]

The extent of delay which may be achieved appears to vary with the length of lag phase in the untreated berry. In Shiraz, a delay of about 14 days has been reported[118] while in Doradillo, a cultivar with a lag phase longer than that of Shiraz, onset of maturity has been delayed by about 28 days.[415]

b. CEPA and Ethylene

Recognition of the role of ethylene (CEPA) as an agent for enhancing fruit quality (color and composition) and advancing maturity has been one of the most important of recent development in the applied use of PGRCs in viticulture. However, we still have a poor understanding of the physiology of the response to this chemical and only the bare bones of the parameters governing its practical use.

The chief cause for concern lies in conflicting reports concerning delay or advance of maturity and the timing of the spray for positive responses. Research on the role of auxins and ethylene in the maturation process demonstrated that CEPA applied directly to the fruit cluster could either advance or delay the onset of maturation and that potential for delay was much greater than that for advancing maturity (see Figure 16).[120] Other studies have reported that timing of application is not as decisive as was at first shown and that an increase in color production and earliness of maturation could be obtained by application of CEPA as early as fruit set, though with reduced effect.[210,302] Jensen et al.[156] found little difference in response of Sultanina fruit treated at veraison ± 7 to 10 days. In these instances, the source of conflict presumably arises from the relative importance in the response of the two sites of action, the leaves and the fruit. In a preliminary study, Weaver and Montgomery[372] showed that CEPA was most effective in stimulating anthocyanin accumulation when applied to the leaves: they obtained no effect when the chemical was applied to the fruit only. This observation is consistent with the role of leaves in synthesis of shikimic acid, the precursor of anthocyanins.[265] Nevertheless, it is difficult to completely resolve the apparent conflict on this basis alone because the effects of CEPA on maturation are most probably located within the fruit itself. Additional studies on this matter are required to resolve the issue and enable clearer guidelines for use to be given.

To a limited extent, CEPA treated fruit may be compared with seedless fruit in that not only is the length of the lag phase of growth shortened somewhat, but apparently, the rate of maturation and extent of accumulation of sugars and anthocyanins is enhanced. Also, accumulation of malic acid is limited. In terms of anthocyanin synthesis, in one case it has been shown that peonidin-3-glucoside synthesis is preferentially enhanced and that this accumulation was associated with improved wine must quality.[270]

The response of a number of cultivars to applied CEPA has been reported (Table 15). The range of use examined has included manipulation of must composition for wine-making purposes,[270] advancement of maturity for table grape production,[93,210,256] improvement of presentation of table grapes through improved color development,[157,256] and improvement of dried fruit (raisin) quality through raised density.[156]

The response to a range of concentrations has not been studied in as much detail as that of other PGRCs, but in most instances the lower end of the range of concentrations reported in Table 15 will probably suffice (i.e., 100 to 300 $mg \cdot \ell^{-1}$). Higher concentrations may cause an excessive reduction in shoot growth.[157,302]

The proper time of application is generally agreed to be when a small proportion (e.g., 10 to 20%) of berries have softened and begun to change color. At this time the CEPA presumably acts to abbreviate the lag phase of the remaining berries, causing a more co-ordinated ripening and may also act to increase the rate of ripening. Studies of time of application also show that in general it is better to err on the early side of the proper time of application than to be late.[372]

c. *Abscisic Acid*

Physiological studies on maturation and anthocyanin synthesis have shown that ABA is intimately related with the onset of ripening and anthocyanin synthesis.[119,204,267,295,415] ABA has been shown to be more effective than CEPA in stimulating anthocyanin synthesis in mature leaf discs[267] and to interact with sucrose in this response. In one experiment ABA was shown to promote the onset of ripening if applied shortly before the onset of the ripening process.[119] Other studies have shown however that ABA content of fruit is correlated with factors which are associated with a delay in ripening; treatment with BTOA and increased number of developed seeds,[415,295] though in the latter case, the rise in concentration was not reported to occur until a time after or concurrent with the onset of ripening.[295] Costs of synthesis at present prevent use of ABA to manipulate maturation processes.

Table 15
SUMMARY OF THE USE AND CONDITIONS OF USE OF CEPA TO MODIFY RIPENING AND COMPOSITION OF FRUIT OF THE GRAPEVINE

Cultivar	Time of treatment	Conc. (mg·ℓ⁻¹)	Responses: Maturation	Responses: Composition	Responses: Col	Responses: Uniformity	Ref.	Comments
Bangalore Purple	H − 28d	300		−A, +S		+	45	
Barbera	15% Col	1000			+		372	
Beauty Seedless	Ver	500	+13d	−A, +S	+	+	210	Earlier treatment less effective
Carignane	?	1000					372	Acts on leaf rather than fruit
Doradillo	Ver −7d	1200	−28d				120	Fruit only treated
	Ver −18d	400	−				415	Cluster only treated
	Ver +3d	400	+				145	Cluster only treated
Emperor	10 to 80% Col	100	+10d				157	Reduced berry firmness
Grenache	15% Col	1000			+		372	
Gulabi	H −28d	300		−A, +S		+	45	
Pinot Noir	Ver	500	+	S	+		270	Improved must quality
Red Malaga	10 to 50% Col	200			+		256	± Girdling, firmness not affected
Ruby Cabernet	20% Col	1000				+	372	
Shiraz	Ver −7d	1200	+4d				120	Cluster only treated
	Ver −14d	1200	−10d				120	
Sultanina	Ver	500	+16d	−A			93	
	Ver ±7d	450	+4 to 8d	−A, +S			156	Improved raisin quality
Tokay	5% Col	200	+14d				157	Reduced shoot growth
Zinfandel	15% Col	1000			+		372	

Note: Ver, veraison; H, harvest; Col, color; A, acids; S, sugars; d, days; +, increases; −, decreases; blank, not reported.

d. Miscellaneous Chemicals

The substituted benzothiadiazole, TH6241 (5-chloro-6-ethoxycarbonylmethoxy-2,1,3-benzothiadiazole) has also been reported to delay maturity when applied close to harvest (4 to 10 days).[104] It may have a special place because it seems to be effective even when applied close to harvest, at a time more suited to last-minute planning decisions or to meet unplanned delays in harvesting. Little is known in detail about the effect of this compound on grape vines and more research is required to establish its usefulness.

Cyclic AMP has been reported to enhance acid content of mature Delaware berries and to reduce the content of soluble solids.[221] However, the effect reported was small in extent and could only be demonstrated in 2 years out of 3.

C. Fruit Abscission

Information on the development physiology and anatomy of abscission of fruit of the grape is not readily available and the lack of such information is limiting the development of reliable methods for controlling the abscission process.

Three potential sites for abscission exist in grape inflorescences, one between the peduncle and rachis and one at each end of the pedicel. In the absence of any fruit set, the whole rachis may abscise at the peduncle, while individual ovaries which fail to set usually abscise at the proximal, rachilla end of the pedicel. The remaining (distal) site for abscission appears to operate during the final stages of ripening, but when fruit are shaken during mechanical harvesting, or when fruit are dipped in an alkaline oil emulsion and dried, separation may occur at the promixal layer, leaving the pedicel attached to the fruit. A more common situation in mechanical harvesting is for the fruit to break away from the pedicel, leaving the central vascular cylinder (brush) attached to the pedicel.

The problem of brush removal during fruit shaking is important and its occurrence suggests that the abscission layer is poorly formed in the region of the xylem fibres and vessels. That development of the abscission layer closest to the fruit can be enhanced is shown by the severe shattering of fruit which may follow rain damage and splitting.[424] Shattering of fruit due to abscission at this site is also a common problem in long-term cold storage of table grapes.

Abscission or retention of fruit at harvest or during storage appears to be subject to two levels of control, one operating through development of the pedicel during the post bloom period of growth, and the other operating through the mechanisms which control the activity of the tissue at harvest.

1. Developmental Control

Plant growth regulating chemicals applied close to bloom have been shown to modify development of the cluster framework during the post-bloom periods of fruit growth (Table 16). Gibberellin, the auxin, 4-CPA, and cincturing each cause enhanced growth in diameter of the pedicel and of the thickness of xylem tissue which is often associated with heightened attachment of the fruit to the pedicel. The authors cited in Table 16 were generally of the opinion that the response to auxin or cincturing was beneficial, not only in increasing degree of attachment, but also in reducing subsequent fruit shatter. Nevertheless there is substantial doubt that the modified growth induced by GA is beneficial, in that the hardening and stiffening of the pedicel and rachis lead to more frequent tearing of the fruit skin during handling and to enhanced shatter in storage.[215,220] This may also be true of the responses reported for 4-CPA and cincturing, but no specific mention of this is made in the reports cited.

Plant growth retarding chemicals have been noted by a number of authors as causing enhanced flexibility of the cluster frame work and thus their use may lead to reduced shatter due to mechanical stresses during handling (see previous sections).

Table 16
SUMMARY OF TREATMENTS USED TO MODIFY DEVELOPMENT AND ACTIVITY OF THE ABSCISSION LAYER IN FRUIT OF *VITIS*

Chemical or treatment	Cultivar	Time	Conc. (mg·mℓ⁻¹)	ABS	ADH	Ref.	Comments
			Modification of Development				
Cincture	Sultanina	bl + 10d			×1.5	348	Increased pedicel diameter
Cincture	Sultanina	bl + 10d			0	368	
4-CPA	Black Corinth	fb	20			113	Increased xylem development in pedicel
4-CPA	Sultanina	bl + 10d	15		×1.5	348, 388	Increased pedicel diameter, effect additive to cincture
4-CPA	Sultanina	bl + 10d	15		0	368	
4-CPA	Black Monukka	bl + 10d	15		0	368	
GA	Black Corinth	bl	20			55	Increased pedicel development
GA	Sultanina	bl,bl +	25,30	+		215	Hardened pedicels, increased brittleness
GA	Sultanina	bl,Ver	40,40		0	306	
GA	Sultanina	bl + 10d	50		0	368	
GA	Kishmish belyi	bl + 10d	200		×2.0	311	Fruit set best time of application
GA	Kishmish chernyi	bl + 10d	200		×1.2	211	
GA	Aneb-e-Shahi	bl − 2d	100	(−)		344	Increased pedicel thickness, reduced shatter
GA	Campbell Early	− bl, + bl	100,100	+		145	Cv. specific effects on xylem different, increased brittleness and mechanical shattering
			Modification of Activity				
CEPA	Alphonse Lavallee	H − 10d	500		×0.87	104	
	Barbera	H − 10d	500		×0.35	104	
	Cabernet Sauvignon	H − 10d	500		×0.62	104	
	Chenin blanc	21.4°B	1600	×6.6		264	Measured 7d after day of application
	Concord	H − 7d	100	+		43	Earlier treatment, less effect, minute conc.
	Concord	H − 6d	2000	100%		53	Temperature increase effect; leaves site of action
	Doradillo	H − 7d	800	13%		264	Effect of wetting agent conc.
	False Tebbiano	17°B	400	×4.0		264	Application at 20°B less effective
	Müller Thurgau	19.4°B	400	60%		264	Effect increased by wetting agent

Table 16 (continued)
SUMMARY OF TREATMENTS USED TO MODIFY DEVELOPMENT AND ACTIVITY OF THE ABSCISSION LAYER IN FRUIT OF *VITIS*

Chemical or treatment	Cultivar	Time	Conc. (mg·mℓ⁻¹)	ABS	ADH	Ref.	Comments
			Modification of Activity				
	Muscadine	H − 3d	2000	46%		266	Reduced harvest damage, leaves recovered
	Purple Cornishon	20°B	500	60%		264	2 Applications on same day give increased response
	Shiraz	H −	800			264	No difference b/n o/a and directed application
	Sultanina	H − 10	500	88%	×0.8	104	2000 More effective, site of abscission
NAA	Sylvaner	20°B	800	40%		264	No difference b/n single and multiple sprays
	Aneb-e-Shahi	H − 3d	100	−		280	Reduced shatter after 6 weeks storage from 9 to 3%
	Cheema Sahebi	H −	50	−	+	83	Increased adherence, reduced shatter
	Muscat Hamburg	H − 4d	20	−		182	Reduced shatter
	Emperor	H	50	0		260	
	James (*V. rotundifolia*)	H	50	0		260	Fruit overripe at application
	Pierce (*V. lubrusca*)	H	50	0		260	Fully ripe at application
	Ribier	H	50	−		260	Mature at application
4-CPA	Sultanina	H	50	0		260	Fruit overripe at application
	Tokay	H	50	0		260	Fully ripe at application
CCC	Aneb-e-Shahi	H − 3d	100	−		280	Reduced shatter in storage, 9 to 3%
CCC	Muscat Hamburg	H − 4d	20	−		182	Reduced berry drop
SADH	Himrod	Ver +	4000	−		35	Reduced shatter in storage (18 to 9%)
	Himrod	Ver +	5000	−		89	Reduced shatter in storage (5 to 2%)
Kinetin	Himrod	H − 28d	4000	−		88	Reduced shatter after 40d storage (16 to 7%)
TH6241	Himrod	Ver +	100	−		89	Reduced shatter after 40d storage (5 to 2%)
	Alphonse Lavalleé	H − 10d	250		×1.21	104	Increased resistance to rupture
	Barbera	H − 10d	250		×1.35	104	Increased resistance to rupture

Table 16 (continued)
SUMMARY OF TREATMENTS USED TO MODIFY DEVELOPMENT AND ACTIVITY OF THE ABSCISSION LAYER IN FRUIT OF *VITIS*

Chemical or treatment	Cultivar	Time	Conc. ($mg \cdot m\ell^{-1}$)	ABS	ADH	Ref.	Comments
			Modification of Activity				
	Cabernet Sauvignon	H − 10d	250		×1.14	104	Increased resistance to rupture
	Sultanina	H − 10d	250		×1.15	104	Increased resistance to rupture

Note: AB, abscission; ADH, adherence; Ver, veraison; H, harvest; Col, color; d, days; +, increases; −, decreases; bl, bloom; °B, °Brix; TH 6241, 5-chloro-6-ethoxycarbonylmethoxy-3,1,3-benzothiadiazole.

2. *Control of Abscission Zone Activity*

Five types of PGRC have been shown to modify the activity of the abscission zone: ethylene, auxins, plant growth retardants, cytokinins, and a substituted benzothiadiazole (Table 16). Ethylene has been shown in many instances to stimulate abscission while auxins, kenetin, and the substituted benzothiadiazole suppress the occurrence of abscission.

a. Ethylene

Ethylene is the only PGRC which has been widely tested for abscission enhancing activity, though it is possible that some other chemicals such as the morphactins may also promote the abscission process. Despite the emphasis given to ethylene (CEPA), many aspects of the technology of use of this compound are inadequately defined (see Section IX).

A sample of the literature on the use of CEPA to promote fruit abscission is documented in Table 16. The responsiveness of different cultivars and species appear to vary widely, though some response was obtained from all the listed cultivars. Undoubtedly, the differences reported are confounded with other factors which influence response to CEPA application (apart from concentration) because response is strongly dependent on environmental conditions prevailing at the time of application (see Section IX).[53] Reliable control of abscission by CEPA application will depend on evaluation of the complex interactions between response and temperature, relative humidity, concentration, pH, and the use of adjuvants such as KI, urea, or polyethylene glycol. Studies on this issue in grapevines have emphasized the necessity of an ambient temperature greater than 17°C (63°F).[53] Fruit removal was easiest at about 25°C (77°F). A temperature higher than 25°C is likely to lead to reduced effectiveness of CEPA.[420]

The list of concentrations applied range over more than an order of magnitude (Table 16), perhaps reflecting the divergence of environmental conditions from those which are optimal. Response to CEPA like that of other PGRCs is well fitted by a linear-log relationship between percentage abscission (y) and concentration of CEPA applied (x, $mg \cdot \ell^{-1}$). In one example[264] the response obtained by CEPA application to Chenin blanc gave the relationship: $y = -67 + 37.6 \log x$, $r^2 = 0.87$. Under different conditions and with another cultivar, both intercept and slope constants might be expected to change. Adjuvants, such as wetting agents, and their concentration have been reported as being an important consideration and use of 2.5% v/v Agral 60® doubled the effectiveness of CEPA applied, compared with a more standard concentration of 0.1%.[264] The effect was associated with dessication of the leaves.

Leaf damage is a common response to CEPA application, with symptoms which may be compared with those of normal senescence.[264] Excessive leaf damage and abscission could lead to reduced productivity.[423] Tolerance has been reported to vary quantitively, with Purple Cornishon and Shiraz being very susceptible (described as being hypersensitive),[264] while

Muscadine grapes tolerated an application of 2000 mg·ℓ⁻¹ without loss of leaves, though with yellowing, followed by regreening.[266] Avoidance of the problem of leaf damage by directed spray application is not practical, because leaves are the site of action.[53] In this regard it is also worth noting that differences in leaf retention of spray between different cultivars is one of the factors determining the responsiveness of particular cultivars.[264]

Timing of the spray application with respect to degree of fruit maturity is an important aspect of the response which has been substantially overlooked in reports examined. In one case, a substantial increase in response was obtained by application at 17° Brix, compared with later application when the fruit had attained 20° Brix (False Trebbiano).[264]

These studies show that scope exists for the use ethylene producing compounds such as CEPA to activate existing abscission layers, but a more concerted approach appears to be required to fully develop their potential.

b. Auxins

A number of reports which describe the use of either NAA or 4-CPA to reduce berry abscission and shatter during harvesting and storage are summarized in Table 16. In general, application of 20 to 100 mg·ℓ⁻¹ of one of these two auxin analogues a few days before harvest increases adhesion of the berry and reduces shatter during storage and transport. The reports do not document the parameters of the response in any detail and substantially more research is required to establish optimum conditions and degree of effect.

c. Other PGRCs

A number of reports have been published which suggest that some other PGRCs such as the plant growth retardants, cytokinins, and a substituted benzothiadiazole (TH 6241) may usefully be used to reduce post-harvest abscission of table grapes (Table 16). Plant growth retardants have been reported to reduce biosynthesis of ethylene (reviewed by Morgan[422]) and may have potential for reducing senescent breakdown of stored fruit, as well as reducing abscission. Kinetin has also been reported to improve storage qualities of fruit but many problems must be overcome before worthwhile and consistent effects could be expected from the application of this class of regulating chemical (see Parthenocarpy, section IV). The remaining chemical, TH 6241, is a curiosity because in other crops it is used to promote fruit abscission.[104] The responses reported were consistent and may justify further study.

In the main part however, reports of the use of PGRCs to prevent berry abscission are not substantial enough to warrant use without more extensive research. Nevertheless, their use for this purpose appears to hold some promise of success.

D. Physical Properties

The physical properties of fruit of the grape have not been as well documented as those of some other fruit.[425] Viticulturally, firmness of fruit is of primary importance for the fresh fruit trade. Strictly speaking, firmness is not a physical property because it is principally a consequence of turgidity, i.e., water content. Strict physical properties are those which reflect the changes in dimension or shape caused by an applied force, and it is not feasible to measure these under compression in a soft fruit. Acknowledgment of this distinction enables a more direct approach to control of either property, firmness or physical properties. The former reflects the interaction of factors such as water status at harvest and rate of water loss during or after harvest, while the second factor reflects the composition, size, and integrity of the structural elements of the fruit, its cuticle, and cell walls. Structural properties are likely to determine such other properties as tolerance or intolerance of handling (resistance to crushing and penetration by stems). The degree of interaction of firmness and physical properties has not been documented to date, although the means exist to distinguish between the two and to obtain some estimate of the bulk deformation moduli.[426]

A range of methods has been used to obtain a measure of physical properties. Most are difficult to interpret, in the sense of obtaining a measure which relates directly to a modulus which is defined in engineering terms or in physical terms such as pressure. Nevertheless they often have a real meaning in practice and are thus justifiable. For example, resistance to crushing,[311] resistance to penetration of a pin,[306] and compressibility.[416]

Ethephon applied to enhance color development and ripening of table grapes has been shown to cause a reduction in fruit firmness after storage[157] of Emperor grapes but not of Red Malaga.[256]

Gibberellin applied to Sultanina grapes at 20 to 40 $mg \cdot \ell^{-1}$ apparently exerts a positive effect on fruit firmness measured either at harvest or following storage, but only if applied towards the end of the first post-bloom growth stage or at veraison (author's unpublished results).[306] The studies of Singh et al.[306] also demonstrated that while the effect on resistance to penetration was relatively small, there was an associated substantial reduction in water loss and a maintenance of almost complete turgidity after 1 month of storage. Higher concentrations (100 to 200 $mg \cdot \ell^{-1}$) applied to Kishmish belyi caused a 2- to 3.4-fold increase in resistance to crushing.[311] The degree of resistance increases with multiple applications (bloom and fruit set) and time (fruit set > bloom).

A number of other plant growth regulators have been examined for effects on physical properties by the author (unpublished data) who found that cincturing 1 month after bloom caused about a 15% decrease in resistance of Sultanina fruit to penetration by a 1/8-in. diameter pin. Earlier treatments had no significant effect. Likewise 4-CPA (20 $mg \cdot \ell^{-1}$) applied either at bloom or fruit set, about 10 days later, caused about a 10 to 15% decrease in resistance to penetration. GA at 30 days after bloom was the only treatment applied which increased resistance to penetration. In all cases the treatment effects were relatively small but they may have important implications for the behaviour of fruit in storage (see results of Singh et al.[306]).

E. Senescence Control/Storage

Plant growth regulators have been widely applied to crops other than grapes to manipulate senescent decline of fruit in storage. Little work has been published on the grape. This fact is unlikely to reflect a lack of interest, rather it probably indicates a general lack of success. Apart from the possible usefulness of GA, discussed in the previous section on fruit firmness and physical properties, there is some indication that plant growth retarding chemicals applied some time during the maturation phase of growth may be beneficial.[35,88,89,94,95,127,276] However, in each of these largely repeated studies, unusually high concentrations of either CCC or Alar® were applied and the effects were not always consistent.

In the matter of control of senescence, the grape is a difficult subject because many of the phenomena of maturation, apart from the accumulation of sugars, resemble senescence in other tissues (see section VI on maturation). In storing the grape, we are presumably attempting to obtain a state of suspended animation or to help repair some of the deleterious effects of maturation such as increased membrane leakiness.[427]

F. Disorders

Fruit cracking and splitting at harvest is a serious problem leading usually to decay and loss of crop. Such fruit cracking may arise through two causes, one associated with the occurrence of rain[428] and the other with a compact bunch structure (see section V on thinning). A compact bunch structure is relatively easy to modify and GA can be used with success even on cultivars which are relatively intolerant of the regulator (see earlier section on thinning for details).[390]

Attempts to modify the physical properties of fruit to produce fruit resistant to rainfall-induced splitting and cracking have been unsuccessful except for one instance, that of a combination of 100 $mg \cdot \ell^{-1}$ CCC plus 1 $mg \cdot \ell^{-1}$ GA applied together to Black Corinth berries a few days after full bloom.[58,96] However, even in this case, the treatment produced fruit which were no more tolerant than untreated fruit, just more tolerant than 4-CPA treated fruit.[58,96]

VII. VEGETATIVE GROWTH

A. Bud Burst

Plant growth regulating chemicals have been applied to vines in attempts to avoid spring frosts by delaying bud burst, to overcome a low percentage bud burst and/or an extended period of bud burst associated with either insufficient winter chilling or with endogenous physiological problems. Timing of bud growth also affects propagation practices in that it is desirable to ensure that shoot growth and the demand created by the shoot for water supply doesn't outstrip the ability of a developing root system or graft union to support the shoot's requirements.

1. Physiology

Bud burst, its occurrence and its timing is known to be influenced by many environmental and physiological factors.[430] Of the environmental factors, temperature is undoubtedly the most important influence acting both to provide a cold requirement[102,171] and to promote the subsequent process of growth.[430] While the terminology relating the cold requirement, whether it constitutes a dormancy or not, has been the subject of debate, there is little doubt that most cultivars respond positively to a cold treatment, the effects of which are reflected in a reduced time to bud burst in temperatures favorable to growth and increased proportion of buds growing.[102,171,430]

Endogenous factors which influence bud burst relate to shoot vigour (proportion of buds to grow varies inversely with vigour[186]) to bud position (basal buds are less likely to burst than distal buds and more likely to be necrotic[186]), and to particular cultivars.[186,342,430]

Treatments other than chemical treatment which have been shown to enhance bud burst of cuttings include dipping in warm water (50°C)[394] and an undefined degree of drying[394] while root pruning has been reported to reduce bud necrosis in field vines.[27]

2. Proportion

Most chemical treatments appear to be detrimental to bud growth, though not many have been examined in sufficient detail to warrant firm recommendations to be made (Table 17). Often the concentrations applied have been excessive.

Compounds with some cytokinin-like activity have, in a number of instances, been shown to promote the process of bud growth and to a substantial degree replace the cold requirement (Table 17). In practice though, they are not effective in the field, presumably because of the difficulty of obtaining sufficient uptake (see previous discussion in section III.F).

A diverse range of other chemicals have been shown to promote bud burst under a range of conditions including SADH at low concentrations[394] (or at higher concentrations if applied in autumn),[399] morphactin[110] (although in some other circumstances it has been reported to delay bud burst in cuttings),[109,394] DNOC,[103] calcium cyanide ($CaCN_2$, calcium carbodiimide, lime nitrogen),[148,261] KNO_3,[103] and ethylene chorohydrin (2-chloroethanol, rhindite).[395] Note that $CaCN_2$ is a very irritating substance and likely to cause severe dermatitis on skin contact and is a defoliant and herbicide and that ethylene chlorohydrin is extremely toxic.[433] Of these, SADH appears to be the most promising because of its low toxicity, availability, and registration status, although CCC has been reported to reduce bud necrosis[27] if applied the previous season. Such an effect may account for some of the beneficial effects on yield in the season following that of application.[211] Further work is required, though, to substantiate or otherwise delineate the reported benefits claimed for these growth retardants.

Chemicals which have been reported to reduce the proportion of buds to grow include

Table 17

SUMMARY OF TREATMENTS WITH PGRCs WHICH MODIFY THE TIME AND/OR PROPORTION OF BUDS TO DEVELOP AND GROW IN *VITIS*

Chemical	Method	Time	Conc. ($mg \cdot \ell^{-1}$)	Cultivar	% BB or proportional effect	Rate/ timing	Ref.	Comments
Auxins								
NAA	Vine	Autumn	400	Sauvignon		D5d	90	Compressed period of BB
NAA	Cuttings	Prechill[a]	100	Sultanina			395	NS effect
2,4-D	Vine	Winter	1000	Tokay		D	363	Sensitivity varied with site of application
BTOA	Cutting		200	St. Emilion	×0		394	Cuttings prechilled
BTOA	Cutting		1000	St. Emilion	×0		399	
Gibberellin								
GA	Vine	bl	100	Black Corinth	×1	?	370	
GA	Vine	±bl	50	Carignane	×0.65	D	370	Only conc. 25 $mg \cdot \ell^{-};1$ effective
GA	Vine	Bloom	>50	Chasselas	×0	D	429	Only epicormic buds grew, others dead
GA	Vine	−bl/+bl	100	Queen of Vineyard	R		405	Time, site, conc. on bud necrosis
GA	Vine	+bl	20	Red Malaga	×0.08		355	Sensitive to conc. > 1 $mg \cdot \ell^{-1}$
GA	Vine	±bl	25	Red Malaga	×0.08	D	370	Effect NS at veraison, recovered next season
GA	Vine	±bl	25	Ribier	×0.34[b]	D	370	Recovered the next season
GA	Vine	+bl	1000	Sultanina	×1	?	370	
GA	Vine	−bl	50	Sylvaner	×0.59	D	160	Lower conc. NS effect
GA	Vine	±bl	25	Tokay	×0.12[b]	D	370	Recovered next season
GA	Vine	bl	20	Waltham Cross	R	D	324	Growth was abnormal
GA	Vine	−bl	250	Zinfandel	×0.06		356	Also inhibited epicormic buds

Table 17 (continued)
SUMMARY OF TREATMENTS WITH PGRCs WHICH MODIFY THE TIME AND/OR PROPORTION OF BUDS TO DEVELOP AND GROW IN *VITIS*

Chemical	Method	Time	Conc. ($mg \cdot \ell^{-1}$)	Cultivar	% BB or proportional effect	Rate/ timing	Ref.	Comments
GA	Vine	±bl	50	Zinfandel	×0.35	D	370	Delay all times, −bl decreased %BB
GA	Cuttings		20 μg	Concord			102	No measurable effect
GA	Cuttings	Prechill	1000	Sultanina		D5d	171	
GA	Cuttings		100	Tokay	×0		356	0.1 $mg \cdot \ell^{-1}$ prevented BB
GA	Cuttings	Postchill	2000	St. Emilion	×0	0	394	
GA	Cuttings	Prechill	100	Sultanina	×0	D	395	Cuttings from summer sprayed vines, NS effect
Cytokinins								
BA	Vine	+BB	500	Palamino			110	No observable effect
BA	Cuttings	Prechill	1000	Sultanina			171	No observable effect
BA	Cuttings	Prechill	1000	St. Emilion	53%	A	357	Untreated failed to grow
BA	Cuttings	Prechill	1000	Sultanina	33%	A	357	Untreated cuttings failed to grow
BA	Cuttings	Prechill	1000	Tokay	31%	A	357	Untreated gave 3% BB
BA	Cuttings	Postchill	1000	St. Emilion		A	394	Substantial increase in rate of BB
BA	Cuttings		1000	St. Emilion		A	399	
Zeatin	Cuttings	Prechill	?	Concord		A3d	102	Applied directly to naked bud
Thiourea	Vine	Winter (×6)	2%	Sultanina	×1.16	A12d	19	Extended BB period ×2
Thiourea	Vine	Winter (×6)	2%	Madelaine Angevine		A12d	19	Extended the period of BB ×2

Thiourea	Vine	PreBB	%	?	+		103	
Thiourea	Vine	BB −77 and −32d	5%	Palamino			110	No measurable effect
Thiourea	Cuttings	Prechill	1000	St. Emilion	23%		357	Untreated gave 0% BB
Thiourea	Cuttings	Prechill	2%	Sultanina		A8d	395	
Abscisic acid								
ABA	Cuttings	Postchill	1000	St. Emilion		D20d	394	No effect on % BB
ABA	Cuttings		30 μg	Concord			102	Applied to naked bud, NS effect
Ethylene								
CEPA	Vines	Harvest	2000	Concord			53	Increased susceptibility to winter frosts
CEPA	Vines	±bloom	2000	Sultanina	R	D	379	Weakened growth
CEPA	Vines	± bloom	2000	Carignane	R	D	379	Weakened growth, reduced cluster count
CEPA	Cuttings	Postchill	2000	St. Emilion	×0.44	D15d	394	200 $mg \cdot \ell^{-1}$ NS effect
Plant Growth Retardants and Inhibitors								
CCC	Vine	?	?	Aneb-e-Shahi			27	Reduced bud death
CCC	Vine	+BB	500	Palamino			110	NS effect on % BB
CCC	Cuttings	Prechill	60 μg	Concord			102	Applied to naked bud, NS effect
CCC	Cuttings	Postchill	2000	St. Emilion		D10d	394	
CCC	Cuttings	Prechill	2000	St. Emilion	×0.23	D	399	200 $mg \cdot \ell^{-1}$ also had a marked effect
SADH	Vine	Autumn	1000	Sauvignon			90	NS effect
SADH	Cuttings	Prechill	2000	St. Emilion		A10d	399	
SADH	Cuttings	Postchill	2000	St. Emilion		D15d	394	
SADH	Cuttings	Postchill	20	St. Emilion	×1.36	A5d	394	
MH	Vine	Aut/Wint	1500	Chasselas		D5d	135	Higher conc. toxic, especially to apical buds
MH	Vine	+BB	3000	Concord			398	Toxic, promoted growth of epicormic buds
IT3456[c]	Vine	Bud swell	1600	Santa Agueda		D6d	109	

Table 17 (continued)
SUMMARY OF TREATMENTS WITH PGRCs WHICH MODIFY THE TIME AND/OR PROPORTION OF BUDS TO DEVELOP AND GROW IN *VITIS*

Chemical	Method	Time	Conc. (mg·ℓ$^{-1}$)	Cultivar	% BB or proportional effect	Rate/ timing	Ref.	Comments
IT3456	Vine	Bud swell	1600	Tempranillo		D6d	109	No effect on % BB
IT3456	Vine	+BB	100	Palamino	×1.32		110	Caused inflorescence abscission
IT3456	Cuttings	Postchill	20	St. Emilion	×0.23	D	394	Higher conc. prevented BB
DNOC	Vine	PreBB	?	?	+		103	
Miscellaneous Chemicals								
EC[d]	Cuttings	Prechill	6000	Sultanina		Alld	395	
Rhindite (EC)	Cuttings	Prechill	1.9%	Sultanina		A7d	395	
$CaCN_2$	Cuttings	Prechill	25%	Zinfandel	95%		148	Untreated buds failed to grow
KNO_3	Vines	PreBB	?	?	+		103	
$CaCN_2$	Vines	Pruning	20%	Niagara-Rosada		+	261	

Note: BB, bud burst; D, delay; A, advance; R, reduced; bl, bloom; d, days; +, increasing; −, decreasing; 0, no effect; blank, not reported.

[a] Prechill, cuttings treated and tested without exposure to natural or artificial chilling treatment to overcome any natural cold requirement for bud burst; Postchill, cuttings either exposed to natural winter chilling or refrigerated before treatment.
[b] Based on degree of yield depression
[c] IT3456, Morphactin, methyl-2-chloro-9-hydroxyfluorene
[d] EC, ethylene chlorhydrin

CCC, CEPA, auxins, but especially GA (Table 17). BTOA, which has been advocated as maturity delaying treatment, appears to be a particularly effective inhibitor of budburst, though no deleterious effects have been reported on bud burst the year following treatment. The effects of GA, however, are known to carry through on a wide range of cultivars, though no detailed survey has been reported. The problem has been examined in detail for one cultivar, Queen of the Vineyard.[405] In this cultivar, as little as 6ng of GA per bud is sufficient to induce bud necrosis. Sensitivity was greatest when the GA was applied during the early phase of shoot growth, before bloom. In this phenomenon, as in the other phenomena discussed previously, cultivars vary in sensitivity and Sultanina and Black Corinth appear to be relatively tolerant.[370] Recovery generally is complete in a further season (i.e., 2 years after the season of application).[370]

Other compounds which have been reported to reduce percentage of bud burst include CEPA (also reported to reduce cold tolerance[53,94]), CCC,[399] and IT3456.[394]

3. Timing

In general, those chemicals which reduce bud burst also delay bud burst and those which enhance the proportion of buds which burst also advance the timing of bud growth (Table 17). Compounds which both reduce and delay include CEPA, GA, CCC (applied to cuttings by soaking), SADH (at high concentrations), MH, and IT 3456. Two PGRCs which have been shown to delay bud burst but not interfere with the proportion of buds which burst are NAA and ABA. Of these two chemicals only NAA is readily available. I have examined only one report which claims such a favorable response[90] and so further work is required to establish the benefit in practical terms. This is especially necessary because other unpublished work[309] has shown that 0.5 to 1% NAA can be applied to vines to prevent growth of epicormic buds.

Of the six chemicals which have been reported to advance the timing of bud burst (Table 17) only SADH at low concentrations holds promise of practical benefit, though $CaCN_2$ may possibly be worth further testing. The cytokinins, while apparently effective on cuttings, do not produce consistent effects on field vines.

Treatments which alter the timing and degree of bud burst also commonly alter the length of period over which bud burst occurs. Thiourea, which has been reported to advance the mean time to bud burst, acted principally by extending the period over which bud burst occurred.[19] The last buds to grow on treated vines emerged at about the same time as those on untreated vines. This broadening of the timing of bud burst may reflect the problem of obtaining uniform entry to the site of action because cytokinin treated cuttings usually emerge over a shorter time span than untreated cuttings. Treatments which delay bud burst may extend the period of emergence,[394,399] though NAA has been reported to compress the period of shoot emergence while retarding the timing of emergence.[90]

B. Shoot Extension and Development

Control of shoot development is assuming a role in viticulture approaching the importance of that of manipulation of fruit development. While often shoot development is inadvertently changed by treatments applied to control some aspect of fruit growth, particular practices have now been proposed to apply the same regulators to specifically regulate shoot growth and development.[185]

Matters for enquiry include restriction of vegetative growth: to enable unimpeded access to the crop; to ensure near optimum partitioning of dry matter between crop, permanent framework, and dispensible portions of the vine; to maximize inflorescence initiation and fruit set; to provide access for air flow and light to minimize the risk of disease and to enhance colour development. Also important is control of physiological status of the shoot, its maturation (to minimize the risk of autumn and winter frost damage), and response to water deficit and other environmental stresses.

Table 18
USE OF ETHYLENE PRODUCING AGENTS TO RESTRICT GROWTH

Agent	Conc. (mg·ℓ^{-1})	Time	Cultivar	Response[a]	Ref.	Comments
CEPA						
	480	bl + 4—6week	Alphonse Lavallée	×0.06	185	Prevented lateral growth for 2 months, no toxic effect
	720		Cabernet Sauvignon	×0.79	302	
	480	bl + 4—6week	Cardinal	×0	185	Prevented lateral growth for 2 months
	960		Cardinal	×0.54	302	
	3000	±bl	Carignane	×0	379	More sensitive than Sultanina to leaf abscission
			Carignane	R	302	
	720		Dabuky	×0.23	302	
	480	bl + 4—6week	Muscat of Hamburg	×0.17	185	
	960		Muscat of Hamburg	×0.67	302	Inhibited lateral bud growth
	750		Muscat of Hamburg	×0.2	129	Reduced node no. and internode length, pH 6.9 > pH 2.2
	10	bl − 14d	Muscat of Alexandria	×1.55	378	Note: enhanced growth
	> 10	bl − 14d	Muscat of Alexandria	nrO	378	No formative effects
	480	bl + 4-6week	Perlette	×0.06	185	
	720	+ bl	Perlette	×0.23	302	Time/topping/vigour: less effect on vigorous shoots
	750		Perlette	×0.11	129	
	3000	±bl	Sultanina	×0	379	Terminal abscission, leaf cupping, −ve carry-over effects
	1000	+ bl	Zinfandel		202	Reduced second crop if applied to leaves
ALSOL						
	2000		Muscat of Hamburg	NS	302	
	2000		Muscat of Hamburg	NS	129	Split the terminal meristem
	2000		Perlette	NS	302	

Note: bl, bloom; d, day; wk, week

[a] Shoot length or growth increment of treated shoots as a factor of untreated shoots

1. Enhancement of Shoot Extension Growth

Of the plant growth promoting chemicals, only GA has been shown to promote shoot extension growth. Auxins and auxin-analogues generally have no discernable effect on shoot extension growth.[184] The effect of GA in enhancing shoot growth is of dubious value and has not been advocated as having any particular merit by the authors who noted its effects.[160,365,434] The effect of GA on shoot extension is generally greatest soon after bud burst.[365] The response may persist for a considerable period of time.[160] Not all cultivars respond equally to applied GA and such comparative studies as have been reported indicate that cultivars differ quantitatively in degree of response.[117,434] Cultivars which are generally unresponsive (<10% increase) included Seeded Thompson, Emperor, and Kober 5BB,[117,434] while responsive cultivars (>30% increase) include Sylvaner and 43-16F.[117,434]

Chlorosis is commonly associated with application of GA and can be severe if especially high concentrations are applied (e.g., 1000 mg·ℓ⁻¹).[160,184,367] Other toxic side effects include excessive lignification and cane splitting.[365]

2. Restriction of Shoot Extension Growth

The traditional method of containing excessive vegetative growth is to top the vines by removing a length of the terminal part of the cane,[402] but this practice suffers from several important disadvantages: its effect is limited in duration,[69] axillary bud growth is promoted, and the proportion of dry matter production allocated to such dispensible portions of growth as lateral shoots may even be increased.[185]

The compound with most promise for overcoming the promotion by topping of lateral shoot growth is CEPA (Table 18). The chemical has been applied at a wide range of stages of shoot development with apparently equal success. However, care should be exercised in applying CEPA at any time before fruit set because it may cause fruit or flower abscission (see sections on fruit set).

CEPA is a very effective inhibitor of shoot growth and need not be used at high concentrations. A concentration as low as 100 mg·ℓ⁻¹ is effective[378] and much higher concentrations may cause formative effects, limiting bud burst and yields in the following season.[379] Restriction of shoot growth is greatest if the chemical is applied directly to the shoot tip and if prepared at a moderately high pH (pH 6.2 more effective than pH 2.2).[129]

While CEPA is effective in restricting excessive production of the dispensible (after harvest) portions of the vine, there is as yet no good evidence that the savings in energy are devoted to additional yield. At present, the best that can be said is that this treatment may be done without prejudicing yield.[185,302]

A range of other PGRCs are also effective as inhibitors of vegetative growth but none seem to have any particular advantage. Some have particular disadvantages (e.g., Morphactin[185,378]). Those chemicals examined include, Dikegulac (2,3:4,6-di-0-isopylidene-L-xylo-2-hexulo-furanosonic acid),[29] Alden (1-allyl-1-(3,7-dimethyloctyl) piperidium bromide),[134] NC 9634 (Fisons),[185] pp413 (ICI),[185] and M7311 (Celamerck).[185]

The plant growth retardants used in viticulture to modify fruit set also affect shoot growth (Figure 17). While the response to CCC is generally predictable, that to SADH is not.[32] Assessment of the response of 23 cultivars of *Vitis* showed that response of vegetative growth to SADH application (500 mg·ℓ⁻¹) varied from +28.3% for cv. G157 to −24.7% for cv. Trollinger. Within this range, there was a continuous distribution of degree of response. Degree of response to CCC also varied but was never positive and was not predictable on the basis of degree of shoot vigor.[32]

While CCC or Alar® are not generally recommended as a means of controlling vegetative growth, the potential exists. An important difference between the action of plant growth retardants and CEPA is that the latter stops growth while CCC merely reduces its extent, having little effect on internode production and thus on leaf production.[32,74] While there is a reduction in leaf area, this is compensated for by increased photosynthetic activity per unit area of leaf.[334]

C. Modification of Tolerance of Stress

Little applied work has been reported in English in this field. CCC has been reported to slightly enhance the cold tolerance of Aligote and Karaburnu.[301] This effect may be related to the effect of CCC in enhancing shoot maturation and early lignification.[274]

The possibility has also been raised of the modification of drought tolerance by application of ABA or phaseic acid.[91,92,130,131,175,176] However, knowledge of the physiology of the

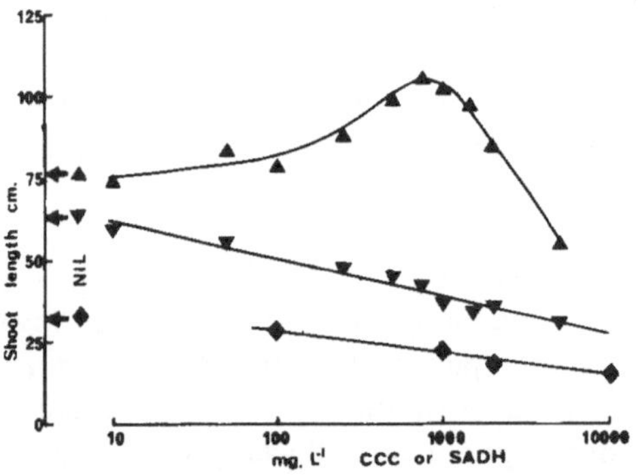

FIGURE 17. Variation in growth of shoots of particular cultivars and species of *Vitis* in response to type of plant growth retardant (CCC or SADH) and concentration. *V. labrusca* cv Concord treated with SADH (▲), *V. vinifera* treated with CCC: Trollinger (▼), and Sylvaner (♦). The regression equation fitted to the Trollinger data was: $y = 74.2 - 11.7 \log (CCC)$, $r^2 = 0.95$; and that to Sylvaner was, $y = 42.0 - 6.8 \log (CCC)$, $r^2 = 0.97$. The Concord data was fitted by eye, but could also be well fitted by two regressions, one covering the concentration range, 100 to 1000 mg·ℓ^{-1} and the other the range from 1000 to 5000 mg·ℓ^{-1} ($y = 12.5 + 32.0 \log (SADH)$, $r^2 = 0.98$, and $y = 297 - 65.0 \log (SADH)$, $r^2 = 0.96$. Note that in each instance that y was measured in terms of centimeters. (Data from Bourquin, H. D. and Alleweldt, A., *Vitis*, 9, 105, 1970. With permission.)

response and its interaction with the presence of a crop has yet to be developed to a practical extent. Such benefits as have been shown are of a short term nature, but may perhaps be usefully applied to reduce transplantation shock for propagated plant material.

VIII. PROPAGATION

Plant growth regulating chemicals are the key to the new technology of plant propagation and improvement, an area which is covered elsewhere.[441] Plant growth regulators also have a role in the less exciting but nevertheless important complementary fields of sexual propagation via seed and the more conventional techniques of vegetative propagation by hardwood cuttings and by grafting.

A. Seed Germination

Satisfactory germination of seeds of grape usually occurs only after a period of after-ripening at about 18°C followed by a cold stratification treatment at about 5°C.[296] Thirty days of each treatment gave a satisfactory germination of about 75% (cv Red Romi). However, the requirement for such treatments can be reduced by treatment with GA,[162,201,255,296,403] though the benefits of cold stratification and GA-treatment may be combined provided that the GA is applied before stratification.[162,201,296] Concentrations recommended have varied from 2000 to 8000 mg·ℓ^{-1} with a period of treatment of up to 24 hr.[201,296,403] The plants produced by these treatments are apparently normal in early growth, although the hypocotyl and first internode are elongated.[403]

A range of other PGRCs have been tested but none appear to be useful (NAA, IAA, ABA, BTOA, Morphactin,® CCC, SADH, BA, Kinetin, Thiourea, and CEPA).[296,301,403] Of the gibberellins tested (GA_3, $GA_{4/7}$, and GA_{13}), only $GA_{4/7}$ wasn't recommended.[255] The $GA_{4/7}$ mixture, though acting with near equal effectiveness of GA_3 at moderate concentrations, caused a marked decrease in germination at higher concentrations (500 mg·ℓ⁻¹).

Grape seeds required for germination should be sorted by immersion in water because seeds that float (empty seeds) are most unlikely to germinate (correlation between % germination and % empty seeds equals −0.95 for 14 cultivars).[435] Empty seededness is considered to be related to embryo abortion at a late stage of seed development, possibly due to an insufficient length of period between the two post-bloom periods of fruit growth.[435] While one solution to this problem is to use embryo culture, a novel use of a fruit maturation delaying treatment has been suggested by Bouquet[31] to enable normal embryo development. Treatment with BTOA at 20 mg·ℓ⁻¹ 4 to 5 weeks after bloom greatly enhanced the percentage germination of seeds of Perle de Csaba. Lesser effects were recorded for Madeleine Angevine and Cardinal. Timing of the application was shown to be quite important and could coincide with the optima shown to apply for maximum extension of fruit growth (Figure 16.)[415]

B. Propagation and Grafting of Hardwood Cuttings

Vegetative propagation by hardwood cuttings is the mainstay of viticulture and is likely to remain so despite the success of mist propagation of young leafy cuttings[440] and the advent of in vitro techniques.[441] Continuity of the use of hardwood cutting techniques of propagation will remain until such time as the newer methods can demonstrate clonal stability and adaptation to mass production methods which provide for grafting. Despite the overall success of hardwood cutting methods of propagation, particular problems and difficulties remain. The most commonly cited problem is that of adequate root initiation, especially in non-vinifera rootstocks, though grafting difficulties nearly match that of root initiation. These problems are minimized in modern practice by stratification, hygiene, and choice of culture conditions, but opportunities still exist for improved efficiency.

1. Root Initiation

Considerable research has been carried out to establish the benefits or other effects of the application of various auxins (Table 19). In an extensive early survey, Harmon[123] showed that about 2/3 of the cultivars he examined showed some positive response, while the remaining cultivars showed negative effects. This general conclusion has been borne out by more recent research which has in general concentrated on a few, economically important rootstock cultivars (Table 19).

Where a positive effect exists, the effect is seen usually as an advance of the time of emergence of roots rather than as an absolute effect.[46] Such an advance can effect important economies in the nursery by enabling a greater throughput and also by reducing the likelihood of failure of a graft due to scion development proceeding ahead of the rootstock.

In particular circumstances, each of the three auxins listed in Table 19, IBA, NAA, and IAA, can produce an enhancement of root initiation. However, in general IAA was only effective if applied in lanolin or if used in low concentrations and applied constantly. Both IBA and NAA have been shown to be effective provided that the concentration applied is not excessive.[8] A mixture of 100 mg·ℓ⁻¹ of each has been recommended,[46] although in a study carried out on 1616, 50 mg·ℓ⁻¹ NAA was shown to be more beneficial than IBA (at 100 mg·ℓ⁻¹).[326]

The hormone may be applied in one of three ways: as a dilute aqueous solution (usually about 50 to 200 mg·ℓ⁻¹ in which the base of cuttings are immersed for a period of up to 24 hr, as a concentrated solution in 50% ethanol (1000 to 5000 mg·ℓ⁻¹) applied by dipping the base of the cuttings in the solution for about 5 to 10 sec, or as a concentrate (2000 to 5000 mg·kg⁻¹) dried onto talc.[46,257] Either of the immersion methods are generally superior to the dry talc method.[46,123,257,438]

Table 19
EFFECT OF APPLIED AUXINS ON ADVENTITIOUS ROOT FORMATION BY A RANGE *VITIS* SPECIES, CULTIVARS AND THEIR HYBRIDS

Name	IBA	NAA	IAA	Ref.
Show increased proportion of cuttings forming roots and increased extent of root initiation				
A × R #2	+			123
Barnes	+			123
Chasselas blanc			+	257[a],439
Constancia	+			123
Dog Ridge	+			9,7,17,46,123
French Columbard		+		8
Harmony	+			7—9,17,46
Isabella (*V. labrusca*)			+	439
Muscat Hamburg			+	257[a]
Mourvedre × Rupestris 1202			+	257[a]
Ramsey (*V. champini*)	+			9
Rupestris du Lot			+	257[a]
Rupestris St. George	+			123
Salt Creek	+			123
Sultanina			+	257[a]
Tisserand	+			123
V. Berlandieri			+	439
V. Monticola × riparia 18815	+			123
V. monticola × rupestris	+			123
V. riparia × cordifolia 107-11	+			123
41B (Berlandieri)			+	257[a]
1613	+	+		8,23
1616	+	+	+	326
Variable or no response				
A × R #1	+	+		7,8
Concord			+	257[a]
Empress			+	257[a]
Harmony	+	+		438
James (*V. rotundifolia*)	+			123
Sultanina	+			7
V. davidi	+			123
Zinfandel	+			7
1613	+	+		438
Show reduced formation of adventitious roots				
Freedom		+		8
Riparia Gloire	+			123
Ruby Cabernet		+		8
Rupestris St. George	+	+		7,8
420 A	+			123

[a] The particular references are cited within this review of early research on the topic.

While basal application of the artificial auxins is generally followed, there are a few reports that the effectiveness of IAA is greatly increased if it is applied to the top of the cutting rather than to the base.[439,442] This practice may not be useful for the artificial auxins because of their secondary effects of stimulating callus production,[436,437] and interfering with the process of bud burst (see Section VII.A).

Studies of the effects of other PGRCs have generally revealed either no benefit or a deleterious effect: the response to GA being one of complete inhibition of root development.[17,326]

2. *Grafting*

Formation of a graft union depends on the process of callus initiation in both stock and scion, on the union of the callus an on the subsequent differentiation of the callus tissue to form the protective and vascular tissues required to form a function unit from the two adjacent plant parts, stock and scion.[443] Difficulties arise when the stock and scion differ in the rate of initiation and proliferation of callus. Differentiation is not usually considered to be a problem.

Differences in rate of callus production by the graft partners is accentuated by the effect of polarity which acts to promote callus formation at the basal end of each, stock and scion. Despite the widespread acknowledgment of the ability of PGRCs and auxins in particular to stimulate callogenesis and the mainly favourable reports of early research workers (reviewed by Pearse[257]) little modern research has been published in this topic. One report[436] claims that treatment with a mixture of 100 $mg \cdot \ell^{-1}$ IBA + NAA + adenine sulphate could promote union formation of grafts which used Ramsey as a stock, but at the time the report was made the research was incomplete. Problems which require evaluation in such programs are the possible interference of the PGRCs used to treat the graft with the normal process of differentiation of vascular and other tissues, and excessive callus formation from a readily callusing scion piece.

C. In Vitro Methods of Propagation and Plant Improvement

Techniques of in vitro propagation of plants is a bourgeoning field of research and cannot be satisfactorily dealt with here. The topic has been reviewed recently in another of this series of books.[441] The technique which appears to have greatest potential for multiplication of plants with minimal somaclonal variation is the fragmented apex method.[23] This technique has been shown to be successful with a range of *V. vinifera* and *V. champini* cultivars but may require modification for *V. rupestris* and its hybrids.[23]

Other techniques which are promising from a plant improvement aspect are the somatic embryo techniques,[197,317] or perhaps the adventitious bud techniques (reviewed elsewhere[275]) which require much further development to attain reliability. Reports of polyembryonic seeds, even though in very low proportions in natural populations, give further encouragement to the use of PGRCs to enhance such developments as a tool in plant improvement.[444] In each of these techniques PGRCs have particular roles but they will not be elaborated here.

IX. PROPERTIES AND USE OF PGRCs

A. Solubility

The problem of solubility is not normally faced by the commercial user of a particular PGRC because commercial formulations usually take account of such problems. These problems are more the concern of the researcher. General guidelines which may be given are as follows:

Organic acid PGRCs (GA, ABA, IAA, IBA, NAA, 4-CPA, etc.) — These may either be dissolved in an organic solvent (usually ethanol) and then diluted with water, or an alternative procedure is to prepare a salt (the potassium salt is usually more soluble than the sodium salt) by dissolving in dilute KOH or in KH_2CO_3. Following dissolution, the solution should be diluted and the pH checked and if necessary brought to a near neutral pH with dilute acid.

Purine bases — These may be dissolved either in isopropanol or more usually by addition of a minimal quantity of dilute HCl or KOH, usually 0.1N is sufficiently strong, then dilute and bring to a more moderate pH. Concentrations higher than about 50 m*M* cannot be maintained in solution at neutral pH and should be kept either at about pH 4 or pH 9.

B. Formulation — Environment Interactions and the Effectiveness of CEPA

One of the major problems facing the user of CEPA as a source of ethylene is the interaction of the activity of the chemical with the environmental conditions applying at the time of spray application.[53] Temperature has a dramatic effect on the rate of ethylene release and a Q_{10} of about 4.2 has been demonstrated.[418] Thus at low temperatures (less than about 17°C (63°F)) insufficient ethylene may be released to elicit a response,[53] while at elevated temperatures (greater than about 32°C — 90°F), the rate of decomposition may be so great that insufficient ethylene penetrates the plant. This problem is compounded by an interaction of temperature with relative humidity which leads to promotion of release at a particular range of relative humidity.[420] The net effect of these interactions is that the user must either carefully choose an appropriate set of conditions or the formulation should be modified to suit the conditions. Within certain limits, the latter approach is probably the most practical. It would require determination of a concentration, pH series to ensure consistent uptake of CEPA and release of ethylene under given conditions. pH is important because rate of decomposition is determined by the proportion of the di-cation existing in the solution. This proportion is approximately 1/10 at pH 6, 5/10 at pH 7, and 9/10 at pH 8.[418] The slow rate of release produced near pH 6 leads to more effective action.[129]

Caution is advisable in considering use of adjuvants for many of the purposes to which CEPA may be put in viticulture. Many of the recommended adjuvants such as KI and urea are defoliants in their own right and act by damaging leaf tissues.[419,425]

C. Effect of Rain Following Spray Application

Few studies have been carried out on this practical and important problem. Studies with GA have shown that application may have to be repeated if rain occurs within 8 to 12 hr of application.[199,231]

D. Toxicology

The following is a list of LD_{50}s ($g \cdot kg^{-1}$) from Sullivan[445] as quoted by Anderson:[10] CCC, 0.5 to 1.0; SADH, 8 to 10; CEPA, 4; maleic hydrazide, 1.4; Dikegulac, 18.0; Morphactins, 6 to 10.

X. CONCLUSION

The overriding conclusion which may be drawn from this review is that while many plant growth regulating chemicals have demonstrably important benefits for viticultural industries, recommendations for their use must be tempered with the knowledge that each cultivar should be treated in the first instance as an individual case and that often it is difficult to create an advantageous balance between the negative and the positive aspects of particular PGRCs. This last point is especially important when considering use of GA or CEPA. I have not attempted to give direct recommendations for these reasons, but rather have attempted to provide sufficient background information to enable the potential user to assess the most likely set of conditions required to optimize benefit. It is also clear that it is an illusion that any general understanding exists of the conditions governing the response of even the most well established application of PGRCs. Where possible I have drawn attention to aspects of problems which if examined could lead to PGRCs having an even greater impact to the benefit of viticulture, viticulturists, and, ultimately, consumers, than presently exists.

ACKNOWLEDGMENTS

I wish to thank Jackie Paterson, Sharyn York, and Heather Kronast who each assisted with typing, Stephen Chambers who prepared the diagrams, and the staff of the Biological Sciences Library, especially Lynn Huddleston and John Lavas. I also wish to acknowledge the Department of Agriculture, Victoria, Sunraysia Horticultural Research Institute, who financed the original literature search.

REFERENCES

1. **Akabane, N. and Yamazaki, T.**, Influence of gibberellic acid on the storage of grapes, *Jpn. Gibberellin Res. Assoc. Proc.*, Fifth meeting, 5, 1961.
2. **Akabane, N. and Yamazaki, T.**, The influence of gibberellic acid on the storage of grapes, *Jpn. Gibberellin Res. Assoc. Proc.*, Sixth Meeting, 106, 1963.
3. **Alleweldt, G.**, Encouraging the growth of grape vine inflorescences by gibberellic acid (GER), *Vitis*, 2, 71, 1959.
4. **Alleweldt, G.**, Inhibition of flower formation by gibberellin in *Vitis rupestris, Naturwissenschaften*, 48, 628, 1961.

4a. **Alleweldt, G.**, The reactions of vines to gibberellin (Ger), *Mitt. Klosterneuburg*, 12A, 67, 1962.

4b. **Alleweldt, G.**, Einfluss von Klimatfaktoren auf die Zahl der Infloreszenzen bei Reben, *Wein Wiss.*, 18, 67, 1963.

5. **Alleweldt, G.**, The relationship between environment and vegetative growth, dormancy and flower formation in vines (Vitis spp.), III. Flower formation (GER), *Vitis*, 4, 240, 1964.
6. **Alleweldt, G.**, Studies on flower initiation in the vine (GER), *Vitis*, 4, 176, 1964.
7. **Alley, C. J.**, Grapevine propagation Part 11, Rooting of cuttings:effects of indolebutyric acid and refrigeration on rooting, *Am. J. Enol. Vitic.*, 30, 28, 1979.
8. **Alley, C. J. and Ferrari, N. L.**, Grapevine propagation 17, Chemical disbudding of cuttings, *Am. J. Enol. Vitic.*, 31, 65, 1980.
9. **Alley, C. J. and Peterson, J. E.**, Grapevine propagation, IX. Effects of temperature, refrigeration, and indole butyric acid on callusing, bud push, and rooting or dormant cuttings, *Am. J. Enol. Vitic.*, 28, 1, 1977.
10. **Anderson, A. S.**, Plant growth retardants, present and future use in food production, in *Plant Regulation and World Agriculture*, Vol. 22, NATO Advanced Institute Series A, Scott, T. K., Ed., Plenum Press, New York, 1978.
11. **Antcliff, A. J.**, 2,3,4,-Trichlorophenoxyacetic acid as a spray to replace cincturing of currants, *J. Aust. Inst. Agric. Sci.*, 23, 242, 1957.
12. **Antcliff, A. J.**, A cincturing experiment on the sultana, *Aust. J. Exp. Agric. Anim. Husb.*, 1, 130, 1961.
13. **Antcliff, A. J.**, A field trial with growth regulators on the Zante Currant (*Vitis vinifera* var.), *Vitis*, 6, 14, 1967.
14. **Antcliff, A. J.**, Increasing the yield dried fruit from the sultana with growth regulators, *Vitis*, 6, 288, 1967.
15. **Arumugam, R. and Madhava Rao, V. N.**, Note on the effect of growth retardants on pollen production and fruit set in grapes, *Indian J. Agric. Sci.*, 43, 619, 1973.
16. **Arutyunyan, E. A., Sklyarova, I. A., and Pogosyan, K. S.**, Effect of synthetic growth preparations on the degree of dormancy, water content, and metabolic change in the grapevine, *Biol. Zh. Arm.*, 29, 90, 1976.
17. **Badr, S. A.**, Root initiation on grape cuttings as influenced by time of collection and treatment with growth regulators, *HortScience*, 8, 274, 1973.
18. **Balasubrahmanyam, V. R. and Khanduja, S. D.**, Effect of foliar sprays of Uracil, Xanthine and Caffeine on the nucleic acid and protein content of leaves and fruiting of Thompson Seedless grapes, *Vitis*, 12, 100, 1973.

19. **Balasubrahmanyam, V. R., Khanduja, S. D., and Abbas, S.,** Effect of thio urea on rest period of grapevine buds, *Am. J. Enol. Vitic.*, 26, 168, 1975.
20. **Baldwin, J. G.,** (deceased), Unpublished data, Sunraysia Horticultural Research Institute, Victoria, Australia.
21. **Bamzai, R. D. and Randhawa, G. S.,** Effects of certain growth substances and boric acid on germination, tube growth and storage of grape pollen (*Vitis* spp.), *Vitis*, 6, 269, 1967.
22. **Barcellos, F. M. and Feliciano, A. J.,** Effect of gibberellic-acid on loosening the clusters and on the characteristics of the grape *Vitis vinifera* cultivar Italia, *Agron. Sulriograndense*, 15, 321, 1979.
23. **Barlass, M. and Skene, K. G. M.,** Studies on the fragmented shoot apex of grapevine. II. Factors affecting growth and differentiation *in vitro J. Exp. Bot.*, 31, 489, 1980.
23a. **Barritt, B. H.,** Ovule development in seeded and seedless grapes, *Vitis*, 9, 7, 1970.
24. **Barritt, B. H.,** Fruit set in seedless grapes treated with growth regulators Alar, 2-chloroethyltrimethyl ammonium chloride and gibberellin, *J. Am. Soc. Hortic. Sci.*, 95, 58, 1970.
25. **Bernon, S.,** Recherches sur la coulure, *Ann. Ec. Agric. Montpellier.*, 24, 57, 1936; *Hort. Abstr.*, 8, 1038, 1938.
25a. **Bertrand, D. E. and Weaver, R. J.,** Effect of exogenous gibberellin on endogenous hormone content and development of Black-Corinth grapes, *Vitis*, 10, 292, 1972.
26. **Bertrand, D. E. and Weaver, R. J.,** Effect of potassium gibberellate on growth and development of Black-Corinth grapes, *J. Am. Soc. Hortic. Sci.*, 97, 659, 1972.
27. **Bindra, A. S., Gupta, K. K., and Chohan, J. S.,** Effect of different fungicidal and cultural treatments on bud killing in Anab-E-Shahi grapevine, *Indian J. Mycol. Plant Pathol.*, 5(1), 44, 1975.
28. **Blaha, J.,** Influence of gibberellic acid on the grapevine and its fruit in Czechoslovakia, *Am. J. Enol. Vitic.*, 14, 161, 1963.
29. **Bocion, P. F., De Silva, W. H., Huppi, G. A., and Szkrybalo, W.,** Group of new chemicals with plant growth regulatory activity, *Nature*, 258, 142, 1975.
30. **Boehm, E. W.,** Cultured Sultanas, Department Agriculture of South Australia, Leaflet no. 3604, 1960.
31. **Bouquet, A.,** Use of benzothiazol-2-oxy-acetic acid to improve the germinability of seeds in early ripening grape varieties, *Vitis*, 16, 100, 1977.
32. **Bourquin, H. D. and Alleweldt, A.,** Der einfluss von CCC and B995 auf das triebwachstum von reben, *Vitis*, 9, 105, 1970.
33. **Branas, J., Bernon, G., and Levadoux, L.,** *Elements de Viticulture Generale*, Montpellier Ecole Nationale d' Agriculture de Montpellier, France, 1946.
34. **Branties, N. B. M.,** Pollinator attraction of *Vitis vinifera* subsp. *silvestris*, *Vitis*, 17, 229, 1978.
35. **Bhullar, J. S., Dhillon, B. S., and Randhawa, J. S.,** Effect of CCC and Alar on the storage of polythene wrapped Himrod grapes, *J. Res. Punjab Agric. Univ.*, 15, 156, 1978.
36. **Bukovac, M. J., Larsen, R. P., and Bell, H. K.,** Effect of gibberellin on berry set and development of Concord grapes, *Q. Bull. Mich. Agric. Exp. St.*, 42, 503, 1960.
37. **Bukavac, M. J., Larson, R. P., and Robb, W. R.,** Effect of N,N-dimenthlaminosuccinamic acid on shoot elongation and nutrient composition of *Vitis labrusca* L cv Concord, *Q. Bull. Mich. Agr. Exp. St.*, 46, 488, 1964.
38. **Buttrose, M. S.,** Fruitfulness in grapevines: effects of light intensity and temperature, *Bot. Gaz.*, 130, 166, 1969.
39. **Buttrose, M. S.,** Fruitfullness in grapevines: the response of different cultivars to light, temperature and daylength, *Vitis*, 9, 121, 1970.
40. **Buttrose, M. S.,** Climatic factors and fruitfulness in grapevines, *Hortic. Abstr.*, 44, 319, 1974.
41. **Cahoon, G. A., Shaulis, N., and Barnard, J.,** Effect of daminozide on Concord grapes, *J. Am. Soc. Hort. Sci.*, 102, 218, 1977.
42. **Carlone, R.,** Effecto della gibberelline e della decorticazione annulare sullar allegagione e lo suiluppo del grappolo di cultivar di uve da tavola normalmento provviste di some, *Ann. Fac. Agron. Torino*, 1, 26, 1962.
43. **Cawthon, D. L. and Morris, J. R.,** Effects of ethephon on juice quality of grapes *Vitis labrusca* cultivar Concord, *HortScience*, 15, 401, 1980.
44. **Celestre, M. R.,** Effecto dell'acido gibberellico e dell'acido 4 parachlorofenoxiacetico sull'uva Ohanez, *Riv. Vitic. Enol.*, 16, 359, 1963.
45. **Chakrawar, V. R. and Rane, D. A.,** Effect of ethrel(2-chloroethyl phosphonic-acid) on uneven ripening and berry characteristics of Gulabi and Banglore purple grapes, *Vitis*, 16, 97, 1977.
46. **Chapman, A. P. and Hussey, E. E.,** The value of plant growth regulators in the propagation of Vitis champini rootstocks, *Am. J. Enol. Vitic.*, 31, 250, 1980.
47. **Chaturvedi, K. N. and Khanduia, S. D.,** Response of Perlette clusters to gibberellic acid applied at different stages of bloom, *Vitis*, 18, 10, 1979.
48. **Christodoulou, A. J., Pool, R. M., and Weaver, R. J.,** Prebloom thinning of Thompson seedless grapes is feasible when followed by bloom spraying with gibberellin, *Calif. Agric.*, 20, 8, 1966.

49. **Chundawat, B. S., Takahashi, E., and Nagasawa, K.,** Effect of gibberellic acid B-nine and kinetin on fruit set parthenocarpy and quality of Kyoho grapes, *J. Jpn. Soc. Hortic. Sci.*, 40, 105, 1971.
51. **Claus, P.,** The action of chlorocholine chloride (CCC) on grapevines (GER), *Wein-Wiss.*, 20, 314, 1965.
52. **Clore, W. J.,** Responses of Delaware grapes to gibberellin, *Proc. Am. Soc. Hortic. Sci.*, 87, 259, 1965.
53. **Clore, W. J. and Fay, R. D.,** The effect of pre-harvest applications of Ethrel on Concord grapes, *HortScience*, 5, 21, 1970.
54. **Considine, J. A.,** Aspects of the Hormonal Physiology of Fruit Development in *Vitis vinifera* L, M. Agric. Sc. Thesis, University of Adelaide, Australia, 1969, 169.
55. **Considine, J. A. and Brown, K.,** Physical aspects of fruit growth: theoretical analysis of distribution of surface growth forces in fruit in relation to cracking and splitting, *Plant. Physiol.*, 68, 371, 1981.
56. **Considine, J. A. and Coombe, B. G.,** The interaction of gibberellic acid and 2-(chloroethyl) trimenthylammonium chloride on fruit cluster development in *Vitis vinifera* L, *Vitis*, 11, 108, 1972.
57. **Considine, J. A. and El-Zeftawi, B. M.,** Gibberellic acid, chlorocholine chloride and yield increases in Zante Currant, *Vitis*, 10, 107, 1971.
58. **Considine, J. A, and Kriedeman, P. E.,** Fruit splitting in grapes, determination of the critical turgor pressure, *Aust. J. Agric. Res.*, 23, 17, 1972.
59. **Considine, J. A. and Knox, R. B.,** Development and histochemistry of the pistil of the grape, *Vitis vinifera*, *Ann. Bot.*, 43, 11, 1979.
60. **Considine, J. A. and Knox, R. B.,** Development and histochemistry of the cells, cell walls and cuticle of the dermal system of the fruit of the grape, *Vitis vinifera*, *Protoplasma*, 99, 347, 1979.
61. **Considine, J. A. and Knox, R. B.,** Tissue origins, cell lineages and patterns of cell division in the developing dermal system of *Vitis vinifera* L., *Planta*, 151, 403, 1981.
62. **Coombe, B. G.,** Artificial parthenocarpy in grape vines, *J. Aust. Inst. Agric. Sci.*, 16, 69, 1950.
63. **Coombe, B. G.,** Setting currants with growth substances, *J. Dep. Agric. S. Aust.*, 56, 186, 196, 1952.
64. **Coombe, B. G.,** Setting currants by spraying with PCPA, *J. Dep. Agric. S. Aust.*, 57, 107, 1953.
65. **Coombe, B. G.,** Last season's experience with PCPA on currants, *J. Dep. Agric. S. Aust.*, 58, 126, 1954.
66. **Coombe, B. G.,** The effect of growth substance sprays on the fruiting of grapevines and in particular for the replacement of cincturing of Zante Currant, unpublished report, 1955.
67. **Coombe, B. G.,** Fruit set and development in seeded grape varieties as affected by defoliation, topping, girdling, and other treatments, *Am. J. Enol. Vitic.*, 10, 85, 1959.
68. **Coombe, B. G.,** Relationship of growth and development to changes in sugars, auxins, and gibberellins in fruit of seeded and seedless varieties of *Vitis vinifera*, *Plant Physiol.*, 35, 241, 1960.
69. **Coombe, B. G.,** The effect of removing leaves, flowers and shoot tips on fruit set in *Vitis vinifera*, *J. Hortic. Sci.*, 37, 1, 1962.
70. **Coombe, B. G.,** The effect of growth substances and leaf number on fruit set and size of Corinth and Sultanina grapes, *J. Hortic. Sci.*, 40, 307, 1965.
71. **Coombe, B. G.,** Unpublished report, Waite Agricultural Research Institute, Glen Osmond, Australia, 1965.
72. **Coombe, B. G.,** Increase in fruit set of *Vitis vinifera* by treatment with growth retardants, *Nature*, 205, 305, 1965.
73. **Coombe, B. G.,** Effects of growth retardants on *Vitis vinifera*, *Vitis*, 6, 278, 1967.
74. **Coombe, B. G.,** Fruit set in grapevines, the mechanism of the CCC, *J. Hortic. Sci.*, 45, 415, 1970.
75. **Coombe, B. G.,** The regulation of set and development of the grape berry, *Acta Hortic.*, 1(34), 261, 1973.
76. **Coombe, B. G.,** The development of fleshy fruits, *Ann. Rev. Plant. Physiol.*, 27, 507, 1976.
77. **Coombe, B. G. and Matile, P.,** Solute accumulation by grape *Vitis vinifera* pericarp cells 1. Sugar uptake by skin segments, *Biochem. Physiol. Pflanz.*, 175, 369, 1980.
78. **Dass, H. C. and Randhawa, G. S.,** Differential response of some seeded grape cultivars by *Vitis vinifera* to gibberellin application, *Vitis*, 6, 385, 1967.
79. **Dass, H. C. and Randhawa, G. S.,** Effect of gizberellin on Golden Queen grape, a Vinifera-lubrusca hybrid, *Am. J. Enol. Vitic.*, 19, 52, 1968.
80. **Dass, H. C. and Randhawa, G. S.,** Response of certain seeded *Vitis vinifera* varieties to gibberellin application at postbloom stage, *Am. J. Enol. Vitic.*, 19, 56, 1968.
81. **Dass, H. C. and Randhawa, G. S.,** Response of Pusa seedless grape to 4-CPA, kinetin, Uracil and GA, *Physiol. Plant*, 21, 298, 1968.
82. **Dass, H. C. and Randhawa, G. S.,** Effect of gibberellin on seeded *Vitis vinifera* with special reference to induction of seedlessness, *Vitis*, 7, 10, 1968.
83. **Dass, H. C., Randhawa, G. S., and Hegi, S. P.,** Effect of growth regulators on post harvest berry drop in Cheema-Sahebi grape *Vitis vinifera*, *Indian J. Hortic.*, 31, 131, 1974.
84. **Dhaliwal, G. S., Dhillon, B. S., and Sharma, K. K.,** Effect of gibberellic-acid and gibberellin 4 plus gibberellin 7 on the compactness of Perlette grapes, *J. Res. Punjab Agric. Univ.*, 14, 41, 1977.
85. **Dhillon, B. S. and Sharma, K. K.,** Regulation of Perlette grapes with 2-chloroethyltrimethyl ammonium chloride gibberellic-acid, 2,4,5-T, and naphthaleneacetic-acid, *J. Res. Punjab Agric. Univ.*, 10, 331, 1973.

86. **Dhillon, B. S. and Singh, L.,** The efficacy of cane and trunk ringing of grapevines, *Proc. Am. Soc. Hort. Sci.*, 53, 259, 1949.
87. **Dhillon, A. S. and Singh, L.,** The influence of thinning and ringing on the cropping and quality of grapes and the vigor of grapevines, *Proc. Am. Soc. Hort. Sci.*, 53, 263, 1949.
88. **Dhillon, B. S., Bhullar, J. S., and Randhawa, J. S.,** Effect of growth regulators on the storage of Himrod grapes, *Indian Fruit Packer*, 31, 36, 1977.
89. **Dhillon, B. S., Randhawa, J. S., and Mann, S. S.,** Effect of growth regulators on the storage of Himrod grapes, *J. Res. Punjab Agric. Univ.*, 12, 241, 1975.
90. **Di Cesare, L.,** Attempt to induce retardation in the budding of grape by means of autumn applications of growth regulating substances, *Phyton*, 25, 129, 1968.
91. **Düring, H.,** Studies on the regulation of stomatal movements on the leaves of *Vitis vinifera, Angew Bot.*, 50, 61, 1976.
92. **Düring, H. and Broquedis, M.,** Effects of abscisic-acid and benzyl adenine on irrigated and nonirrigated grapevines *Vitis vinifera, Sci. Hortic.*, 13, 253, 1980.
93. **El-Banna, G. I. and Weaver, R. J.,** Effect of ethephon and gibberellin on maturation of ungirdled Thompson seedless grapes, *Am. J. Enol. Vitic.*, 30, 11, 1979.
94. **El-Latief, F. I. A.,** Effect of pre-harvest sprays of some growth retardants on the quality of banati grapes during cold storage, *Acta Agron. Acad. Sci. Hung.*, 25, 156, 1976.
95. **El-Latief, F. I. A.,** The effect of N di methylamino succinamic-acid on the quality of Banati and Gharibi grapes during cold storage, *Acta Agron. Acad. Sci. Hung.*, 26, 151, 1977.
96. **El-Zeftawi, B. M.,** Effects of GA and CCC on setting, splitting, yield and quality of Zante currant (*Vitis vinifera* var), *Vitis*, 10, 27, 1971.
97. **El-Zeftawi, B. M. and Weste, H. L.,** Some effects of gibberellic acid and Cycocel on grapevines, *J. Aust. Inst. Agric. Sci.*, 9, 274, 1969.
98. **El-Zeftawi, B. M. and Weste, H. L.,** Effects of some growth regulators on the fresh and dry yield of Zante currant *Vitis vinifera, Vitis*, 9, 47, 1970.
99. **El-Zeftawi, B. M. and Weste, H. L.,** Effects of girdling and parachlorophenoxyacetic acid and gibberellic acid sprays on yields and quality of Zante Currant (*Vitis vinifera* L), *Hortic. Res.*, 10, 74, 1970.
100. **El-Zeftawi, B. M. and Weste, H. L.,** Effect of topping, pinching, cincturing and PCPA on the yield of Zante currant (*Vitis vinifera* var), *Vitis*, 9, 184, 1970.
101. **El-Zeftawi, B. M. and Weste, H. L.,** Time and level of pruning with cincturing or PCPA in relation to yield and quality of Zante Currant, *Aust. J. Exp. Agric. Anim. Husb.*, 10, 484, 1970.
102. **Emmerson, J. G. and Powell, L. E.,** Endogenous abscisic-acid in relation to rest and bud burst in three *Vitis* species, *J. Am. Soc. Hortic. Sci.*, 103, 677, 1978.
103. **Erez, A., Lavee, S., and Samish, R. M.,** Improved methods to break rest in deciduous fruit species, *Proc. Int. Hortic. Congr.*, 1, 82, 1970.
104. **Eynard, I.,** Effects of pre-harvest application of TH6241 and CEPA on *Vitis vinifera, Vitis*, 13, 303, 1975.
105. **Fallot, J.,** Callogenesis self grafting culture of tissues, *Bull. O. I. V.*, 43 (475), 908, 1970.
106. **Farmahan, H. L. and Jawanda, J. S.,** Response of seeded grape *Vitis vinifera* cultivar Khalili to growth regulators, *J. Res. Punjab Agric. Univ.*, 12, 236, 1975.
107. **Farmahan, H. L. and Pandey, R. M.,** Effect of alar and cycocel on the storage life of Pusa seedless grapes, *Indian J. Plant Physiol.*, 22, 275, 1979.
108. **Flore, J. A. and Bukovac, M. J.,** A species effect on conjugation of foliar absorbed naphthalene acetic acid, *HortScience*, 9, 289, 1974.
109. **Frances, V., Chinchetru, G., and Esteban, P.,** Combating spring frosts in vine growing by the use of morphactins spanish communication, *Bull. O. I. V.*, 47 (524), 758, 1974.
110. **Free, J. B.,** *Insect Pollination of Crops*, Academic Press, London, 1970.
111. **Funt, R. C. and Tukey, L. D.,** Influence of exogenous daminozide and gibberellic acid on cluster development and yield of the Concord grape, *J. Am. Soc. Hortic. Sci.*, 102, 509, 1977.
112. **Galzy, P. and Nigond, J.,** Essai d'obtention d'un retard a'la maturation des raisins de table, *Prog. Agric. Vitic.*, 148, 86—92, 125—130, 141—150, 1957.
113. **Gifford, E. M., Jr. and Weaver, R. J.,** Effects of 4-chlorophenoxyacetic acid and girdling on the anatomy of Black Corinth grapes, *Am. J. Enol. Vitic.*, 11, 140, 1960.
114. **Guelfat-Reich, S. and Safran, B.,** Maturity responses of Sultanina grapes to gibberellic acid treatments, *Vitis*, 12, 33, 1973,
115. **Haeseler, C. W.,** Responses of mature Concord grapevines to succinic-acid 2,2-dimethy hydrazide in Pennsylvania, *HortScience*, 11, 265, 1976.
116. **Haeseler, C. W. and Fleming, H. K.,** Response of Concord grapevines (*Vitis lubrusca* L.) to various controlled day temperatures, *Pa. Agr. Exp. Sta. Bull.*, 739, 1967.
117. **Hagiwara, K., Ryugo, K., and Olmo, H. P.,** Comparison between responsiveness of selected grape clones to gibberellin applications and their endogenous levels in breaking buds and maturing berries, *Am. J. Enol. Vitic.*, 31, 309, 1980.

118. **Hale, C. R.,** Growth and senescence of the grape berry, *Aust. J. Agric. Res.*, 19, 939, 1968.
119. **Hale, C. R., and Coombe, B. G.,** Abscisic acid an effect on the ripening of grapes, *R. Soc. N. Z. Bull.*, 12, 831, 1974.
120. **Hale, C. R., Coombe, B. G., and Hawker, J. S.,** Effects of ethylene and 2-chloroethylphosphoric acid on the ripening of grapes, *Plant Physiol.*, 45, 620, 1970.
121. **Handre, L. and Hascoet, M.,** Dosage des residus de CCC dans les raisins *Compte rendus de la Premiere Journee d'Etude sur les applications du CCC sur vigne,* Nice, 1968.
122. **Hardie, W. J. and Considine, J. A.,** The response of grapes to water deficit stress during particular stages of development, *Am. J. Enol. Vitic.*, 27, 55, 1976.
123. **Harmon, F. N.,** Influence of indolebutyric acid on the rooting of grape cuttings, *Proc. Am. Soc. Hortic. Sci.*, 42, 383, 1943.
124. **Harris, J. M., Kriedemann, P. E., and Possingham, J. V.,** Anatomical aspects of grape berry development, *Vitis,* 7, 106, 1968.
125. **Haudenard, M. Van,** The possibility of chemical thinning in vine glasshouses with the product PRB 200-E-50, *Tuinbouw,* 38, 201, 1974.
126. **Hedberg, P. R. and Goodwin, P. B.,** Factors affecting natural and ethephon induced grape *Vitis vinifera* berry abscission, *Am. J. Enol. Vitic.*, 31, 109, 1980.
127. **Hifney, H. A. A. and Abdel-All, R. S.,** Effect of GA and CCC on physical and chemical changes in seedless grapes under cold storage conditions, *Vitis,* 16, 27, 1977.
128. **Hidalgo, L. and Candela, M. R.,** Efectos inducidos par el acido giberelico (Berelex) en tratamiento unico, sobre la *Vitis vinifera* L, *Bol. Inst. Nac. Invest. Agron.*, Madrid, 25, 1, 1965; *Hort. Abs.*, 36, 2690, 1966.
129. **Hirschfield, G. and Lavee, S.,** Control of vegetative growth of grapevine shoots by ethylene releasing substances, conditions and sites of action, *Vitis,* 19, 308, 1980.
130. **Hoad, G. V., Loveys, B. R., and Skene, K. G. M.,** Studies on the hormonal physiology of Vitis vinifera leaf tissue, *J. Sci. Food. Agric.*, 27, 797, 1976.
131. **Hoad, G. V., Loveys, B. R., and Skene, K. G. M.,** The effect of fruit removal on cytokinins and gibberellin-like substances in grape leaves, *Planta,* 136, 25, 1977.
132. **Hopping, M. E.,** Effect of bloom applications of gibberellic acid on yield and bunch rot of the wine grape Seibel-5455, *N.Z. J. Exp. Agric.*, 4, 103, 1976.
133. **Hopping, M. E.,** Effect of growth regulators and dormancy breaking chemicals on bud break and yield of Palomino grape vines, *N.Z. J. Exp. Agric.*, 5, 339, 1977.
134. **Hueppi, G. A., Bocion, P. F., and De Silva, W. H.,** A new quaternary ammonium compound with plant growth regulatory activity, *Experientia (Basel),* 32, 37, 1976.
135. **Huglin, P. and Julliard, B.,** Action de l'Hydrazide Meleique sur la vigne, *Vitis,* 2, 65, 1959.
136. **Huglin, P. and Julliard, B.,** Reactions de la vigne a' quelques reducteurs de croissance, *CRJ d'Etudes sur les applications du CCC sur vigne,* Nice, 1968, 65.
137. **Hull, J., Jr.,** Grape fruit set increased by Alar, *Hortic. Rep. Mich. St. Univ.*, 30, 15, 1966.
138. **Iizuka, M. and Hashizume, T.,** Induction of female organs in stominate grape by 6-substituted adenine derivatives, *Jpn. J. Genet.*, 43, 393, 1968.
139. **Iizuka, T., Kim, R., Rao, L. M., and Kozaki, I.,** *Jpn. J. Breed.*, 18, 217, 1968; cited in Ref. 219.
140. **Inoue, S.,** Gibberellin treatment on grapes, *Jpn. Gibberellin Res. Assoc., Fifth Meet.*, 32, 1961.
141. **Inoue, S. and Fujiwara, Y.,** Studies on gibberellin treatment of grapes, *Jpn. Gibberellin Res. Assoc., Sixth Meeting,* 87, 1963.
142. **Isoda, R.,** Effect of 2 chloroethyltrimethyl ammonium chloride on promotion of flower initiation in some cultivars of grapevines, *Bull. Hiroshima Agric. Coll.*, 5, 419, 1977.
143. **Itakura, T. and Kozaki, I.,** Studies with gibberellin application on grapes, *Jpn. Gibberellin Res. Assoc., Fifth Meeting,* 27, 1961.
144. **Itakuru, T. and Kozaki, I.,** Experiments on gibberellin application for grapes, *Jpn. Gibberellin Res. Assoc., Sixth Meeting,* 83, 1963.
145. **Itakura, T., Kozaki, I., and Machida, Y.,** Studies on the action of gibberellin on grapes, *Jpn. Gibberellin Res. Assoc., Sixth Meeting,* 5, 1963.
146. **Itakura, T., Kozaki, I., and Machida, Y.,** Studies on gibberellin application in relation to the response of certain vine varieties, *Bull. Hortic. Res. Sta. Hiratsuka, Ser. A,* No. 4, 67, 1965.
147. **Ito, H., Motomura, Y., Konno, Y., and Hatayama, T.,** Exogenous gibberellin as responsible for the seedless berry development of grapes, Part 1, Physiological studies on the development of seedless Delaware grapes, *Tohoku J. Agr. Res.*, 20, 1, 1969.
148. **Iwasaki, K. and Weaver, R. J.,** Effects of chilling calcium cyanamide and bud scale removal on bud break, rooting, and inhibitor content of buds of Zinfandel grape *Vitis vinifera, J. Am. Soc. Hortic. Sci.*, 102, 584, 1977.
149. **Iyer, C. P. A. and Randawa, A. S.,** Chemical induction of pollen sterility in grapes, *Curr. Sci.*, 34, 411, 1965.

150. **Jacob, H. E.,** Some responses of the seedless varieties of *Vitis vinifera* to girdling, *Proc. Am. Soc. Hortic. Sci.*, 25, 223, 1928.
151. **Jacob, H. E.,** The response of the Hunisa grape to girdling, *Proc. Am. Soc. Hortic. Sci.*, 32, 386, 1934.
152. **Jako, N.,** Influence of treatment with adenine and uracil upon the ribose and deoxyribose content of vine leaves, *Mitt. Klosternenburg*, 18, 411, 1968.
153. **Jacquinet, A.,** A method of controlling the vigour of growth of grape vines (FR), *Vitis*, 12, 291, 1974.
154. **Jawanda, J. S., Singh, R., and Pal, R. N.,** Effects of growth regulators on floral bud drop fruit characters and quality of Thompson Seedless grape (*Vitis vinifera* L.), *Vitis*, 13, 215, 1974.
155. **Jensen, F. L.,** Effects of timing gibberellin sprays for berry sizing on maturity of table Thompson seedless grape, *Calif. Agric.*, 23, 13, 1969.
156. **Jensen, F., Christensen, J. P., Andris, H., Swanson, F., Leavitt, G., and Peacock, W. L.,** The effects of ethephon on Thompson seedless grapes and raisins, *Am. J. Enol. Vitic.*, 31, 257, 1980.
157. **Jensen, F. L., Kissler, J. J., Peacock, W. L., and Leavitt, G. M.,** Effect of ethephon on color and fruit characteristics of Tokay and Emperor Table grapes, *Am. J. Enol. Vitic.*, 26, 79, 1975.
158. **Jona, R. and Webb, K. J.,** Callus and axillary bud culture of *Vitis vinifera* cultivar Sylvaner-Riesling, *Sci. Hortic.*, 9, 55, 1978.
159. **Jones, L. T.,** Hormone sprays and their effects on the setting, yield and vigour of currant grape vines, *Bull. Dep. Agric. West Aust.*, No. 2289, 1955.
160. **Julliard, B. and Balthazard, J.,** Physiological effects of gibberellic acid on some varieties of grape, *Ann. Amelior.*, 15, 61, 1965.
161. **Kachru, R. B., Chacko, E. K., and Singh, R. N.,** Physiological studies on dormancy in grape seeds (*Vitis vinifera*), I. On the naturally occurring growth substances in grape seeds and their changes during low temperature after-ripening, *Vitis*, 8, 12, 1969.
162. **Kachru, R. B., Singh, R. N., and Yadav, I. S.,** Physiological studies on dormancy in grape seeds (*Vitis vinifera* var. Black Muscat), II. On the effect of exogenous application of growth substances, low chilling temperature and subjection of the seeds to running water, *Vitis*, 11, 289, 1972.
163. **Kasimatis, A. N. and Halsey, D.,** Caution advised when using 4-CPA on Thompson seedless, *West. Fruit Grower.*, 17, 26, 1963.
164. **Karanov, E. N.,** Structure activity relationships of some new growth substances, in *Plant Regulation and World Agriculture*, Scott, T. K., Ed., Nato Advanced Study Institutes Series, Series A, Life Sciences, Vol. 22, Plenum Press, New York, London, 237—249, 1979.
165. **Kasimatis, A. N., Swanson, F. H., and Vilas, E. P., Jr.,** Effects of bloom-applied gibberellic acid on soluble solids and berry weight of Thompson Seedless grapes and on raisin grades, *Am. J. Enol. Vitic.*, 29, 263, 1978.
166. **Kasimatis, A. N., Weaver, R. J., and Pool, R. M.,** Effects of 2,4-D and 2,4-DB on the vegetative development of 'Tokay' grapevines, *Am. J. Enol. Vitic.*, 20, 194, 1969.
167. **Kasimatis, A. N., Swanson, F. H., Vilas, E. P., Peacock, W. L., and Leavitt, G. M.,** The relation of bloom-applied gibberellic acid to the yield and quality of Thompson Seedless raisins, *Am. J. Enol. Vitic.*, 30, 224, 1979.
168. **Kasimatis, A. N., Weaver, R. J., Pool, R. M., and Halsey, D. D.,** Response of 'Perlette' grape berries to gibberellic acid applied during bloom or at fruit set, *Am. J. Enol. Vitic.*, 22, 19, 1971.
169. **Kessler, B., Bak, R., and Cohen, A.,** Flowering in fruit trees and annual plants as affected by purines, pyrimidines and triiodobenzoic acid, *Plant Physiol.*, 34, 605, 1959.
170. **Kessler, B. and Lavee, S.,** Effect of purines, pyrimidines and metals upon the flowering of olive trees and grape vines, *Isr. J. Agric. Res.*, 9, 261, 1959.
171. **Kliewer, W. M. and Soleimani, A.,** Effect of chilling on budbreak in Thompson seedless and Carignane grapevines, *Am. J. Enol. Vitic.*, 23, 31, 1972.
172. **Kobayashi, A. H., Fukushema, T., Nii, N., and Harada, K.,** Effect of day and night temperatures on yield and quality of Delaware grapes, *J. Jpn. Soc. Hortic. Sci.*, 36, 1, 1967.
173. **Koblet, W.,** Fruchtansatz bei Reben in Abhangigkeit von Triebbehandlung und Klimafactoren, *Weinwiss.*, 7, 297, 1966.
174. **Koblet, W. and Vetsch, U.,** Entwicklung der Rebblute und Fruchtansatz, *Schweizer Z. Obst. Weinbau.*, 104, 383, 1968.
175. **Kriedmann, P. E. and Loveys, B. R.,** Hormonal mediation of plant responses to environmental stress, *R. Soc. N. Z. Bull.*, 12, 461, 1974.
176. **Kriedemann, P. E. and Loveys, B. R.,** Hormonal influences on stomatal physiology and photosynthesis, in *Environmental and Biological Control of Photosynthesis*, Marcelle, R., Ed., Dr. W. Junk, B. V., Hague, The Netherlands, 1975, 227.
177. **Kriedemann, P. E., Loveys, B. R., Possingham, J. V., and Satoh, M.,** Sink effects on stomatal physiology and photosynthesis, in *Transport and Transfer Processes in Plants*, Wardlaw, I. F. and Passioura, J. B., Ed., Academic Press, New York, 1976, 401.

178. **Krishnamurthi, S., Randhawa, G. S., and Singh, J. P.,** Effect of gibberellic acid on fruit set, size and quality in the Pusa seedless variety of grapes (*Vitis vinifera* L), *Indian J. Hortic.*, 16, 1, 1959; *Hortic. Abstr.*, 30, 1861.
179. **Krul, W. R. and Worley, J. F.,** Formation of adventitious embryos in callus cultures of Seyval, a French hybrid grape, *J. Am. Soc. Hortic. Sci.*, 102, 360, 1977.
180. **Kurennoj, N. M. and Barbas, I. P.,** An effective method of pollinating vines and increasing yields, *Sadovod. Vinograd. Vinodel. Mold.*, 10, 30, 1966.
181. **Larsen, R. P. and Bauovac, M. J.,** The effects of several foliar applied nutrient and plant growth substances on a "shelling" disorder of Concord grapes, *Q. Bull. Mich. Agric. Exp. St.*, 44, 608, 1962.
182. **Lavee, S.,** Physiological aspects of post-harvest berry drop in certain grape varieties, *Vitis*, 2, 34, 1959
183. **Lavee, S.,** Effect of gibberellic acid in seeded grapes, *Nature*, 185, 395, 1960.
184. **Lavee, S.,** IAA reversible growth inhibition of grape shoots (*Vitis vinifera*) by maleic hydrazide compared to gibberellic acid induced growth, *Vitis*, 19, 207, 1980.
185. **Lavee, S., Erez, A., and Shulman, Y.,** Control of vegetative growth of grapevines *Vitis vinifera* with chloroethyl phosphonic acid (ethephon) and other growth inhibitors, *Vitis*, 16, 89, 1977.
186. **Lavee, S., Melamud, H., Ziu, M., and Bernstein, Z.,** Necrosis in grapevine buds (*Vitis cinifera* cv Queen of Vineyard), I. Relation to vegetative vigor, *Vitis*, 20, 8, 1981.
187. **Leonard, O. A. and Weaver, R. J.,** Absorption and translocation of 2,4-D and amitrole in shoots of the Tokay grape, *Hilgardia*, 31, 327, 1961.
188. **Lider, L. A. and Sanderson, G. W.,** Effects of girdling and rootstock and crop production with the grade variety Chardonnay, *Proc. Am. Soc. Hortic. Sci.*, 74, 383, 1959.
189. **Lilov, D. and Andonova, T.,** Cytokinins, growth, flower and fruit formation in *Vitis vinifera*, *Vitis*, 15, 160, 1976.
190. **Linser, H., Kuhn, M., and Bohring, J.,** Zur Frage der Nachwirkung von chlorocholinchlorid, *Bodendultur*, 14, 111, 1963.
191. **Looney, N. E.,** Control of fruit maturation and ripening with growth regulators, *Acta Hortic.*, 1, 397, 1973.
192. **Looney, N. E.,** Some growth regulator effects on berry set yield and quality of Himrod and De-chaunac grapes, *Can. J. Plant Sci.*, 55, 117, 1975.
193. **Looney, N. E.,** Some growth regulator and cluster thinning effects on berry set, berry quality and annual productivity of de-Chaunac grapes, *Vitis*, 20, 22, 1981.
194. **Looney, N. E. and Wood, D. F.,** Some cluster thinning and gibberellic acid effects on fruit set berry size vine growth and yield of de-Chaunac grapes, *Can. J. Plant Sci.*, 57, 653, 1977.
195. **Loreti, F. and Natali, S.,** Effect of (2-chloroethyl) trimethylammonium chloride on growth and fruiting of Ciliegliolo grape variety, *Am. J. Enol. Vitic.*, 25, 21, 1974.
196. **Luckwill, L. C.,** *Growth Regulators in Crop Production*, The Institute of Biology's Studies in Biology, No. 129, Edward Arnold, 1981.
197. **Lynn, C. D. and Jensen, F.,** Grape thinning bonus from bloomtime gib, *West. Fruit Grower*, Feb, 1966.
198. **Lynn, C. D. and Jensen, F. L.,** Thinning effects of bloomtime gibberellin sprays on Thompson Seedless table grapes, *Am. J. Enol. Vitic.*, 17, 283, 1966.
199. **Manankov, M. K.,** Determining the optimum concentrations, application times and techniques for applying gibberellic acid to vines, (in Russian), *Gibberelliny i ikh Deistvie na Rasteniya*, 226, 1963.
200. **Manankov, M. K.,** Effect of gibberellin on different varieties of grape, *Sov. Plant Physiol.*, 17, 607, 1970.
201. **Manivel, L. and Weaver, R. J.,** Effect of growth regulators and heat on germination of Tokay grape seeds, *Vitis*, 12, 286, 1974.
202. **Mannini, F., Weaver, R. J., and Johnson, J. O.,** Effects of early bloom sprays of ethephon on irrigated and nonirrigated vines of Zinfandel grapes, *Am. J. Enol. Vitic.*, 32, 277, 1981.
203. **Marro, M.,** Parthenocarpy in the vine variety Trebbiano treated with gibberellin (Italian), *Frutticoltura*, 22, 123, 1960.
204. **Matsui, H., Yuda, E., Nakagawa, S., and Yonemori, K.,** Physiological studies on the ripening of Delaware grapes, 2. Effect of light intensity to the cluster on sugar accumulation and carbon dioxide fixation in the berries under light and dark conditions, *J. Jpn. Soc. Hortic. Sci.*, 48, 405, 1980.
205. **Maxwell, N. P.,** Effect of gibberellin on size of Thompson seedless and Perlette grapes in South Texas, *J. Rio Grande Valley Hortic. Soc.*, 13, 117, 1959.
206. **May, P.,** Reducing inflorescence formation by shading individual Sultana buds, *Aust. J. Biol. Sci.*, 18, 463, 1965.
207. **May, P. and Antcliff, A. J.,** The effect of shading on fruitfulness and yield in the Sultana, *J. Hortic. Sci.*, 38, 85, 1963.
208. **May, P. and Cellier, K. M.,** The fruitfulness of grape buds, II. The variability of bud fruitfulness in ten cultivars over four seasons, *Ann. Amelior. Plant.*, 23, 13, 1973.

209. **McEachern, G. R. and Storey, J. B.,** The influence of photoperiod gibberellic acid thio urea and uran on the emergence of Pecans grapes and peaches from rest, *HortScience,* 9, 296, 1974.
210. **Mehta, P. K. and Chundawat, B. S.,** Effect of ethephon (2-chloroethyl phosphoric acid) on quality and ripening of Beauty seedless grape, *Vitis,* 18, 117, 1979.
211. **Menary, R. C.,** Effect of CCC ((2-chloroethyl) - trimethyl ammonium chloride) on fruiting behaviour of Cabernet Sauvignon, *Vitis,* 18, 17, 1979.
212. **Mičurin, V. G.,** Concerning the effects of succinic acid (Russian), *Sadovod. Vinograd. Vinodel. Mold.,* 8, 30, 1964; *Hort. Abstr.,* 35, 643, 1965.
213. **Miele, A., Weaver, R. J., and Johnson, J. O.,** Effect of potassium gibberellate on fruit-set and development of Thompson seedless and Zinfandel grapes, *Am. J. Enol. Vitic.,* 29, 79, 1978.
214. **Miele, A., Weaver, R. J., and Johnson, J. O.,** Effect of application of naphthaleneacetic acid on berry thinning of Carignane grapes, *Vitis,* 17, 369, 1978.
215. **Mosesian, R. M. and Nelson, K. Z.,** Effect on Thompson seedless fruit of gibberellic acid bloom sprays and double girdling, *Am. J. Enol. Vitic.,* 19, 37, 1968.
216. **Motomura, Y.,** Exogenous gibberellin as responsible for the seedless berry development of grapes, VII. Change in the activity of GA applied to the inflorescence, *Tohoku J. Agric. Res.,* 32, 87, 1981.
217. **Motomura, Y. and Hori, Y.,** Exogenous gibberellin as responsible for the seedless berry development of grapes, IV. Effects of temperature on the activity of applied gibberellin on the seedlessness and the seedless berry development in Delaware and Campbell early grapes, *Tohoku J. Agric. Res.,* 28, 8, 1977.
218. **Motomura, Y. and Hori, Y.,** Effects of cyclic AMP applied with gibberellin on the induction of seedless fleshy berries in Delaware and Campbell early grapes, *Tohoku J. Agric. Res.,* 28, 72, 1977.
219. **Motomura, Y. and Hori, Y.,** Exogenous gibberellin as responsible for the seedless berry development of grapes, V. The effects of 6 benzyl adenine applied with gibberellin on the induction of seedless fleshy berries in Delaware and Campbell early grapes, *Tohoku J. Agric. Res.,* 29, 1, 1978.
220. **Motomura, Y. and Hori, Y.,** Exogenous gibberellin as responsible for the seedless berry development of grapes, VI. Explanation of GA effects on the induction of seedlessness and seedless berry development varying with cultivars, *Tohoku J. Agric. Res.,* 29, 111, 1978.
221. **Motomura, Y., Hori, Y., and Ishiyama, J.,** Increase of the acid contents in grape berries by treatment with cyclic AMP, *Vitis,* 18, 301, 1979.
222. **Motomura, Y. and Ito, H.,** Exogenous gibberellin as responsible for the seedless berry development of grapes, II. Role and effects of the pre-bloom gibberellin application as concerned with the flowering seedlessness and seedless berry development of Delaware and Campbell early grapes, *Tohoku J. Agric. Res.,* 23, 15, 1972.
223. **Motomura, Y. and Ito, H.,** Exogenous gibberellin as responsible for the seedless berry development of grapes, III. Role and effects of the pre-bloom gibberellin application on the blossom bud development to anthesis, *Tohoku J. Agric. Res.,* 24, 151, 1973.
224. **Müller-Thurgau, H.,** Uber das abfallen der Rebenblüten und die Enstehung bernloun Tranbenbeesen, *Weinban,* no. 22, 1883.
225. **Mullins, M. G.,** Regulation of fruit set in the grapevine, *Aust. J. Biol. Sci.,* 20, 1141, 1967.
226. **Mullins, M. G.,** Morphogenetic effects of roots and of some synthetic cytokinins in *Vitis vinifera* L, *J. Exp. Bot.,* 18, 206, 1967.
227. **Mullins, M. G.,** Regulation of inflorescence growth in cuttings of the grapevine (*Vitis vinifera* L), *J. Exp. Bot.,* 19, 532, 1968.
228. **Mullins, M. G. and Osborne, D. J.,** Effect of abscisic acid on growth correlation in *Vitis vinifera, Aust. J. Biol. Sci.,* 23, 479, 1970.
229. **Muranishi, S.,** On the action of gibberellins towards the fruiting of grapes, I. On the formation of seedless berries, *Jpn. Gibberellin Res. Assoc., Fifth Meeting,* 30, 1961.
230. **Muranishi, S.,** The effect of gibberellins on the fruiting habit of grapes, *Jpn. Gibberellin Res. Assoc., Sixth Meeting,* 95, 1963.
231. **Muranishi, S.,** The action of gibberellins on the fruiting of grapes, *Sci. Bull. Fac. Agric. Kyushu,* 23, 225, 1968.
232. **Naito, R. and Hayashi, T.,** Promotion of berry set in grapes by growth retardants, III. Effects of the pre-bloom application of SADH and CCC on gibberellin and cytokinin activity in florets of grape varieties, Kyoho and Muscat of Alexandria, *J. Jpn. Soc. Hortic. Sci.,* 45, 135, 1976.
233. **Naito, R. and Kawashima, T.,** Promotion of berry set in grapes by growth retardants, 4. Comparison of succinic acid 2,2-dimethyl hydrazide cluster dipping shoot pinching and flower thinning with regards to their effects on berry set in grape *Vitis labrusca* × *Vitis vinifera* cultivar Kyoho, *J. Jpn. Soc. Hortic. Sci.,* 49, 297, 1980.
234. **Naito, R. and Moriyama, S.,** Effects of the prebloom application of gibberelic-acid combined with Benzyl adenine and urea on the set and growth of seedless berries of Delaware Grape, Part 2. Influence of Urea concentration, *Bull. Fac. Agric. Shimane Univ.,* 10, 1, 1976.

234a. **Naito, R. and Nakano, M.,** Gibberellin-like substances in immature berries of seeded and gibberellin induced seedless Delaware grapes, *J. Jap. Soc. Hortic. Sci.*, 40, 1, 1971.

235. **Naito, R., Kawashima, T., and Fijimoto, J.,** Promotion of berry set in grapes by growth retardants, 5. Effects of cluster dipping with succinic acid-2, 2-Di Methyl Hydrazide or 2-Chloroethyltrimethyl ammonium chloride at different times before anthesis on berry set in cultivars Kyoho and Muscat-of-Alexandria, *J. Jpn. Soc. Hortic. Sci.*, 49, 539, 1981.

236. **Naito, R., Miura, K., and Matsuda, K.,** Effects of the prebloom application of GA combined with BA and urea on the set and growth of seedless berries in Delaware grapes, I. Enhanced berry growth by the combined application, *J. Jpn. Soc. Hortic. Sci.*, 43, 215, 1974.

237. **Naito, R., Ueda, H., and Hayashi, T.,** Promotion of berry set in grapes by growth retardants, II. Effects of SADH and CCC applied directly to clusters on berry set and shoot growth in Kyoho and Muscat of Alexandria grapes, *J. Jpn. Soc. Hortic. Sci.*, 43, 109, 1974.

238. **Naito, R., Ueda, H., and Ishihana, Y.,** Promotion of berry set in grapes by growth retardants, I. Comparison of the effects of B-zine and CCC applied as shoot spray and cluster dip on berry set and shoot growth in 'Kyoho' grapes, *Bull. Fac. Agric. Shimane Univ.*, 6, 10, 1972.

239. **Negi, S. S. and Olmo, H. P.,** Conversion and determination of sex in *Vitis vinifera* L (*sylvestris*), *Vitis*, 9, 265, 1971.

240. **Negi, S. S. and Olmo, H. P.,** Induction of sex conversion in male *Vitis*, *Vitis*, 10, 1, 1971.

241. **Negral, A. M.,** Variabilitat und Verebung des Geschlechts bei der Rebe, *Gartenbauwissenschaft*, 10, 215, 1936.

242. **Nelson, J. M. and Sharples, G. C.,** Influence of chlormequat succinic- acid-2, 2-di methyl hydrazide and a cytokinin on fruit set in the seeded cardinal grape, *HortScience*, 9, 598, 1974.

242a. **Niimi Y., Ohkawa, M., and Torikata, H.,** Changes in auxin and abscisic acid-like activities in grape berries, *J. Jpn. Soc. Hortic. Sci.*, 46, 139, 1977.

243. **Nijjar, G. S. and Bhatia, C. L.,** Effect of gibberellic acid and para-chlorophenoxy-acetic acid on the cropping and quality of Anab-e-Shahi grapes, *J. Hortic. Sci.*, 44, 91, 1969.

244. **Nil, G. and Lavee, S.,** Persistence, uptake and translocation of ^{14}C ethephon (2-chloroethyl phosphonic acid) in Perlette and Cardinal grape vines, *Aust. J. Plant Physiol.*, 8, 57, 1981.

245. **Nito, N. and Kuraishi, S.,** Abnormal auxin distribution and poor berry setting coulure in grapes, *Sci. Hortic.*, 10, 63, 1979.

246. **Nitsch, J. P., Pratt, C., Nitsch, C., and Shaulis, N. J.,** Natural growth substances in Concord and Concord Seedless grapes in relation to berry development, *Am. J. Bot.*, 47, 566, 1960.

247. **Oberle, G. D.,** A genetic study of variations in floral morphology and function in cultivated forms of Vitis, *N.Y. State Agric. Exp. Sta. Tech. Bull.*, 250, 1938.

248. **Oinoue, Y.,** Effect of girdling on rate of ripening of grapes (Japanese, French summary), *Bull. Inst. Oinoue, Japan*, 2, 1, 1938.

249. **Oinoue, Y.,** Effect of boron on the setting of grapes in the Muscat of Alexandria, *J. Hort. Res. Japan*, 9, 141, 1938.

250. **Oinoue, Y.,** Artificial parthenocarpy by use of auxin, *Agric. Hortic.*, 13, 2213, 1938.

251. **Olmo, H. P.,** Pollination and setting of fruit in the Black Corinth grape, *Proc. Am. Soc. Hortic. Sci.*, 34, 402, 1936.

252. **Olmo, H. P.,** Pollination in the Almeria grape, *Proc. Amer. Soc. Hortic. Sci.*, 41, 219, 1942.

253. **Olmo, H. P.,** Correlations between seed and berry development in some seeded varieties of *Vitis vinifera*, *Proc. Am. Soc. Hortic. Sci.*, 48, 291, 1946.

254. **Overcash, J. P.,** Some effects of certain growth regulating substances on the ripening of Concord grapes, *Proc. Am. Soc. Hortic. Sci.*, 65, 54, 1955.

255. **Pal, R. N., Singh, R., Vij, U. K., and Sharma, J. N.,** Effect of gibberellins GA_3, GA_{4+7} and GA_{13} on seed germination and subsequent seedling growth in Early Muscat grape (*Vitis vinifera*), *Vitis*, 17, 265, 1976.

256. **Peacock, W. L., Jensen, F., Else, J., and Leavitt, G.,** The effects of girdling and ethephon treatments on fruit characteristics of Red Malaga, *Am. J. Enol. Vitic.*, 28, 228, 1977.

257. **Pearse, H. L.,** Growth substances and their practical importance in horticulture, *Commonw. Bur. Hortic. Plant. Crops, Tech. Commun.*, No. 20, 1948.

258. **Pearson, H. M.,** Parthenocarpy and seedlessness in *Vitis vinifera*, *Science*, 76, 594, 1932.

259. **Pearson, H. M.,** Parthenocarpy and seed abortion in *Vitis vinifera*, *Proc. Am. Soc. Hortic. Sci.*, 29, 169, 1932.

260. **Pentyer, W. T.,** Studies on the shatter of grapes with special reference to the use of solutions of naphthalene acetic acid to prevent it, *Proc. Am. Soc. Hortic. Sci.*, 38, 397, 1941.

261. **Pereira, F. M., Simao, S., Martins, F. P., Lowler, P., and Igue, T.,** Effects of lime nitrogen and gibberellic-acid on the anticipation of harvest time of the grape cultivar Niagra-rosada, *Cientifica (Jaboticabal)*, 7, 411, 1979.

262. **Peterson, J. R.,** A bunch position effect on response to 2-chloroethyltrimethyl ammonium chloride in Cabernet Sauvignon grapevines, *Aust. J. Exp. Agric. Anim. Husb.*, 14, 122, 1974.
263. **Peterson, J. R.,** Effect of N-dimethylamino succinamic acid on Cabernet Sauvignon grapevines, *Aust. J. Exp. Agric. Anim. Husb.*, 14, 126, 1974.
264. **Peterson, J. R., and Hedberg, P. R.,** Some factors affecting the response of grapevines to ethephone, *Sci. Hortic.*, 3, 275, 1975.
265. **Peynaud, E. and Ribereau-Gayon, P.,** The grape, in *The Biochemistry of Fruits and Their Products,* Vol. 2, Hulme, A. C., Ed., Academic Press, New York, 1971, 171.
266. **Phatak, S. C., Austin, M. E., and Mason, J. S.,** Ethephon as harvest aid for muscadine grapes *Vitis rotundifolia* cultivar unt, *HortScience,* 15, 267, 1980.
267. **Pirie, A. and Mullins, M. G.,** Changes in anthocyanin and phenolics content of grapevine leaf and fruit tissues treated with sucrose nitrate and abscisic acid, *Plant Physiol.*, 58, 468, 1976.
268. **Pool, R. M.,** Effect of cytokinin on in vitro development of 'Concord' flowers, *Am. J. Enol. Vitic.*, 26, 43, 1975.
269. **Pool, R. M., Weaver, R. J., Kliewer, W. M.,** The effect of growth regulators on changes in fruits of Thompson Seedless grapes during cold storage, *J. Am. Soc. Hortic. Sci.*, 97, 67, 1972.
270. **Powers, J. R., Shively, E. A., and Nagel, C. W.,** Effect of Ethephon on colour of Pinot-noir fruit and wine, *Am. J. Enol. Vitic.*, 31, 203, 1980.
271. **Pratt, C. and Shaulis, N. J.,** Gibberellin induced parthenocarpy in grapes, *Proc. Am. Soc. Hortic. Sci.*, 77, 322, 1961.
272. **Pratt, R., Dufrenoy, J., and Sah, P. P. T.,** Growth regulating properties of some thiosemicarbazones, *Plant Physiol.*, 27, 622, 1952.
273. **Proebsting, E. L., Jr.,** Value of growth regulators as aids to mechanical harvesting of apples cherries and grapes, *Acta Hortic.*, 1, 363, 1973.
274. **Pugliano, G.,** Alcuni effetti del cloruro di (2 cloroetil) trimetilammonio (CCC) sul vitigno Trebbiano bianco, *Ann. Fac. Sci. Agar. Napoli, Ser. 4,* 2, 233, 1967.
275. **Rajasekaran, K. and Mullins, M. G.,** Organogenesis in internode explants of grapevines, *Vitis,* 20, 218, 1981.
276. **Randhawa, J. S., Dhillon, B. S., and Mann, S. S.,** Effect of the pre-harvest application of 2-chloroethyltrimethyl ammonium chloride and kinetin on the cold storage life of perlette grapes, *J. Res. Punjab Agric. Univ.*, 13, 267, 1976.
277. **Randhawa, G. S. and Pal, N. C.,** Further studies on seed germination and subsequent seedling growth in grape *Vitis, Indian J. Hortic.*, 25, 148, 1968.
278. **Rane, D. A. and Tukey, L. D.,** Effect of various growth retardants on photosynthesis of Concord grape leaves, *HortScience,* 8, 265, 1973.
279. **Rane, D. A. and Tukey, L. D.,** Effect of various growth retardants on shoot development of Concord grapes, *HortScience,* 8, 265, 1973.
280. **Rao, M. M., Navasimham, P., Nagaraja, N., and Anandaswamy, B.,** Effect of pre-harvest spray of alpha-Naphthalene acetic acid and para-chlorophenoxyacetic acid on control of berry drop in Aneb-3-Shahi grapes (*Vitis vinifera* L.), *J. Food. Sci. Technol.*, 5, 127, 1968.
281. **Rapp, A. and Klenert, M.,** Influence of seeds on maturation of *Vitis vinifera* L. berries (GER), *Vitis,* 13, 222, 1974.
282. **Reid, D. M. and Carr, D. J.,** Effects of the dwarfing compound CCC on the production and export of gibberellin-like substances by root systems, *Planta,* 73, 1, 1967.
283. **Rives, M. and Pouget, R.,** Action de la gibberelline sur la compacite des grappes de deux varieties of vigne, *C. R. Seances Acad. Agric. Fr.*, 45, 343, 1959.
284. **Roubelakis, K. A. and Kliewer, W. M.,** Influence of light intensity and growth regulators on fruit set and ovule fertilization in grape cultivars under low temperature conditions, *Am. J. Enol. Vitic.*, 27, 163, 1976.
285. **Saeed, M., Haq, A., and Khan, A. H.,** Some studies on the effect of gibberellic acid on size and quality of seedless grapes (Shudokhani) in Quetta/Kulat region, *West Pak. J. Agric. Res.*, 6 (2), 85, 1968.
286. **Sachs, R. M. and Weaver, R. J.,** Gibberellin and toxin-induced berry enlargement in *Vitis vinifera* L, *J. Hortic. Sci.*, 43, 185, 1968.
287. **Samish, R. M. and Lavee, S.,** Spray thinning of grapes with growth regulators, *Ktavim,* 8, 273, 1958.
288. **Sarooshi, R. A. and Roberts, E. A.,** Effect of trellising crown bunch handling and the number of oil emulsion and gibberellic-acid sprays on harvest pruned sultanas, *Aust. J. Exp. Agric. Anim. Husb.*, 19, 122, 1979.
289. **Sato, K.,** A certain problem for commercial production of "Seedless Delaware" by gibberellin application, *Jpn. Gibberellin Res. Assoc.*, 73, 1961.
290. **Sato, K.,** Field experiment on the commercial production of "Seedless Delaware" by gibberellin application, *Jpn. Gibberellin Res. Assoc.*, 119, 1963.

291. **Sato, T. and Yamabe, K.,** Studies on gibberellin application for grapes, *Jpn. Gibberellin Res. Assoc.*, 29, 1961.
292. **Sato, T. and Yamabe, K.,** Studies on the application of gibberellin on grapes, *Jpn. Gibberellin Res. Assoc.*, 94, 1963.
293. **Schwabe, W. W. and Mills, J. J.,** Hormones and parthocarpic fruit set: a literature survey, *Hortic. Abstr.*, 51, 661, 1981.
294. **Scienza, A. and Düring, H.,** Nitrogen supply and water relations in grapevines, *Vitis*, 19, 301, 1980.
295. **Scienza, A., Miravalle, R., Visai, C., and Fregoni, M.,** Relationships between seed number, gibberellin and abscisic acid levels and ripening in Cabernet Sauvignon grape berries, *Vitis*, 17, 361, 1978.
296. **Selim, H. H., Ibrahim, F. A., Fayek, M. A., El-Deen, S. A. S., and Gamal, N. M.,** Effect of different treatments on germination of Romi red grape seeds, *Vitis*, 20, 115, 1981.
297. **Sharples, G. C., Hilgeman, R. H., and Milne, R. L.,** The relation of cluster thinning and trunk girdling of cardmal grapes to yield and quality of fruit in Arizona, *Proc. Am. Soc. Hortic. Sci.*, 66, 225, 1963.
298. **Shaulis, N. and Oberle, G. D.,** Some effects of pruning severity and training on Fredonia and Concord grapes, *Proc. Am. Soc. Hortic. Sci.*, 51, 263, 1949.
299. **Shaulis, N. J. and Smart, R. E.,** Grapevine canopies: management, microclimate and yield responses, *Proc. XIXth Int. Hortic. Congr.*, Warsaw III, 255, 1974.
300. **Shaulis, N., Amberg, H. and Crowe, D.,** Response of Concord grapes to light, exposure and Geneva double curtain trimming, *Proc. Am. Soc. Hortic. Sci.*, 89, 268, 1966.
301. **Sherer, V. A. and Kucher, G. M.,** Frost resistance of grapevine canes in relation to treatment with TUR, *Sadovod. Vinograd. Vinodel. Mold.*, 30, 1978.
302. **Shulman, Y., Hirschfeld, G., and Lavee, S.,** Vegetative growth control of 6 grapevine *Vitis vinifera* cultivars by spray application of 2-chloroethyl phosphonic acid (Ethephon), *Am. J. Enol. Vitic.*, 31, 288, 1980.
303. **Sidahmed, O. A. and Kliewer, W. M.,** Effects of defoliation, gibberellic acid and 4-chlorophenoxy acetic acid on growth and composition of Thompson seedless grape berries, *Am. J. Enol. Vitic.*, 31, 149, 1980.
304. **Singh, R. K. N. and Campbell, R. W.,** Some effects of 4-thianaphthenacetic acid on ripening of Concord grapes, *Proc. Am. Soc. Hortic. Sci.*, 84, 259, 1964.
305. **Singh, I. S., Weaver, R. J., and Chundawat, B. S.,** Timing of ethephon application on ripening and quality of raisin grape cv. Black Corinth, *Haryana Agric. Univ. J.*, 7, 97, 1977.
306. **Singh, K., Weaver, R. J., and Johnson, J. O.,** Effect of applications of gibberellic acid on berry size shatter and texture of Thompson seedless grapes, *Am. J. Enol. Vitic.*, 29, 258, 1978.
307. **Skene, K. G. M.,** A comparison of the effects of "Cycocel" and tipping on fruit set in *Vitis vinifera* L., *Aust. J. Biol. Sci.*, 22, 1305, 1969.
308. **Skene, K. G. M.,** The relationship between the effects of CCC on root growth and cytokinin levels in the bleeding sap of *Vitis vinifera* L., *J. Exp. Bot.*, 21, 418, 1970.
309. **Soderland, R.,** Personal communication.
310. **Smart, R. E.,** The grape shoot system, a review in preparation.
311. **Smirnov, K. V. and Perepelitsyna, E. P.,** On the effect of gibberellin on seedless varieties of vine and other processing products, *Sov. Plant Physiol.*, 12, 259, 1965.
312. **Sood, S. K.,** Effect of gibberellic acid on the fruit-set, and partheno-carpic development of fruit in Bhakri grapes (*Vitis vinifera*, L.), *Lal-Baugh*, 12, 25, 1968.
313. **Sperling, D. and Eaton, G. W.,** Some effects of gibberellic acid applications on two french hybrid grape varieties, *Fruit Var. Hortic. Dig.*, 26, 67, 1972.
314. **Sproule, R. S. and Stannard, M. C.,** Effects of growth regulators and bunch trimming on size maturity and yield of sultana grapes, *Aust. J. Exp. Agric. Anim. Husb.*, 10, 474, 1970.
315. **Srinivasan, C. and Mullins, M. G.,** Control of flowering in the grapevine *Vitis vinifera*: formation of inflorescences *in-vitro* by isolated tendrils, *Plant Physiol.*, 61, 127, 1978.
316. **Srinivasan, C. and Mullins, M. G.,** Flowering in *Vitis*: conversion of tendrils into inflorescences and bunches of grapes, *Planta*, 145, 187, 1979.
317. **Srinivasan, C. and Mullins, M. G.,** High frequency somatic embryo production from unfertilized ovules of grapes, *Sci. Hortic.*, 13, 245, 1980.
318. **Srinivasan, C. and Mullins, M. G.,** Effects of temperature and growth regulators on formation of Anlagen tendrils and inflorescences in *Vitis vinifera* cultivar Muscat of Alexandria, *Ann. Bot.*, 45, 439, 1980.
319. **Srinivasan, C. and Mullins, M. G.,** Flowering in *Vitis*: effects of genotype on cytokinin-induced conversion of tendrils into inflorescences, *Vitis*, 19, 293, 1980.
320. **Srinivasan, C. and Mullins, M. G.,** Induction of precocious flowering in grapevine seedlings by growth regulators, *Agronomie*, 1, 1, 1981.
321. **Srinivasan, C. and Mullins, M. G.,** Modification of leaf formation of cytokinin and chlormequat (CCC) in *Vitis*, *Ann. Bot.*, 48, 529, 1981.

322. **Srinivasan, C. and Mullins, M. G.,** Physiology of flowering in the grapevine *Vitis vinifera*: a review, *Am. J. Enol. Vitic.*, 32, 47, 1981.
323. **Staalduine, Van D.,** The growing and setting of fruit of Muscat grapes (Dutch), *Meded. Dir. Tuinb.*, 25, 786, 1962; *Hortic. Abstr.*, 33, 2584, 1963.
324. **Stannard, M. C., Peterson, J. R., and Sproule, R. L.,** Effects of gibberellic acid on fruit, "hen and chicken" condition, and vine development in the following season of Waltham Cross grapes, *Aust. J. Exp. Agric. Anim. Husb.*, 14, 256, 1974.
325. **Stewart, W. S., Halsey, D., and Ching, F. T.,** Effect of the potassium salt of Gibberellic acid on fruit growth of Thompson Seedless grapes, *Proc. Am. Soc. Hortic. Sci.*, 72, 165, 1958.
326. **Stino, G. R., Fayek, M. A., and Mikhail, N. M.,** Effect of various treatments on the production of first grade grafts of Thompson Seedless grapevine on *Vitis solonis* × *V. riparia* 1616 rootstock, *Vitis*, 16, 20, 1977.
327. **Stout, A. B.,** Types of flowers and intersexes in grapes with reference to fruit development, *N.Y. Agric. Exp. Stn. Geneva, Tech. Bull.*, No. 82, 1921.
328. **Sugiura, A., Utsunomiya, N., and Tomana, T.,** Induction of inflorescence by CCC application on primary shoots of grapevines *Vitis*, 15, 88, 1976.
329. **Sullivan, E. F.,** *Plant Growth Regulator Handbook*, Plant Growth Regulation Working Group, Great Western Sugar Co. Longmont, Colorado, 1977.
330. **Tafazoli, E.,** Increasing fruit set in *Vitis vinifera*, *Sci. Hortic.*, 6, 121, 1977.
331. **Takagi, T., Fukukawa, Y., and Soejima, T.,** Jpn. Soc. Hortic. Sci., 46, 173, 1977; Cited in Reference 219.
332. **Tamamura, K., Miyoshi, T., Shiba, H., and Hirata, K.,** Studies on the treatment of gibberellin in grape, *Jpn. Gibberellin Res. Assoc., Fifth Meeting*, 89, 1961.
333. **Tamponi, G. C.,** The effects of 2,4,5-TP on Cardinal and Moscato dell'Adda grapes, *Riv. Vitic. Enol.*, 22, 174, 1969.
334. **Tezuka, T., Sekiya, H., and Ohno, H.,** Physiological studies on the action of chloroethyltrimethyl ammonium chloride in Kyoho grapes, *Plant Cell Physiol.*, 21, 969, 1980.
335. **Tkačenko, G. V.,** The effect of gibberellin on fruiting in the vine variety Cauš, *Fiziol. Rast.*, 7, 348, 1960.
336. **Tripathi, S. N.,** Studies on the effect of gibberellic acid on fruit thinning, branch size and chemical composition of Perlette grape (*Vitis vinifera* L.), *Punjab Hortic. J.*, 7, 60, 1967.
337. **Tripathi, S. N.,** Effect of gibberellic acid on branch and berry size and quality of Pusa seedless grape (*Vitis vinifera* L.), *Punjab Hortic. J.*, 8, 152, 1968.
338. **Tukey, L. D.,** Effects of controlled temperatures following bloom on berry development of the Concord grape *(Vitis lubrusca)*, *Proc. Am. Soc. Hortic.*, 71, 157, 1958.
339. **Tukey, L. D. and Fleming, H. K.,** Fruiting and vegetative effects of N-dimethylaminosuccinamic acid on "Concord" grapes, *Vitis lubrusca* L, *Proc. Am. Soc. Hortic. Sci.*, 93, 300, 1968.
340. **Ueda, H. and Naito, R.,** Effects of shoot vigour on the induction of seedless berries by G.A. application in Muscat Bailey A. grape, *J. Jpn. Soc. Hortic. Sci.*, 50, 92, 1981.
341. **Eda, H., Sainoto, Y., and Naito, R.,** Production of seedless berries with GA in Muscat Bailey A grapes, I. Effects of the timing, concentration and frequency of GA application on the seedlessness ratio and bunch quality, *Bull. Fac. Agric. Shimane Univ.*, 6, 16, 1972.
342. **Ueda, H., Miyamoto, T., and Naito, R.,** Production of seedless berries with gibberellic-acid in Muscat-Bailey A grapes, Part 2. Effects of the repeated applications of gibberellic acid and combined application of gibberellic acid and benzyl adenine before full bloom on the seedlessness ratio and bunch quality, *Bull. Fac. Agric. Shimane Univ.*, 8, 1, 1974.
343. **Ueda, H., Miyamoto, T., and Naito, R.,** Production of seedless berries with gibberellic acid in Muscat Bailey A grapes, Part 3. Effects of benzyl adenine applied at different concentrations with gibberellic acid before full bloom on compactness and other bunch qualities, *Bull. Fac. Agric. Shimane Univ.*, 11, 1, 1977.
344. **Venkataratnam, L.,** Effect of gibberellic acid on Anab-e-Shahi grape *(Vitis vinifera)*, *Proc. Am. Soc. Hortic. Sci.*, 84, 255, 1964.
345. **Vidal, J. P. and Marcelin, H.,** Action de regulateurs de croissance sur la floraison du Grenache noir, *Prog. Agric. Vitic.*, 169, 128, 1969.
346. **Weaver, R. J.,** Response of Black Corinth grapes to applications of 4-chlorophenoxy-acetic acid, *Bot. Gaz.*, 144, 107, 1952.
347. **Weaver, R. J.,** Thinning and girdling of Red Malaga grapes in relation to size of berry color, and percentage of total soluble solids of fruit, *Proc. Am. Soc. Hortic. Sci.*, 60, 132, 1952.
347a. **Weaver, R. J.,** Berry size of seedless grapes, *Calif. Agric.*, 7, 15, 1953.
348. **Weaver, R. J.,** Further studies on effects of 4-chlorophenoxyacetic acid on development of Thompson seedless and Black Corinth grapes, *Proc. Am. Soc. Hortic. Sci.*, 61, 135, 1953.

349. **Weaver, R. J.,** Effect of benzothiayol-2-oxyacetic acid in delaying maturity of grapes, *Science,* 119, 287, 1954.
350. **Weaver, R. J.,** Preliminary report on thinning grapes with chemical sprays, *Proc. Am. Soc. Hortic. Sci.,* 63, 194, 1954.
351. **Weaver, R. J.,** Effect of benzothiazol-2-oxyacetic acid on development of Black Corinth grapes, *Bot. Gaz.,* 115, 365, 1954.
352. **Weaver, R. J.,** Use of benzothiazol-2-oxyacetic acid to delay maturity grapes, *Bot. Gaz.,* 116, 266, 1955.
353. **Weaver, R. J.,** Relation of time of girdling to ripening of fruit of Red Malaga and Ribi grapes, *Proc. Am. Soc. Hortic. Sci.,* 65, 183, 1955.
354. **Weaver, R. J.,** Effect of gibberellic acid on fruit set and berry enlargement in seedless grapes of *Vitis vinifera, Nature,* 181, 851, 1958.
355. **Weaver, R. J.,** Toxicity of gibberellin to seedless and seeded varieties of *Vitis vinifera, Nature,* 187, 1135, 1960.
356. **Weaver, R. J.,** Growth of grapes in relation to gibberellin, in *Gibberellins* Advances in Chemistry Series, No. 28, American Chemical Society, Applied Publications, Washington, D.C., 1961, 89.
357. **Weaver, R. J.,** Use of a kinin in breaking rest in buds of *Vitis vinifera, Nature,* 198, 207, 1963.
358. **Weaver, R. J.,** Experiments on thinning grapes with alpha-naphthalene acetic acid and dinitro-sec-butyl-phenol, *Vitis,* 4, 1, 1963.
359. **Weaver, R. J.,** Effect of chlormequat ((2-chloroethyl)-trimethylammonium chloride) on small-berried wine grapes, *Am. J. Enol. Vitic.,* 24, 69, 1973,
360. **Weaver, R. J.,** Effect of growth retardant sprays on fruitfulness and cluster development of cultivar Thompson-Seedless grapes, *Am. J. Enol. Vitic.,* 26, 47, 1975.
361. **Weaver, R. J.,** Effect of time of application of potassium gibberellate on cluster development of 'Zinfandel' grapes, *Vitis,* 14, 97, 1975.
362. **Weaver, R. J. and Ibrahim, I. M.,** Effect of thinning and seededness on maturation of *Vitis vinifera* grapes, *Proc. Amer. Soc. Hortic. Sci.,* 92, 311, 1968.
363. **Weaver, R. J. and Leonard, O. A.,** Effect of dormant-season applications of 2,4-D on pruned and nonpruned Tokay grapes, *Hilgardia,* 37, 661, 1967.
364. **Weaver, R. J. and McCune, S. B.,** Response of Thompson Seedless grapes to 4-Chlorophenoxyacetic acid and benzothiazol-2-oxyacetic acid, *Hilgardia,* 27, 189, 1957.
365. **Weaver, R. J. and McCune, S. B.,** Response of certain varieties of *Vitis vinifera* to gibberellin, *Hilgardia,* 28, 297, 1959.
366. **Weaver, R. J. and McCune, S. B.,** Girdling; its relation to carbohydrate nutrition and development of Thompson Seedless, Red Malaga and Ribier grapes, *Hilgardia,* 28, 421, 1959.
367. **Weaver, R. J. and McCune, S. B.,** Effect of gribberellin on seeded *Vitis vinifera* and its translocation within the vine, *Hilgardia,* 28, 625, 1959.
368. **Weaver, R. J. and McCune, S. B.,** Effect of gibberellin on seedless *Vitis vinifera, Hilgardia,* 29, 247, 1959.
369. **Weaver, R. J. and McCune, S. B.,** Further studies with gibberellin on *Vitis vinifera* grapes, *Bot. Gaz.,* 121, 155, 1960.
370. **Weaver, R. J. and McCune, S. B.,** Effect of gibberellin on vine behaviour and crop production in seeded and seedless *Vitis vinifera, Hilgardia,* 30, 325, 1961.
371. **Weaver, R. J. and McCune, S. B.,** Studies on prebloom sprays of gibberellin to elongate and loosen clusters of Thompson Seedless grapes, *Am. J. Enol. Vitic.,* 13, 15, 1962.
372. **Weaver, R. J. and Montgomery, R.,** Effect of Ethephon on coloration and maturation of wine grapes, *Am. J. Enol. Vitic.,* 25, 39, 1974.
373. **Weaver, R. J. and Nelson, K. E.,** Improving grape quality by thinning, girdling, plant growth regulators, *Calif. Agric. Exp. St.,* Leaflet 120, 1959.
374. **Weaver, R. J. and Pool, R. M.,** Morphactins induce berry abscission in grapes, *Calif. Agric.,* 22, 10, 1968.
375. **Weaver, R. J. and Pool, R. M.,** Induction of berry abscission in *Vitis vinifera* by morphactin, *Am. J. Enol. Vitic.,* 19, 121, 1968.
376. **Weaver, R. J. and Pool, R. M.,** Effect of various levels of cropping on *Vitis vinifera* grapevines, *Am. J. Enol. Vitic.,* 19, 185, 1968.
377. **Weaver, R. J. and Pool, R. M.,** Berry response of 'Thompson Seedless' and 'Perlette' grapes to application of gibberellic acid, *J. Am. Soc. Hortic. Sci.,* 96, 162, 1972.
378. **Weaver, R. J. and Pool, R. M.,** Effect of Ethrel, abscisic acid, and a morphactin on flower and berry abscission and shoot growth in *Vitis vinifera, J. Am. Soc. Hortic. Sci.,* 94, 474, 1969.
379. **Weaver, R. J. and Pool, B. M.,** Effect of Ethephon and morphactin on growth and fruiting of Thompson Seedless and Cariganme grapes, *Am. J. Enol. Vitic.,* 22, 234, 1971.
380. **Weaver, R. J. and Pool, R. M.,** Effect of succinic acid-2, 2-di-methyl hydrazide and 2-chloroethyltrimethyl ammonium chloride on shoot growth of Tokay grapes, *Am. J. Enol. Vitic.,* 22, 223, 1971.

381. **Weaver, R. J. and Pool, B. M.,** Effect of (2-chloroethyl) phosphoric acid (Ethephon) on maturation of *Vitis vinifera, J. Am. Soc. Hortic. Sci.*, 96, 725, 1971.
382. **Weaver, R. J. and Pool, R. M.,** Thinning Tokay and Zinfandel grapes by bloom sprays of gibberellin, *J. Am. Soc. Hortic. Sci.*, 96, 820, 1971.
383. **Weaver, R. J. and Pool, R. M.,** Chemical thinning of grape clusters, *Vitis*, 10, 201, 1971.
384. **Weaver, R. J. and Pool, R. M.,** Effect of time of thinning on berry size of girdled, gibberellin-treated Thompson Seedless grapes, *Vitis*, 12, 97, 1973.
385. **Weaver, R. J. and Pool, R. M.,** Bloom spraying with gibberellin loosens clusters of Thompson Seedless grapes, *Calif. Agric.*, 19, 14, 1965.
386. **Weaver, R. J. and Williams, W. J.,** Response of flowers of Black Corinth and fruit of Thompson Seedless grapes to applications of plant growth regulators, *Bot. Gaz.*, 111, 477, 1950.
387. **Weaver, R. J. and Williams, W. O.,** Response of certain varieties of grapes to plant growth regulators, *Bot. Gaz.*, 113, 75, 1951.
388. **Weaver, R. J. and Wrinkler, A. J.,** Increasing the size of Thompson Seedless grapes by means of 4-chlorophenoxy-acetic acid, berry thinning and girdling, *Plant Physiol.*, 27, 626, 1952.
389. **Weaver, R. J., Abdel-Gawad, H. A., and Martin, G. C.,** Translocations and persistence of 1,2-^{14}C-(2-chloroethyl)-phosphonic acid (Ethephon) in Thompson Seedless, *Physiol. Plant*, 26, 13, 1972.
390. **Weaver, R. J., Kasimatis, A. N., and McCune, S. B.,** Studies with gibberellin on wine grapes to decrease bunch rot, *Am. J. Enol. Vitic.*, 13, 78, 1962.
391. **Weaver, R. J., Kasimatis, A. N., and Pool, R. M.,** Effect of wetting agents on Thompson Seedless and Perlette grapes, *Am. J. Enol. Vitic.*, 22, 40, 1971.
392. **Weaver, R. J., Leonard, O. A., and McCune, S. B.,** Response of clusters of *Vitis vinifera* grapes to 2,4-dichlorophenoxyacetic acid and related compounds, *Hilgardia*, 31, 113, 1961.
393. **Weaver, R. J., Leonard, O. A., and McCune, S. B.,** Response of Tokay grapes to spray applications of 2,4-D, *Hilgardia*, 31, 419, 1961.
394. **Weaver, R. J., Manivel, L., and Jensen, F. L.,** The effects of growth regulators, temperature, and drying on *Vitis vinifera* buds, *Vitis*, 13, 23, 1974.
395. **Weaver, R. J., McCune, S. B., and Coombe B. G.,** Effects of various chemicals and treatments on rest period of grape buds, *Am. J. Enol. Vitic.*, 12, 131, 1961.
396. **Weaver, R. J., McCune, S. B., and Hale, C. R.,** Effect of plant regulators on set and berry development in certain seedless and seeded varieties of *Vitis vinifera* L., *Vitis*, 3, 84, 1964.
397. **Weaver, R. J., Van Overbeek, J., and Pool, R. M.,** Effect of kinins on fruit set and development in *Vitis vinifera, Hilgardia*, 37, 181, 1966.
398. **White, I. G. and Kennard, W. C.,** A preliminary report on the use of maleic hydrazide to delay blossoming of fruits, *Proc. Am. Soc. Hortic. Sci.*, 55, 147, 1950.
399. **Weaver, R. J., Yeou-Der, K., and Pool, R. M.,** Relation of plant regulators to bud rest in *Vitis vinifera* grapes, *Vitis*, 7, 206, 1968.
400. **Weinhaus, H.,** Responses in vigour of the grape vine to the application of metabolism inhibiting factors and uncouplers during ripening process, *Vitis*, 12, 105, 1973.
401. **Winkler, A. J.,** Effects of overcropping, *Am. J. Enol.*, 5, 4, 1954.
402. **Winkler, A. J.,** *General Viticulture*, The Jacaranda Press, 1963; University of California Press, Berkeley, Cal., 1962.
403. **Yeou-Der, K., Weaver, R. J., and Pool, R. M.,** Effect of low temperature and growth regulators on germination of seeds of Tokay grapes, *Proc. Am. Soc. Hortic. Sci.*, 92, 323, 1968.
404. **Yokozawa, Y. and Fukunaga, S.,** Studies on gibberellin availability to grapes, *Jpn. Gibberellin Res. Assoc.*, 33, 1961.
405. **Ziv, M., Melamud, H., Bernstein, Z., and Lavee, S.,** Necrosis in grapevine buds (*Vitis vinifera* cv. Queen of Vineyard), II. Effect of gibberellic acid (GA_3) application, *Vitis*, 20, 105, 1981.
406. **Zuluaga, P. A. and Lumelli, J.,** Empeleo de reguladores de crecimiento en vid, *Oeste (Mendosa)*, 5 (15), 39, 1966.
407. **Zuluaga, P. A.,** Métodos fisiológicos para la obtención de uvas sin semilla a partir de variedados de rid con semillas normales, *Oeste*, 6, 29, 1967.
408. **Zuluaga, E. M., Lumelli, J., and Christensen, J. H.,** Influence of growth regulators on the characteristics of berries of *Vitis vinifera* L. *Phyton*, 25, 35, 1968.
409. **Zuluaga, P. A., Zuluaga, E. M., and De La Iglesia, F. J.,** Induction of stimulative parthenocarpy in *Vitis vinifera* L., *Vitis*, 7, 97, 1968.
410. **Zuluaga, P. A., Zuluaga, E. M., and De La Iglesia, F. J.,** Regulation of parthenocarpy in berries of *Vitis vinifera, Phyton*, 28, 137, 1971.
411. **Gardner, F. E. and Marth, P. C.,** Parthenocarpic fruits induced by spraying with promoting compounds, *Bot. Gaz.*, 99, 184, 1937.
412. **Čigrin, V. N.,** The effect of gibberellin on the yield of the vine variety Madeleine Angevine, *Sb. Nauchn. Rab. Vses. Nauchn. Issled. Inst. Sadovod.*, 12, 247, 1967.

413. **Alleweldt, G.,** Studies on flower initiation in the vine (in Ger), *Vitis,* 4, 176, 1964.
415. **Coombe, B. G. and Hale, C. R.,** The hormone content of ripening grape berries and the effects of growth substance treatments, *Plant Physiol.,* 51, 629, 1973.
416. **Coombe, B. G. and Bishop, G. R.,** Development of the grape berry, II. Changes in diameter and deformability during veraison, *Aust. J. Agric. Res.,* 31, 499, 1980.
417. **Kondrja, S. M.,** Ringing of vines, *Sadovod Vinograd. Vinodel. Mold.,* 12, 24, 1967.
418. **Biddle, E., Kerfoot, D. G. S., Kho, Y., Russell, K. E.,** Kinetic studies of the thermal decomposition of 2-chloroethyl-phosphonic acid in aqueous solution, *Plant Physiol.,* 58, 700, 1976.
419. **De Wilde, R. C.,** Practical application of Ethrel in agricultural production, Paper presented at symposium on ethylene, 67th Meet. Am. Soc. Hortic. Sci., Miami Beach, Fla., (cited in Reference 422), 1970.
420. **Klein, I., Lavee, S., Ben-Tal, Y.,** Effect of water vapour pressure on the thermal decomposition of 2-chloroethylphosphonic acid, *Plant Physiol.,* 63, 474, 1979.
421. **Luckwill, L. C. and Child, R. D.,** The meadow orchard — a new concept of apple production based on growth regulations, *Acta Hortic.,* 34(1), 213, 1973.
422. **Morgan, P. W.,** Regulation of ethylene as an agricultural practice, *Acta Hortic.,* 34(1), 41, 1973.
423. **Scholefield, P. B., Neales, T. F., and May, P.,** Carbon balance of the Sultana vine (*Vitis vinifera* L.) and the effects of autumn defoliation by harvest pruning, *Aust. J. Plant Physiol.,* 5, 561, 1978.
424. **Swift, J. G., May, P., and Lawton, E. A.,** Concentric cracking of grape berries, *Vitis,* 13, 30, 1974.
425. **Mohsenin, N. N.,** *Physical Properties of Plant and Animal Materials, Vol. 1, Structure, Physical Characterists and Mechanical Properties,* Gordon and Breach, London, 1970.
426. **Bernstein, Z. and Lustig, I.,** A new method of firmness measurement of grape berries and other juicy fruits, *Vitis,* 20, 15, 1981.
427. **Simon, E. W.,** Leakage from fruit cells in water, *J. Exp. Bot.,* 28, 1147, 1977.
428. **Considine, J. A.,** A statistical study of rain damage of grapes grown for drying in Victoria, *Aust. J. Exp. Agr. Anim. Husb.,* 13, 604, 1973.
429. **Branas, J. and Vergnes, A.,** New observations on the effects of gibberellin on vines (in French), *Prog. Agric. Vitic.,* 80, 75, 1963.
430. **Bugnon, F. and Besis, R.,** *Biologie de la Vigne,* Masson & Cie, Paris, 1968.
431. **May, P.,** Uber die Knospen-und Infloreszenzentwicklung der Rebe, *Weinwiss.,* 19, 457, 1964.
432. **Hopping, M. R.,** The effect of light intensity during cane development on subsequent bud break and yield of 'Palamino' grape vines, *N.Z. J. Exp. Agric.,* 5, 287, 1977.
433. **Stecher, P. G., Windholz, M., and Leahy, D. S., Eds.,** *The Merck Index,* 8th ed., Merck & Co., Rahway, N.J., 1968.
434. **Alleweldt, G.,** The relationship between environment and vegetative growth, dormancy and flower formation in vines, II. Reactions to gibberellin and growth phases of buds (in German), *Vitis,* 4, 152, 1964.
435. **Olmo, H. P.,** Empty-seededness in Varieties of *Vitis vinifera, Proc. Am. Soc. Hortic. Sci.,* 32, 376, 1934.
436. **Goussard, P. G. and Oreffer, C. J.,** The propagation of Salt Creek, *Decid. Fruit Grow.,* 29, 56, 1979.
437. **Goussard, P. G., and Orffer, C. J.,** Callusing ability of cane segments of rootstock cultivars (Vitis) and the effect of stimulants on callus formation, *Proc. S. A. Soc. Enol. Vitic.,* 24, 1977.
438. **Harmon, F. N. and Weinberger, J. H.,** Studies to improve the bench grafting of vinifera grapes, *Proc. Am. Soc. Hortic. Sci.,* 90, 149, 1967.
439. **Julliard, B.,** Sur la rhizogenese chez la vigne, *Vitis,* 6, 375, 1967.
440. **Goheen, A. C. and Nyland, G.,** The big breakthrough: mist propagation, *Wines and Vines,* 52, 25, 1971.
441. **Skirvin, R. M.,** Fruit crops, in *Cloning Agricultural Plants via In Vitro Techniques,* Conger, B. V., Ed., CRC Press, Boca Raton, Fla., 1980, 51.
442. **Evenari, N. and Konis, E.,** The effect of heteroauxin on root formation by cuttings and on grafting, *Palestine J. Bot.,* 1, 13, 1938, (Cited in Reference 257).
443. **Schenk, Von W.,** Untersuchungen über die Verwachsungvorgänge bei Pforpf-reben *Weinberg Keller,* 22, 55, 1975.
444. **Bouquet, A.,** Effect of some genetic and environmental factors on spontaneous polyembryony in grape *(Vitis vinifera), Vitis,* 19, 134, 1980.
445. **Sullivan, E. F.,** *Plant Growth Regulator Handbook,* Plant Growth Regulator Working Group, Great Western Sugar Co. Longmont, Colorado, 1977.
446. **Considine, J. A.,** unpublished data.

Chapter 7

SUGARCANE

Louis G. Nickell

TABLE OF CONTENTS

I. INTRODUCTION

Sugarcane (*Saccharum officinarum* L.) has been known from ancient times, long before the Christian era. The original home of sugarcane, for many years in dispute, is now believed to be New Guinea. The dispersal of cultivated forms of sugarcane from New Guinea is closely related to the ancient migrations which covered a large part of the world. Sugarcane was one of the first tropical crops to be adapted to large-scale farming. The length of the growing season varies from less than 9 to 10 months in Louisiana (frost dates setting the time limits), to 2 years in Peru and South Africa, and 2 years or more in Hawaii. Most of the rest of the sugarcane is grown in 14- to 18-month plant crops followed by 12-month ratoon crops. Cane sugar is produced commercially in over 70 countries, territories, and island groups, generally within a band around the world bounded by 35°N and 35°S latitudes.

Plant growth regulators have been used in the sugarcane industry worldwide for over two decades to increase the recoverable yield of sucrose in sugarcane.[1-5] The first commercial success was in the prevention of flowering, followed by the application of gibberellic acid for the increase of stalk elongation which ultimately resulted in increased sugar production. Currently, the primary interest centers around the use of chemicals for the control of maturation — the so called chemical "ripeners, . These uses are now on a commercial basis and will be discussed along with the on/off use of paraquat as a desiccant. Research has been successful also in affecting both germination of the vegetative "seed pieces" used for propagation and on the tillering of young plants of sugarcane. Because of the importance of and the success with ripening control, these latter two uses have not been pursued rigorously until the last few years. Consequently, they are not yet at the commercial stage. The various uses of plant growth regulators will be presented sequentially, as they would be used in the development of the crop, i.e., from planting through harvest.

II. GERMINATION

Because of the problems involved in emergence of sugarcane after planting (including such factors as depth of planting, angle of the bud, adverse weather conditions — particularly excess moisture and cold — and fungal or bacterial infection),[6] it has always been the aim of sugarcane planters to try to obtain a "good stand". In order to get a good stand, one has to have germination and rapid early growth, i.e., emergence of the first green leaf. In a search for compounds which might stimulate the rate of germination or early growth of the young sugarcane plant, it was found that the amino acid arginine caused a considerable stimulation of germination.[7] This amino acid was also shown to cause an increase in the growth rate of the sugarcane plant at later stages, resulting in increased production of both cane and sugar.[6] Additional studies showed that this effect occurs at the cellular level, i.e., cells of sugarcane grown in suspension culture respond rather dramatically to this amino acid.[8,9] Subsequently, considerable work has been done with arginine and its effects on sugarcane growth and metabolism.[10-14] The dramatic results obtained with arginine suggest that if ways to exploit these findings commercially should be devised they could easily lead to one of the greatest increases in sugar production yet achieved by the application of chemicals. In order to achieve this, much more whole plant physiology and field work must be conducted.

III. TILLERING

A tiller is a branch (stalk) from the base of the plant or from the axil of a lower leaf. Tillering is a general characteristic of grasses. The manner of tillering provides a means for dividing grasses into two groups: tufted grasses and sod formers. In sod formers, there is intense underground branching which permeates the soil. These underground branches combined with the root system form a coherent mat which is found in lawns and permanent pastures. In tufted grasses, the underground branching is limited. This is followed by the formation of a number of erect stalks. The individual plants (in clumps) are easily distinguishable. The cereals and sugarcane belong to the tufted grasses.

Of the many variables involved in the production of sugar from the cane plant, probably the most significantly related factor is the number of stalks per acre of land at harvest.[6,15] In order to obtain the maximum number of stalks supportable by an acre of cane, it is necessary to induce branching at an early stage. A search for compounds that would induce tillering in sugarcane was started several years ago. Several compounds, such as ethephon, were found to be active in Hawaii.[16,17] Recent results from Jamaica confirm the Hawaiian results in response to ethephon.[18] Two of the three cultivars studied tillered freely in response to ethephon treatment; the third did not tiller — regardless of treatment. Results from Brazil show no effect of ethephon in the plant crop but a significant increase in the number of millable stalks in the ratoon.[19-21] In the Philippines, both ethephon at high concentrations and Bualta at low concentrations significantly stimulated tillering,[22,23] while in Taiwan, ethephon and chlormequat were effective.[24] Subsequent studies in Taiwan have shown that both indolebutyric acid and ethephon increase tillering in ratoon crops, although the effect varies with cultivars.[25-27] Recent greenhouse studies in Louisiana emphasize that the tillering process in sugarcane is a function of genotype. Ethephon was found to be the most effective treatment for those cultivars which did respond.[28,29]

A uniform lack of effect on cultivar NCo-310, one of the most widely used cultivars of the world, has been reported in all countries.

IV. INCREASED GROWTH OF STALK

A. Gibberellins

The control of plant and organ size can be of great importance in agriculture. If maximum weight, length, or diameter affects final yield, then an increase in size is desirable. On the other hand, if it can be of commercial benefit, it may be important to be able to reduce the overall size of the plant.

The elongated "foolish seedling" effect in rice, caused by infection with the fungus *Gibberella fujikuroa*, has been known for many decades. It was not until 1938, however, that a metabolite of this fungus was isolated and shown to be the causative agent of the disease. The isolation, crystallization, and structural determination of this material led to the discovery of a new class of hormones, the gibberellins. In most plants, the outstanding effect of the gibberellins is to elongate the primary stalk. This effect occurs in the young tissues and growth centers and is caused by an increase in cell length, an increase in the rate of cell division, or a combination of both, depending on the specific types of plant treated. Gibberellins have remarkable effects on many dwarf plants including peas, corn, and beans; when treated wth gibberellins, these plants grow to full size. Also, gibberellins (1) affect the extent to which the plant develops side branches and (2) increase the size of many young fruits, especially grapes. Because gibberellins induce the production of the enzyme amylase in barley, they are commonly used in the malting of this grain.

Although the gibberellins can induce flowering in many plant species, their greatest commercial uses have been in increasing the size of grapes and in stimulating the growth of sugarcane to increase the length of the primary stalk.

During the 1950s when experimental amounts of gibberellic acid and related gibberellins became available, they were applied to sugarcane, and stalk elongation was demonstrated.[30,31] Because of the high cost of crystalline gibberellic acid and its unavailability in quantity, more than a decade passed before such materials were evaluated to determine whether or not they would increase crop production on a commercial scale. During the early studies of this class of plant hormones, more and more potential uses became apparent. Fermentation companies consequently started making larger amounts of gibberellins, and eventually unrefined fermentation liquids became available for evaluation, at greatly reduced costs. These products were shown to increase cane tonnage per unit area, particularly when applied during the cold or winter months when growth is slowed down considerably.[32] Later, field work with several types of products including broths and semipurified and crystalline gibberellins showed on average, under Hawaiian plantation conditions, a gain of about 5 ton/acre at harvest, with a gain in sugar from 0.2 to 0.5 ton/acre acre from the application of 2 to 3 oz/acre of gibberellin.[2,33] Most of the early work was done by application during the second season of the 2-year crop cycle in Hawaii. Subsequent to the registration of gibberellin products with regulatory agencies, additional studies have been carried out with first season applications and these also have proved to be effective, indicating that maximum response is obtained with applications made to young sugarcane plants.[34] More recent studies have shown that the maximum response is a function of the cultivar treated; the amount of gibberellic acid applied; the number of applications; the time interval between applications; and the allowance of sufficient time before harvest for maturation of the gibberellin-produced growth.[35-41]

Gibberellic acid has been evaluated on sugarcane since the mid-1950s and evaluated under commercial field conditions since the mid-1960s. As the increases are relatively small compared to the total crop yield, they are difficult to measure. As a result, there has been a trend toward evaluating gibberellic acid effects on the basis of individual stalks or on those portions of the stalk showing the greatest responses. Taking into consideration the high variability among plots under most sugarcane cultural conditions, it has been calculated that at least 60 plots would be necessary to give an 85% probability of measuring a 5% yield difference at the 5% level of significance. Recently such a yield test was run in Hawaii on sugarcane cultivar H59-3775. In this experiment one half of 120 plots, each 150 m^2, was sprayed with two applications of 70 g/ha of gibberellic acid. The results show that gibberellic acid significantly increased the yield of sugarcane by 3.7% (10 ton/ha). Sugar yields were increased by 2.8% (1.1 ton/ha). There was no effect of gibberellic acid on the quality of the cane or juice.[41]

The latest summary of gibberellin effects in Hawaii showed consistent increases in cane growth and sugarcane yields over untreated cane when the following certain conditions were met:

1. Utilizing best responding varieties
2. Timing properly to coincide with winter stunting
3. Allowing sufficient time before harvest
4. Applying in sequential doses at 2- to 4-week intervals.

When such conditions were met significant gains in fresh weight were obtained without loss of cane quality. These gains were of the order of 1.75 metric tons of sugar per hectare.[42]

B. Ethephon

It has been known since the early days of ripening evaluations that ethephon can be active as a ripener. The fact that the results obtained over a period of time have been variable between tests (particularly under different conditions such as season) and that the activity usually was not as great as that of other available materials throughout the testing time suggest either that ethephon is not a competitive ripener or that it has other activity in addition to ripening which complicates the picture. The first suggestion of what this complication might be was presented by Jaramillo and co-workers,[43] who showed increased cane weights in tests initiated to evaluate ethephon as a ripener. This work was conducted in Colombia and reported in 1977. Ethephon was included in tests over several years in Hawaii comparing it with other potential ripeners. The results showed, under the conditions of the tests, that the quality of the cane following ethephon treatment was depressed, which is an indication of growth rather than ripening. The harvest results confirmed this, showing consistently increased stalk fresh weight between 8 and 10 weeks following treatment.[44] This led to serious consideration of ethephon as an enhancer of cane weight to be followed by a different ripener as a sequential treatment. The results of these tests over a 2-year period using glyphosate following ethephon treatment showed consistent increases in cane growth attributable to ethephon. Although the reason for this ethephon-induced growth stimulation has not been determined, it has been likened to that postulated for gibberellic acid. In addition to increasing the distance between photosynthetically active leaves, the leaf length is reduced as a result of ethephon treatment. These investigators suspect that the shorter, more erect, more widely spaced leaves may allow for greater light penetration into the canopy and account for the increased level of dry matter production.[44,45]

Studies confirming the increased growth effects from ethephon application and comparing the results to those obtained with similar treatment with gibberellic acid have been reported from South Africa.[46] The similarity to results obtained with Gramineae[47] is pointed out.

V. FLOWERING

Control of flowering is one of the most important practical aspects of horticulture and agriculture. With many horticultural crops, the key to financial success is the capability to induce flowering and more importantly, to induce it on command in order to meet certain major market and holiday dates. Conversely, the ability to prevent flowering is extremely important in agricultural crops when flowering causes a decreased economic benefit.

To improve yields, it is commercially worthwhile to prevent flowering in some crops, such as sugarcane. In other crops, among them almond, peach, and tall oil trees, delaying the onset of flowering may be useful to avoid adverse weather conditions such as extremes in temperature and moisture. Furthermore, such a delay can (1) bring two plant varieties with different flowering dates into synchronization for breeding purposes (as in the case of varieties of almond trees) or (2) control the timing of flowering plants such as carnation or poinsettia to coincide with major holidays when selling prices are higher.[48]

Flowering in plants is the end product of the cumulative effect of many subtle metabolic changes resulting in the initiation of flower buds. In 1920, Gardner and Allard[49] established that many plants show a peculiar sensitivity to light, in that flower buds will be initiated in such plants only under certain day lengths. Subsequent work established that the night length is a critical period. Sugarcane is considered to be among the most sensitive of plants to light. Sugarcane was classified by Allard in 1938[49] as an intermediate type, but later work showed that sugarcane is a short day plant, initiating flowers only within a critical range of day lengths. Members of this class must have an uninterrupted dark period in order to flower. The briefest interruption, for sugarcane as little as 50-ft-candle minutes[2,4,5,48,49] of incandescent light is usually sufficient to prevent flowering.

The flowering process in sugarcane is extremely sensitive to the environment. This is true of flower initiation, emergence, and pollen fertility. The optimum photoperiod appears to be about 12.5 hr, and most commercial varieties respond to this throughout the world. The 12.5 hr-day occurs about September 2 in Hawaii, and night interruption experiments indicate this is the approximate date on which initiation begins in that state. For most commercial varieties studied thus far, initiation is usually complete by September 20. The highly reproducible short period of floral initiation makes possible the suppression of flowering in young cane on a commercial scale. Experiments in Hawaii have established that flower inhibition by night lights, chemical sprays, or the withdrawal of water can be expected to increase yields by 10 to 20% under conditions of heavy flowering.

After it had been determined that night interruption from September 1 to 20 would inhibit flowering (commonly referred to as tasseling or arrowing in the sugarcane industry) for the varieties worked with in Hawaii at the time, field experiments were conducted to determine what quantitative effect flowering might have on the yield of sucrose. The 1949 results of a replicated field test of ten paired plots using a current commercial variety supported the belief, long held by sugar growers, that flowering reduces sugar yields. In this field test, the average gain from suppression of flowering was 15% — the equivalent of 1.3 tons of sucrose per acre.[49]

In the years immediately following these studies, the factors affecting flowering were extensively studied as were methods for preventing its occurrence. Effective ways found to prevent flowering included night interruption with light, lower temperature, leaf and spindle trimming, withdrawal of water, and application of chemicals. Because temperature cannot be controlled in the field and because leaf trimming and night interruption on a commercial scale were not operationally feasible, emphasis was placed on water withdrawal and application of chemicals. Withdrawal of water was possible only on irrigated plantations and, because such a practice had other operational problems, using chemicals eventually became standard practice in Hawaii.

The first potentially useful commercial chemical was maleic hydrazide, which gave about 60% control at best. Rapid developments led to the establishment of first monuron, and then diuron as the chemicals of choice. Properly applied, 4 lb/acre of either chemical gave virtually complete control of flowering in the heavy flowering varieties used in Hawaii during the 1950s and 1960s. Continued testing for active chemicals to prevent flowering led to the discovery that diquat applied from the air at rates as low as 0.125 lb/acre of the cation form was as active as monuron at 4 lb/acre, making diquat one of the most active compounds yet evaluated for this purpose. The resultant cut in cost per unit acre for control was substantial. Tests included both the heavy flowering varieties in use at that time as well as the lighter flowering varieties that were emerging as the preferred ones in the late 1960s. Positive effects with chemical control of flowering — with diquat being the compound of choice — have been obtained in Guyana, Mexico, the Philippines,[50] and Taiwan,[51] as well as in Hawaii.

With the introduction and increasingly widespread use of drip irrigation, even fields that previously used water withdrawal to control flowering are now shifting to the use of diquat.[52]

VI. RIPENING

Compounds that affect crop metabolism, particularly those that regulate crop maturity, are especially likely to have a dramatic impact on agriculture in the years ahead. Several such compounds already are used commercially and their success is in large part responsible for increased interest in plant growth regulators in agriculture. Most of the compounds used on economic crops have a direct or indirect effect on final yield, on quality, or on both.

Ripening is considered one of the most important aspects of sugarcane production from both a research and an operational point of view. To say that the phenomenon of cane ripening is extremely complex would be, at best, a gross understatement. Studies on the use of plant growth regulators have appeared in the literature spasmodically since 1949. The first material reported to be effective was 2,4-D.[53] This was followed by studies with maleic hydrazide, triiodobenzoic acid, dalapon, CMU, DCMU, EDTA, Trysben, Pesco 1850 (a mixture of MCPA and Trysben), as well as a number of enzyme inhibitors and metabolic inhibitors. No large-scale program was launched, however, until basic studies on the effects of defoliation on translocation in sugarcane had furnished a solid basis for such a program.[54,55]

The screening test used is a very simple one, consisting of adding the test material by pipette or needle and syringe into the whorl of leaves at the top of the sugarcane stalk, which is field grown and almost at the stage of normal maturation. At a specified time or times (4, 5, and/or 8 weeks) after application of the test material, 5 to 10 stalks are harvested, analyzed, and compared with an untreated group of stalks. The effectiveness of a test compound as a ripener is based on its ability to increase the quality of the treated stalks in two major parameters for sugar production (juice purity and sugar as a percent of field cane weight).[56]

Although attempts were made for several decades to control the ripening of sugarcane by the use of chemicals, no concerted effort was made until the start of a research program in Hawaii in the early 1960s.[2,56] This effort was soon joined by investigators in Australia[57] and Trinidad.[58] The initial success resulted in extensive field testing throughout the sugarcane world,[59] especially in Mauritius,[60-64] South Africa[65-67] India,[68-73] Brazil,[74,75] the Philippines,[76-80] Taiwan,[81-85] Guyana,[86,87] Colombia,[43] Puerto Rico,[88-91] Australia,[92-95] Jamaica,[96,97] Trinidad,[58,98] and the mainland U.S.A.[99-104] Originally, very few companies were involved; chemicals used were primarily those available from chemical supply houses together with the few materials synthesized by research organizations in the sugarcane industry around the globe. The initial success led a number of companies to become interested in supplying chemicals for evaluation. This total effort has resulted in a surprising number of chemicals that increase the sucrose content of sugarcane at harvest. Most of those compounds that have met with sufficient success to have information published about them are given in Table 1. Some compounds were never developed beyond the initial screening stages; others are too new to have been reported other than through an initial publication or an issued patent.

The first material seriously considered as a candidate ripener for increasing sucrose yields of sugarcane was the dimethylamine salt of 2,3,6-trichlorobenzoic acid. Because of a number of technical, environmental, and legal problems, this material did not prove successful commercially.[153] Nevertheless, it served as a standard for comparison in screening tests aimed at finding better sugarcane ripeners. It continued to be the standard for comparison until the registration of the first ripener for sugarcane in the U.S. This compound is N,N-bis(phosphonomethyl)glycine, known generically as glyphosine and marketed by Monsanto as the product Polaris®.[115,213]

Polaris has been evaluated over a period of several years on close to 100,000 acres in Hawaii and other sugar producing areas,[115] and has given substantial gains — about 10 to 15% increase in yields — which is an increase of 1 ton/acre or more when applied to certain varieties grown on the rainy coasts of the Island of Hawaii. More recent work has shown that varieties previously thought to be nonresponsive to this ripener have been found to respond positively when surfactants are added to the formulation.[1] Similarly, it has been found to be effective on irrigated lands when surfactants are added. Glyphosine treatment results in a reduced rate of terminal cane growth, but how this relates to its mode of action has not yet been established.[115]

Until late 1980 this was the only compound registered for this use in the U.S. In the fall of 1980 phosphonomethyl glycine was also registered as a ripener for sugarcane. This compound, known generically as glyphosate,[118,197] is marketed by Monsanto as the product Polado®; it is the sodium salt of the same compound that is the active ingredient of the herbicide Round-Up®. Glyphosate is almost an order of magnitude more active than glyphosine. Glyphosate formulations improve the sucrose content over a wide range of climatic conditions, are less cultivar specific, and the ripening response they induce in sugarcane is more consistent and rapid than that obtained with glyphosine.

Three other chemicals — Ripenthol®, chlormequat, and mefluidide — have been registered under experimental labels in the U.S. for field evaluation as commercial ripeners.

Ripenthol, the monoamine salt of Endothall®, was one of the first materials found to have

Table 1
SUGARCANE RIPENING COMPOUNDS

Compound	Common name	Code designation	Trade name	Ref.
p-Aminobenzenesulfonyl urea	—	—	—	105
2-Amino-6-methyl benzoic acid	6-Amino-o-toluic acid	ARC-1308	—	106, 107
Aminomethylphosphonic acid	—	AMPA	—	108
6-Aminopenicillanic acid	—	6-APA	—	3, 109
4-Amino-3,5,6-trichloropicolinic acid	Picloram	—	Tordon	3, 56
Ammonium ethyl carbamoyl phosphonate	—	—	—	110
Ammonium isobutyrate	Ammonium isobutyrate	AIB	—	111
N-benzoyl-N-(3,4-dichlorophenyl)aminopropionic acid	—	—	—	112
N,N-bis(phosphonomethyl) glycine	Glyphosine	CP-41845	Polaris	1, 2, 4, 31, 60—63, 69—72, 74—77, 79—82, 89, 91, 92, 94, 97, 99—101, 103, 113—140
bis-(N,O-trifluoracetyl)-N-phosphonomethyl glycine	—	—	—	141
2-Bromobenzylphosphonic acid	—	—	—	142
5-Bromo-3-*sec*-butyl-6-methyluracil	Bromacil	—	Hyvar X	143
2-Chlorobenzoic acid	—	—	—	2, 3, 144
2-(Chloro-2,4-dimethoxyphenyl)thioureido-3-cyano-4,5-tetramethylenethiophene	—	—	—	145
2-Chloroethylaminodi(methyl phosphonic acid)	—	—	—	146
2-Chloroethylphosphonic acid	Ethephon	Amchem 66-329	Ethrel Cepha	2—4, 43, 65—68, 80, 83, 88, 91, 93, 97, 116, 120, 122, 132, 136, 138, 147—151
2-Chloroethyltrimethyl ammonium chloride	Chlormequat	CCC	Cycocel	1, 3, 4, 56, 63, 68, 69, 73, 85, 91, 116, 117, 131, 152—155
5-Chloro-2-thenyl-tri-n-butylphosphonium chloride	—	CHE-8728	—	3, 116
3-Cyclohexene-1-carboxylic acid	Tetrahydrobenzoic acid	—	—	3
2,6-Dichlorobenzylphosphonic acid	—	—	—	142
2,3-Dichloro-6-methylbenzoic acid	—	—	—	156

2,4-Dichlorophenoxyacetic acid	2,4-D	2,4-D	Brush Killer Weedone Super-D Weed-B-Gon Weedtrol	53, 113, 157
2,2-Dichloropropionic acid	Dalapon	—	Dowpon Basfapon	85, 87, 158
N-(2,3-dihydroxy-1-propyl)-N-phosphonomethyl glycine, disodium salt	—	—	—	159
Diisobutylphenoxyethoxy-ethyldimethylbenzyl ammonium chloride	—	—	Hyamine 1622	1—3, 70, 72, 83, 116, 117, 160
2-(β-Dimethylamino-ethoxy)-4-(3′,4′-dichlorophenyl)thiazole hydrochloride	—	—	—	161
Dimethylarsenic acid	Cacodylic acid	—	Phytar 138	2, 3, 162
N,N-dimethylglycine	—	—	—	163
3-(2-[3,5-Dimethyl-2-oxocyclohexyl]-2-hydroxy-ethyl) glutarimide	Cycloheximide	—	Actidione	2, 3, 115
N-[2,4-dimethyl-5-[[(trifluoromethyl)sulfonyl]amino] phenyl]acetamide	Mefluidide	MBR-12325	Embark	2—4, 78, 80, 120, 122, 129, 135, 147, 148, 150, 164
Diphenylchlorophosphate	—	—	—	165
Ethyl N-(2-cyanomethyl)-N-ethoxy phosphonomethylglycinate, monosodium salt	—	—	—	166
2-Formyl-4-chlorophenoxyacetic acid	—	SAEI-517	—	147, 167
Hexadecyltrimethylammonium bromide	Cetyltrimethyl ammonium bromide	CTAB	Cetrimide	1—3, 70, 72, 116, 117, 168
Furfuryl 2-methoxy-3,6-dichlorobenzoate	—	—	—	169
Ethyl 2-[4-(5-bromo-2-pyridyloxy) phenoxy]propionate	—	—	—	170
1-Hydroxy-1,1-ethane diphosphonic acid	—	—	—	171
4-Hydroxy-3-methoxy benzaldehyde	Vanillin	—	—	2, 3, 172
Imidodicarbonic diamide	Carbamyl urea	—	Biuret	173
Isochlortetracycline	Isoaureomycin	—	—	2, 3, 116, 152, 153, 174, 175
Laurylmercaptotetrahydropyrimidine	—	—	—	2, 3, 56, 116, 152, 153, 176
β-Mercaptovaline	Penicillamine	—	Cuprimine	2, 109
N-[4-methoxy-6-methyl-1,3,5-triazin-2-yl)-amino carbonyl]-2-chlorobenzene sulfonamide	—	—	—	187
N-[(4-methoxy-6-methylamino-1,3,5-triazine-2-yl)-aminocarbonyl]benzene sulfonamide	—	—	—	185, 186
Methyl-3,6-dichloro-o-anisate	Disugran	60-CS-16	Racuza	3, 68, 80, 88, 89, 11, 117, 177

Table 1 (continued)
SUGARCANE RIPENING COMPOUNDS

Compound	Common name	Code designation	Trade name	Ref.
Methyl-2-[4-(2,4-dichlorophenoxy)phenoxy]propionate	—	—	—	178
7-Methyl indole	—	PP-757	—	83, 179
3-(2-Methylphenoxy)pyridazine	—	H-722	Credazine	2, 3, 116, 180
2-Methyl-1-propanol	Isobutanol	—	—	181
Methylsulfanil-yl-carbamate	Asulam	MB-9057 FR-600/1	Asulox	2, 3, 83, 132, 135, 147, 182—184
Methyl-2-(ureidooxy)propionate	—	DA-5	—	64, 92
2-(p-Methoxybenzyl)3,4-pyrolidine-diol-3-acetate	Anisomycin	—	Flagecidin	2, 3, 152
7-Oxabicyclo-(2,2,1)-heptane-2,3-dicarboxylic acid, monoalkylamine salt	Endothall	TD-191	Ripenthol	1, 2, 4, 56, 116, 117, 152, 153, 188
n-Pentanoic acid	n-Valeric acid	—	—	2, 3, 189
6-Phenoxyacetamide-penicillanic acid	Penicillin V	Pen V	Pen Vee	2, 3, 190
N-(2-phenoxyethyl)-N-propyl-1H-imidazole-1-carboxamide	—	BTS 34-273	—	191, 192
N-phenylphosphinylmethyliminodiacetic acid-N-oxide	—	—	—	193
N-phenylsulfonamido-N-phosphonomethyl glycine	—	—	—	194
Phosphonic acid, (2,2,2-trichloro-1-hydroxy-ethyl)-bis-2-(2-hydroxy-propoxy)-1-methylethyl ester	—	—	—	195
N-phosphonomethylglycine	Glyphosate	MON-8000	Polado	2—4, 65, 76, 93, 95, 100, 102, 103, 115, 116, 118—120, 132, 133, 135, 139, 147—149, 178, 198—202
Poly[oxyethylene (dimethyliminio) ethylene (dimethyliminio) ethylenedichloride]	—	—	Bualta	80, 83, 93, 203, 204
Sucrose ester of 2-methoxy-3,6-dichlorobenzoic acid	—	—	—	205
Tetrahydrofuroic acid hydrazide	—	—	—	2, 3, 206
1,2,4-triazine-3,5(2H,4H)-dione	6-Azauracil	—	—	2, 3
N-trichloroacetylaminomethylenephosphonic acid	—	—	—	207
2,3,6-Trichlorobenzoic acid, dimethylamine salt	2,3,6-TBA	TBA	Trysben	2—4, 56, 58, 80, 85, 89, 97, 138, 152, 153, 174, 208

3-(Trifluoromethyl-sulfonamido)-p-aceto-toluidide	Fluoridamid	MBR-6033	Sustar	1—3, 68, 80, 116, 117
	Bacitracin	—	Bacitracin	2, 3, 209
	Mineral oil	—	—	58, 210
	—	—	Tergitol NPX	2, 3, 117, 211
	—	—	Tween-20	2, 3, 117, 136, 212
	—	LFA 2129	—	128, 135
	—	ACR 1093	—	129

significant activity on sugarcane in Hawaii.[56] Numerous relatives of this compound were tested in the early screening stages, and it was found that, while the acid itself had very low activity, amine salts were more active than disubstituted amines. Ripenthol (also known as Hydrothol®) has considerable phytotoxic activity and, because of this, care must be taken in its application, especially to avoid drift when applied by air.

Chlormequat (2-chloroethyl-trimethylammonium chloride), also known as Cycocel®, is among the most widely used plant growth regulators in the world on crops other than cane. It has been evaluated on more than 1000 acres of sugarcane in Hawaii, but preliminary results suggest that its activity might be too low to be commercially successful.

Mefluidide®, also known as Embark®, is being tested at the present time under an experimental label in Hawaii, the Philippines, and certain other countries.

The ethylene producing compound, ethephon, is used commercially in sugarcane in South Africa[65,66] and Rhodesia.[67] The effectiveness of ethephon as a sugarcane ripener has not been comparable to that of glyphosine in some areas of the world, although its growth effects enable it to result in an increase in yield. These effects are being evaluated in research programs in the sugar industry.

The number, and very chemical nature, of compounds that are active as sugarcane ripeners suggest that there are several modes of action to enhance the ripening of sugarcane and that the active compounds might fall into any of these several classifications. There are also varietal differences, as well as differences due to (1) fertilizer status (particularly nitrogen); (2) age of the crop and its condition; (3) climate (both during the growth of the crop and prior to harvest); (4) physiological state of the cane; and (5) purity of the juice in the young growing tops. These variables, and probably many others, suggest that there is room for a number of ripeners on sugarcane. Additional variables to be considered are (1) phytotoxicity of the ripeners; (2) cost effectiveness of the compound under consideration; and (3) effects on the processing of sugarcane.

The effect(s) of ripeners on the growth of subsequent ratoon crops is an important factor because of potential adverse effects on tillering and regrowth of the ratoon crop. If this occurs, it also raises the question of residues in the stubble.

Studies in Florida showed both glyphosine and glyphosate to reduce the regrowth of two test cultivars in commercial-scale tests, with glyphosate causing a greater reduction than glyphosine.[214] Ratoon regrowth studies in Hawaii showed poorer growth at higher glyphosate rates (1 and 2 lb/acre), but no adverse effect at the recommended rate (0.5 lb/acre).[118] Results in Louisiana were similar to those in Hawaii.[102]

Early work with glyphosine in Louisiana produced numerous nonresponsive fields.[215] Studies to determine the agent(s) or factor(s) which might reduce the level of response showed that cane infected with ratoon stunt disease did not respond to glyphosine treatment.[216]

Recent comparative studies with four active sugarcane ripeners — glyphosine, glyphosate, ethephon, and mefluidide — have demonstrated two models for increasing sucrose per stalk in sugarcane. Glyphosine and glyphosate increase sucrose per stalk by increasing the partitioning of dry matter toward sucrose storage and away from fiber production. Both mefluidide and, to a much larger extent, ethephon increased the production of sucrose per stalk by increasing the total amount of dry matter produced, with a greater portion incorporated as fiber and a lesser amount as sucrose.[120,217]

Probably because glyphosine was the first registered sugarcane ripener, the use of this compound had a meteoric rise, at least in the sugar areas of the U.S. From its registration in 1972 and experimental use on a few hundred acres, it reached over 60,000 acres in Hawaii by 1977. A similar situation occurred in Florida, increasing from 178 acres in 1972 to more than 46,000 acres in 1979.[104] Because of the greater activity and considerably lower costs with glyphosate, it is expected that glyphosine will be replaced very quickly. Recommended dosage for glyphosine is about 4 lb active per acre whereas for glyphosate it is about 0.5 lb active per acre.

Rostron[66] has found ethephon to be much more effective than Polaris in southern Africa, whereas the reverse was found to be true in Hawaii and other places where the two have been compared. Glyphosate has been found to be quite active in the sugarcane growing regions of southern Africa and it is expected that it will be an effective competitor for ethephon.

Interest seems to be declining for three materials that received early consideration: chlormequat, disugran, and Ripenthol.® In several major sugar growing areas, mefluidide appears to be losing popularity, but it is being more seriously evaluted in tropical countries such as the Philippines.

The financial return to the grower is substantial through the use of ripeners; increased sugar yield produced by such compounds can be as much as 20% depending on the variety of sugarcane treated as well as on prevailing weather and soil conditions. Chemical control of maturation in sugarcane is now an established practice. In fact, it is so established that many research organizations in the sugar industry are shifting part of their efforts to looking at other stages in the development of the sugarcane crop for additional potential uses and times of chemical treatment. It is thought ethephon might fall into this sort of economic use not unlike that found for gibberellic acid.

VII. DESICCATION

Preharvest burning or "detrashing" is a common practice in sugarcane culture designed to lower the percentage of extraneous material shipped to the factory as well as to facilitate harvest procedures. When leaves accompany stalks during the milling operations, the recovery of sucrose is decreased. In addition to (1) extra costs of harvesting and transportation due to trash, (2) extra fiber contributing to the milling operation without sugar, and (3) impurities in the trash which contribute to low juice purity, the presence of trash leads to greater insect infestation, larger rat populations, and reduced effectiveness of preemergence herbicide treatments.

The importance of eliminating trash before milling operations is widely recognized, but is seldom accomplished to the satisfaction of either growers or factory superintendents. Burning has been the traditional means of eliminating trash. However, burns are seldom uniform, seldom completely effective, and, under humid conditions, quite ineffective.

One means of accomplishing trash removal would be chemical defoliation. Such compounds are commonly used on other crops and there the process is helped immensely by the sensitive abscission layer found at the base of the petiole. No such structure exists in sugarcane. As in most grasses, leaf fall is usually prompted by mechanical disturbance after the death or partial decomposition of the leaf sheath which clings tenaciously to the stalk. A more practical means of leaf removal in sugarcane would be by desiccation of the attached leaf and its removal by fire. Long ago chemicals were evaluated for desiccation purposes. Systematic searches were conducted during the 1940s and 1950s with a number of compounds. Several decades of testing show that the herbicides diquat and paraquat are the most effective chemicals for this purpose,[218,219] In recent years these two compounds have become almost synonymous with the term "desiccant" in sugarcane growing regions. The desirable properties of these two herbicides are (1) low mammalian toxicity; (2) solubility in water; (3) rapid absorption; and (4) inactivation upon contact with the soil. Now, after more than two decades of evaluation of these quaternary ammonium herbicides, paraquat has emerged as the preferable compound. Unfortunately, the use of paraquat has failed to produce consistent results as a desiccant. Reports vary from no effect on sucrose production and yield to slight negative effects.[86,220-223] However, as higher tonnage varieties are developed and with the persistent losses in both harvesting and milling, there has been renewed interest, particularly in areas using mechanical harvesters, in preharvest desiccation with paraquat.

Although these dipyridyl herbicides are the best chemical desiccants so far evaluated, their limited systemic action in sugarcane is a major shortcoming. It is virtually impossible to contact more than a fraction of the green canopy under field conditions by aircraft application. Another disadvantage is the rather rapid decrease in cane quality after treatment with these desiccants necessitating rather immediate harvest.

VIII. CONCLUDING REMARKS

Traditionally, the sugar industry worldwide has supported a multipronged investigation of the activity of chemicals on most of the steps of cane development, from germination through ripening and harvest. While the use of plant growth regulators is still in its infancy, success to date with ripening alone in terms of yield increases greater than 10% substantiates the belief that the regulation of crop growth and metabolism may result in one of the most important quantitative gains yet achieved in agriculture. The monumental task of producing raw materials to supply the food of the world and to supplement its energy requirements may depend to a large degree on achievements of this magnitude in a wide range of crops.

Hawaii, with its high costs of operation and high yields of sugarcane, nonseasonal environmental conditions, and the necessity to harvest the year round, can afford high-priced chemicals. In other cane producing countries this may not be the case. For example, Australian investigators were among the first to study the use of chemical ripeners. In fact, during the early 1960s there was a cooperative program between Hawaiian and Australian workers on this subject. This was not pursued when it was realized that Australian conditions, with cool and dry weather at harvest, were for the most part conducive to excellent natural ripening. In other cane growing areas in the world, however, as more is learned about the relationship of a given chemical to the process which it affects, as sugarcane agronomy improves and its economy becomes more favorable, the use of chemical ripeners will undoubtedly become more widespread; essentially the same can be said for the use of gibberellins and chemicals for flower control and other uses.

Historically, agricultural research has been primarily concerned with improvement of total crop yield by the removal of obstacles to optimize production. Now that many of these obstacles can be overcome with herbicides, pesticides, fertilizers, irrigation, and improved management practices, the stage is set for further yield increases by the use of sophisticated techniques of physiological manipulation of the plant and its metabolism.

REFERENCES

1. **Nickell, L. G.,** Plant growth regulants in sugarcane, *Bull. Plant Growth Regul.*, 2(4), 51, 1974.
2. **Nickell, L. G.,** Chemical growth regulation in sugarcane, *Outlook Agric.*, 9, 57, 1976.
3. **Nickell, L. G.,** Chemical enhancement of sucrose accumulation in sugarcane, *Am. Chem. Soc. Adv. Chem. Ser.*, 159, 6, 1977.
4. **Nickell, L. G.,** Uses of plant growth substances in the production of sugarcane: a practical case history, in *Plant Growth Substances 1979,* Skoog, F., Ed., Springer-Verlag, New York, 1980, 419.
5. **Nickell, L. G.,** Plant growth regulators in the sugarcane industry, in *Chemical Manipulation of Crop Growth and Development,* McLaren, J. S., Ed., Butterworths, London, 1982, chap. 13.
6. **Nickell, L. G.,** Sugarcane, in *Ecophysiology of Tropical Crops,* Alvim, P. de T. and Kozlowski, T. T., Eds., Academic Press, New York, 1977, 89.
7. **Nickell, L. G. and Kortschak, H. P.,** Arginine: its role in sugarcane growth, *Haw. Plant. Rec.*, 57, 230, 1964.
8. **Maretzki, A. and Nickell, L. G.,** Dependence on arginine of sugarcane cells grown in submerged culture, *Proc. Int. Congr. Biochem.*, 7 (Abstr. Vol. J-207), 1968.
9. **Nickell, L. G. and Maretzki, A.,** Growth of suspension cultures of sugarcane cells in chemically defined media, *Physiol. Plant.*, 22, 117, 1969.

10. **Maretzki, A., Nickell, L. G., and Thom, M.,** Arginine in growing cells of sugarcane. Nutritional effects, uptake, and incorporation into proteins, *Physiol. Plant.*, 22, 827, 1969.
11. **Nickell, L. G. and Maretzki, A.,** Developments and biochemical studies with cultured sugarcane cell suspensions, *Proc. Int. Ferment. Symp.*, 4, 681, 1972.
12. **Glenn, E. P.,** Growth effects and metabolism of arginine in stationary and rapidly dividing sugarcane cell suspensions, *Physiol. Plant*, 52, 59, 1981.
13. **Glenn, E. P. and Maretzki, A.,** Properties and subcellular distribution of two partially purified ornithine transcarbamyolases in cell suspensions of sugarcane, *Plant Physiol.*, 60, 122, 1977.
14. **Komor, E., Thom, M., and Maretzki, A.,** Mechanism of arginine uptake by sugarcane cells, *Annu. Rep. Exp. Stn. Haw. Sugar Plant. Assoc.*, 36, 1980.
15. **Nickell, L. G.,** Agricultural aspects of transplanting and spacing, *Rep. Haw. Sugar Technol. 1967*, 147, 1968.
16. **Takahashi, D. T.,** Chemical effects on tillering, *Annu. Rep. Exp. Stn. Haw. Sugar Plant. Assoc.*, 50, 1969.
17. **Teshima, A., Osgood, R. V., and Olende, C.,** Tillering response to ethephon, *Annu. Rep. Exp. Stn. Haw. Sugar Plant. Assoc.*, 32, 1980.
18. **Eastwood, D.,** Tillering and early growth of sugarcane sets in response to preplant treatment with (2-chloroethyl)phosphonic acid, *Trop. Agric.*, 56, 11, 1979.
19. **Lucchesi, A. A., Florencio, A. C., Godoy, O. P., and Stupiello, J. P.,** Influence of 2-chloroethyl-phosphonic acid on the tillering of sugarcane variety N.A. 56-79 (Portuguese), *Bras. Acuc.*, 93, 19, 1979.
20. **Lucchesi, A. A., Florencio, A. C., Godoy, O. P., and Stupiello, J. P.,** Influence of 2-chloroethyl-phosphonic acid in the induction of tillering in sugarcane (*Saccharum* spp.) variety N.A. 56-79 (Portuguese), *Bras. Acuc.*, 94, 209, 1979.
21. **Lucchesi, A. A., Godoy, O. P., Florencio, A. C., and de Araujo, V. B.,** Effect of 2-chloroethylphos-phonic acid and its mixture with urea on the induction of tillering in ratooning of sugarcane (*Saccharum* spp.) variety CB 41-14, *Bras. Acuc.*, 95, 232, 1980.
22. **Madrid, P. V. and Rosario, E. L.,** Comparative effect of some chemicals on the tillering ability of sugarcane variety Phil. 56-226, *Phil. J. Crop Sci.*, 2, 168, 1977.
23. **Madrid, P. V. and Rosario, E. L.,** Effects of chemicals on the tillering ability of variety Phil. 56-226, *Sugarland*, 16(2), 11, 1979.
24. **Pao, T. P.,** Response of sugarcane to plant growth regulators, *Annu. Rep. Taiwan Sugar Exp. Stn.*, 5, 1973.
25. **Yang, P. C., Ho, F. W., and Wei, C. C.,** Application of plant growth regulators for promoting sprouting and growth of ratoon cane, *Annu. Rep. Taiwan Sugar Res. Inst.*, 9, 1978/1979.
26. **Yang, P. C., Ho, F. W., and Wei, C. C.,** Application of plant growth regulators for promoting sprouting, growth and yield of ratoon canes, *Annu. Rep. Taiwan Sugar Res. Inst.*, 8, 1979/1980.
27. **Yang, P. C., Ho, F. W., and Wei, C. C.,** Application of plant growth regulators for promoting sprouting and growth of ratoon cane, *Taiwan Sugar*, 27(4), 131, 1980.
28. **Wong-Chong, J. and Martin, F. A.,** Greenhouse studies on the interaction of genotype and plant growth regulators with regard to early tillering in sugar cane, *Sugar Azucar*, 76(6), 28, 1981.
29. **Wong-Chong, J. and Martin, F. A.,** Greenhouse studies on the interaction of genotype and plant growth regulators, *Sugar J.*, 44(3), 13, 1981.
30. **Coleman, R. E.,** The effect of gibberellic acid on growth of sugarcane, *Sugar J.*, 20, 23, 1958.
31. **Coleman, R. E., Todd, E. H., Stokes, I. E., and Coleman, O. H.,** Some responses of sugarcane to gibberellic acid, *Sugar J.*, 23, 11, 1960.
32. **Tanimoto, T. and Nicell, L. G.,** Re-evaluation of gibberellin for field use in Hawaii, *Rep. Haw. Sugar Technol. 1966*, 184, 1967.
33. **Tanimoto, T. and Nickell, L. G.,** Effects of gibberellin on sugarcane growth and sucrose production, *Rep. Haw. Sugar Technol. 1967*, 137, 1968.
34. **Buren, L. L.,** Projections for gibberellic acid, *Rep. Haw. Sugar Technol. 1971*, 104, 1972.
35. **Moore, P. H.,** Use of gibberellic acid to increase sugarcane yields in Hawaii, *Proc. Plant Growth Regul. Working Group*, 4, 173, 1977.
36. **Moore, P. H.,** Effect of gibberellic acid on sugarcane yields, *Rep. 36th Annu. Conf. Haw. Sugar Technol.*, 67, 1977.
37. **Moore, P. H.,** Sugarcane growth response to serial applications of gibberellic acid, *Proc. Plant Growth Regul. Working Group*, 5, 158, 1978.
38. **Moore, P. H.,** Additive and non-additive effects of serial applications of gibberellic acid on sugarcane internode growth, *Physiol. Plant.*, 9, 271, 1980.
39. **Moore, P. H. and Buren, L. L.,** Gibberellin studies with sugarcane. I. Cultivar differences in growth responses to gibberellic acid, *Crop Sci.*, 17, 443, 1978.
40. **Moore, P. H. and Ginoza, H.,** Gibberellic studies with sugarcane. III. Effects of rate and frequency of gibberellic acid applications on stalk length and fresh weight, *Crop Sci.*, 20, 78, 1980.

41. **Moore, P. H., Osgood, R. V., and Ginoza, H. S.,** Small paired plots to evaluate the effect of GA_3 on tons cane per acre and tons sugar per acre, *Proc. Plant Growth Regul. Working Group,* 7, 146, 1980.
42. **Moore, P. H.,** Use of gibberellic acid to increase sugarcane yields in Hawaii, *Proc. Int. Soc. Sugar Cane Technol.,* 17(1), 556, 1980.
43. **Jaramillo, H., Schuitemaker, F., and Garcia, C.,** Ripening sugarcane with Ethrel plant growth regulator in Colombia, *Proc. Int. Soc. Sugar Cane Technol.,* 16(2), 1931, 1977.
44. **Osgood, R. V. and Teshima, A.,** Response of unirrigated 59-3775 to summer-applied ethephon (Ethrel), *Annu. Rep. Exp. Stn. Haw. Sugar Plant. Assoc.,* 40, 1979.
45. **Osgood, R. V. and Teshima, A.,** Growth stimulation studies with ethephon continue, *Annu. Rep. Exp. Stn. Haw. Sugar Plant. Assoc.,* 33, 1980.
46. **Clowes, M. S. J.,** Growth stimulation from Ethrel and the effects of gibberellic acid when applied to sugarcane, *Proc. S. Afr. Sugar Technol.,* 54, 146, 1980.
47. **van Andel, O. M. and Verkerke, D. R.,** Stimulation and inhibition by ethephon of stem and leaf growth of some gramineae at different stages of development, *J. Exp. Bot.,* 29, 639, 1978.
48. **Nickell, L. G.,** Plant growth regulators. Controlling biological behavior with chemicals, *Chem. Eng. News,* 56(41), 18, 1978.
49. **Tanimoto, T. and Nickell, L. G.,** Field control of sugarcane flowering in Hawaii with diquat, *Proc. Int. Soc. Sugar Cane Technol.,* 12, 113, 1965.
50. **Benedicto, F.,** Diquat controls tasselling in Victorias, *Victorias Milling Co. Exp. Stn. Bull.,* 3, March/April 1967.
51. **Yang, P. C., Pao, T. P., and Ho, F. W.,** Studies on the chemical control of sugarcane flowering in Taiwan, *Taiwan Sugar,* 19(1), 21, 1972.
52. **Smith, B.,** Tassel control in drip irrigated fields at Wailuku Sugar Company, *Rep. 36th Annu. Conf. Haw. Sugar Technol.,* 81, 1977.
53. **Beauchamp, C. E.,** A new method of increasing the sugar content of sugarcane, *Proc. 23rd Annu. Meet. Assoc. Tecn. Azucareros Cuba,* 55, 1949.
54. **Hartt, C. E.,** Translocation of sugar in the cane plant, *Rep. Haw. Sugar Technol. 1963,* 151, 1964.
55. **Hartt, C. E., Kortschak, H. P., and Burr, G. O.,** Effects of defoliation, eradication, and darkening the blade upon translocation of C^{14} in sugarcane, *Plant Physiol.,* 39, 15, 1964.
56. **Nickell, L. G. and Tanimoto, T. T.,** Effects of chemicals on ripening of sugarcane, *Rep. Haw. Sugar Technol. 1965,* 152, 1966.
57. **Glasziou, K. T.,** Chemical control of growth and ripening, *Rep. David North Res. Cent.,* 48, 1964.
58. **Vlitos, A. J. and Lawrie, I. D.,** Chemical ripening of sugarcane. A review of field studies carried out in Trinidad over a five year period, *Proc. Int. Soc. Sugar Cane Technol.,* 12, 429, 1965.
59. **Humbert, R. P.,** Sugarcane production: growing interest in chemical ripeners, *World Farming,* 16(12), 25, 1974.
60. **Julien, M. H. R.,** Studies of ripeners on sugarcane. I. Effects of MON-045 on growth and sucrose content, *Exp. Agric.,* 10, 113, 1974.
61. **Julien, M. H. R.,** Studies of ripeners on sugarcane. II. The distribution of dry matter and sucrose in the sugarcane stalk following treatment with ripener MON-045, *Exp. Agric.,* 10, 123, 1974.
62. **Julien, M. H. R. and Goolambossen, M.,** Results of industrial trials with ripener "Polaris", *Rev. Agric. Sucr. Maurice,* 55, 389, 1976.
63. **Julien, M. H. R., Soopramanien, G. C., Martine, J. F., and Medan, H.,** Growth, sucrose content and regrowth in sugarcane treated with ripener "Polaris", *Rev. Agric. Sucr. Maurice,* 57, 172, 1978.
64. **Lalouette, J. A., Mazery, G., and Ng Ying, R.,** Notes on experiments conducted with ripener DA5, *Annu. Rep. Mauritius Sugar Ind. Res. Inst.,* 17, 107, 1970.
65. **Clowes, M. S. J. and Wood, R. A.,** Post-harvest deterioration of whole stalk sugarcane treated with chemical ripeners, *Proc. S. Afr. Sugar Technol. Assoc.,* 52, 166, 1978.
66. **Rostron, H.,** A review of chemical ripening of sugarcane with Ethrel in southern Africa, *Proc. Int. Soc. Sugar Cane Technol.,* 16(2), 1605, 1977.
67. **Sweet, C. P. M.,** Ethrel as an early season cane ripener in Rhodesia, *Proc. Int. Soc. Sugar Cane Technol.,* 16(2), 1619, 1977.
68. **Chacravarti, A. S., Sarkar, A. K., and Thakur, A. K.,** Effects of some chemical ripeners on maturation of cane and maintenance of quality in late season, *Proc. Jt. Conv. Ind. Sugar Technol. Assoc.,* 5, Ag67, 1975.
69. **Chacravarti, A. S., Thakur, A. K., and Sarkar, A. K.,** Biochemical responses of sugarcane to some chemical ripener treatments, *Proc. Jt. Conv. Ind. Sugar Technol. Assoc.,* 6, Ag167, 1977.
70. **Kumar, A. and Narasimhan,** Ripening in sugarcane with "Polaris", "Cetrimide", and "Hyamine 1622", *Ind. Sugar,* 26, 817, 1977.
71. **Rao, K. C. and Asokan, S.,** Chemical ripener studies — big mill tests (BMT) at sugar factories in Tamil Nadu, *Proc. Jt. Conv. Ind. Sugar Technol. Assoc.,* 6, Ag133, 1977.

72. **Sharma, R. A., Sharma, R. K., and Sharma, S. R.,** Effect of different ripening agents on quality of sugarcane juice, *Ind. Sugar Crops J.*, 4, 35, 1977.
73. **Srivastava, S. C., Singh, B., and Singh, K.,** Effect of Cycocel on ripening of sugarcane, *Proc. Jt. Conv. Ind. Sugar Technol. Assoc.*, 4, A1, 1971.
74. **Alves, A. S., Azzi, G. M., and Kumar, A.,** Effect of "Polaris" application on the juice quality of sugarcane in the decline phase of the maturity curve, *Proc. Int. Soc. Sugar Cane Technol.*, 16(2), 1713, 1977.
75. **Azzi, G. M., Alves, A. S., and Kumar, A.,** Chemical ripener studies with Polaris in sugarcane in Northeast Brazil, *Proc. Int. Sugar Cane Technol.*, 16(2), 1653, 1977.
76. **Gonzales, M. Y. and Tianco, A. P.,** Chemical ripening of sugarcane with glyphosate ripeners, *Victorias Agric. Res. Rep.*, 18, 16, 1978.
77. **Porquez, P. H., Panol, F. Y., and Gibe, J. N.,** Cane ripening effect of Polaris, *Proc. Conv. Philsutech*, 21, 73, 1973.
78. **Rosario, S. B. and Javier, R. Q.,** Performance of Embark as cane ripener in the Philippines, *Proc. Conv. Philsutech*, 25, 152, 1977.
79. **Tianco, A. P. and Escober, T. R.,** Chemical ripening of sugarcane with CP-41845, *Proc. Conv. Philsutech*, 18, 93, 1970.
80. **Zamora, O. B. and Rosario, E. L.,** Physiological and morphological responses of sugarcane to some selected chemical ripeners, *Phil. J. Crop Sci.*, 2, 133, 1977.
81. **Yang, P. C. and Ho, F. W.,** Chemical ripeners, *Annu. Rep. Taiwan Sugar Res. Inst.*, 8, 1977.
82. **Yang, P. C. and Ho, F. W.,** Effects of aerial spraying with "Polaris" on sucrose enhancement, growth and yield of sugarcane, *Proc. Int. Soc. Sugar Cane Technol.*, 16(2), 1701, 1977.
83. **Yang, P. C. and Ho, F. W.,** Effect of chemical ripeners on cane quality and cane yield in subsequent ratoon, *Taiwan Sugar*, 25(3), 101, 1978.
84. **Yang, P. C. and Ho, F. W.,** Effect of chemical ripeners on varieties of F160 and F177, *Taiwan Sugar*, 26(1), 30, 1979.
85. **Yang, P. C., Hsu, C. J., and Ho, F. W.,** Artificial ripening of sugarcane by foliar application of Cycocel, dalapon, and TBA, *Annu. Rep. Taiwan Sugar Exp. Stn.*, 48, 1969.
86. **Evans, H. and Bates, J. F.,** A summary of investigations on the possibility of artificially ripening sugarcane with various chemicals, *Proc. Int. Soc. Sugar Cane Technol.*, 11, 298, 1963.
87. **Yates, R. A. and Bates, J. F.,** Preliminary experiments on the effects of chemicals on the ripening of sugarcane, *Proc. Br. West Indies Sugar Technol. 1957*, 174, 1957.
88. **Alexander, A. G. and Montalvo-Zapata, R.,** Evaluation of chemical ripeners for sugarcane having constant nitrogen and water regimes. I. Growth, quality and enzymic responses of nine potential ripeners, *Trop. Agric.*, 50, 35, 1973.
89. **Alexander, A. G. and Montalvo-Zapata, R.,** Evaluation of chemical ripeners for sugarcane having constant nitrogen and water regimes. II. Superior activity of C.P. 41845 (Monsanto), *Trop. Agric.*, 50, 307, 1973.
50. **Delfel, N. E., Ortiz-Torres, E., Colberg, C., and Samuels, G.,** Evaluation of phosfon and maleic hydrazide as late season yield stimulants on sugarcane, *Trop. Agric.*, 43, 199, 1966.
91. **Samuels, G., Velez, A., Yates, R. A., and Walker, B.,** Evaluation of chemical ripeners for sugarcane, *J. Agric. Univ. Puerto Rico*, 56, 370, 1972.
92. **Bieske, G. C.,** Chemical ripening of sugarcane, *Proc. Queensland Soc. Sugar Cane Technol.*, 37, 117, 1970.
93. **Chapman, L. S. and Kingston, G.,** Cane ripeners, *Proc. Queensland Soc. Sugar Cane Technol.*, 44, 143, 1977.
94. **Hurney, A. P. and Schmalzl, K.,** Chemical ripening with "Polaris" under commercial conditions in north Queensland, *Proc. Queensland Soc. Sugar Cane Technol.*, 45, 139, 1978.
95. **Kingston, G., Chapman, L. S., and Hurney, A. P.,** Chemical ripening of cane — BSES experiments during 1977, *Proc. Queensland Soc. Sugar Cane Technol.*, 45, 37, 1978.
96. **Eastwood, D.,** Assessment of "Polaris", a sugarcane-ripening chemical, on field scale, *J. Assoc. Sugar Technol. Jamaica*, 35, 34, 1974.
97. **Yates, R. A.,** Field experiments on the chemical ripening of sugarcane in Jamaica and Belize in 1970, *Trop. Agric.*, 49, 235, 1972.
98. **Vlitos, A. J. and Fewkes, D. W.,** Sugar research: biological and agricultural, *Sugar Azucar*, 64(9), 27, 1969.
99. **Andreis, H. J. and DeStefano, R. P.,** Chemical ripening of sugarcane suckers and of variety CL 41-91, *Sugar J.*, 41(11), 21, 1979.
100. **Andreis, H. J. and DeStefano, R. P.,** Chemical ripening of sugarcane suckers, *Sugar J.*, 43(1), 26, 1980.
101. **Holder, D. G. and DeStefano, R. P.,** Response of Clewiston (CL) varieties of Polaris in Florida during the 1978 to 1979 season, *Sugar J.*, 41(9), 21, 1979.
102. **Legendre, B. L., Martin, F. A., and Dill, G. M.,** Preliminary investigations on the effects of Polado on regrowth of sugarcane in Louisiana, *Proc. Plant Growth Regul. Working Group*, 7, 148, 1980.

103. **Martin, F. A., Legender, B. L., Dill, G. M., and Steib, R. J.**, Chemical "ripening" of Louisiana sugarcane, *La. Agric.*, 24(1), 4, 1980.
104. **Rice, E. R., Holder, D. G., and DeStefano, R. P.**, Response of several Florida sugarcane varieties tested during the 1979—1980 harvest, *Sugar J.*, 43(5), 23, 1980.
105. **Hamm, P. C.**, Method of increasing the sugar content of sugarcane, U.S. Patent 3,525,603, August 25, 1970.
106. **de Silva, W. H., Bocion, P. F., Eggenberg, E., and Mur, A.**, The plant growth regulating and herbicidal activity of 2-amino-6-methylbenzoic acid, *Z. Pflanzenkrankh. Pflanzenschutz*, 86, 546, 1979.
107. **Thomas, G. J.**, Plant growth regulating agents, U.S. Patent 4,094,664, June 13, 1978.
108. **Otten, G. G.**, Method of increasing sucrose yield of sugarcane, U.S. Patent 4,120,688, October 17, 1978.
109. **Nickell, L. G.**, Aminopenicillanic acid or penicillamine as ripener for sugarcane, U.S. Patent 3,992,187, November 16, 1976.
110. **Quebedeaux, B.**, Method of increasing sugar content of crops, U.S. Patent 3,619,166, November 9, 1971.
111. **Nickell, L. G.**, Ripening of sugarcane by use of ammonium isobutyrate, U.S. Patent 4,033,755, July 5, 1977.
112. **Leach, R. W. A.**, Alanine derivatives as sugarcane ripeners, U.S. Patent 4,056,385, November 1, 1977.
113. **Chacravarti, A. S., Srivastava, D. P., and Khanna, K. L.**, Foliar application of 2,4-D to increase sugar in cane, *Sugar J.*, 18(6), 23, 1955.
114. **Julien, M. H. R. and McIntyre, G.**, Notes on experiments conducted with ripener CP-41845, *Annu. Rep. Mauritius Sugar Ind. Res. Inst.*, 18, 130, 1970.
115. **Nickell, L. G. and Takahashi, D. T.**, Ripening studies in Hawaii with CP-41845, *Rep. Haw. Sugar Technol. 1971*, 73, 1972.
116. **Nickell, L. G. and Takahashi, D. T.**, A review of chemical ripening studies with sugarcane in Hawaii, *Rep. Haw. Sugar Technol. 1972*, 47, 1973.
117. **Nickell, L. G. and Takahashi, D. T.**, Sugarcane ripeners in Hawaii — 1973, *Rep. Haw. Sugar Technol. 1973*, 76, 1974.
118. **Osgood, R. V.**, The effect of glyphosate salts on sugarcane quality and ratoon regrowth, *Rep. 36th Annu. Conf. Haw. Sugar Technol.*, 60, 1977.
119. **Osgood, R. V. and Teshima, A.**, Sucrose accumulation in sugarcane treated with glyphosate and glyphosine, *Proc. Plant Growth Regul. Working Group*, 6, 29, 1979.
120. **Osgood, R. V. and Teshima, A.**, The effect of several growth regulators on dry matter production and partitioning in sugarcane cv. H59-3775, *Proc. Plant Growth Regul. Working Group*, 7, 150, 1980.
121. **Pan. Y. C. and Lee, Y. P.**, Field trials with "Polaris" as a sugarcane ripener. II. The effect on cane yield and sugar yield, *Proc. Int. Soc. Sugar Cane Technol.*, 16(2), 1693, 1977.
122. **Teshima, A. and Osgood, R. V.**, Comparison of chemical ripeners in unirrigated sugarcane, *Rep. 36th Annu. Conf. Haw. Sugar Technol.*, 63, 1977.
123. **Zschoche, W. C.**, "Polaris" sugarcane ripener field performance in Hawaii, *Sugar Azucar*, 72(4), 21, 1977.
124. **Legendre, B. L. and Benda, G. T. A.**, Single plants for testing glyphosine effects in sugarcane, *Proc. Plant Growth Regul. Soc. Am.*, 8, 169, 1981.
125. **Oudman, L.**, Polaris response in 1975, *Rep. 34th Annu. Conf. Haw. Sugar Technol.*, 87, 1975.
126. **Mason, G. F.**, Chemical ripening in Trinidad of variety B-41227 with glyphosine, *Proc. Meet. West Indies Sugar Technol.*, 130, 1976.
127. **Jackson, N. E.**, Commercial application of Polaris chemical ripener in Guyana, *Proc. Meet. West Indies Sugar Technol.*, 139, 1976.
128. **Eastwood, D.**, Chemical ripening in Jamaica — progress and prospects, *Proc. Meet. West Indies Sugar Technol.*, 143, 1976.
129. **Anon.**, Chemical ripeners, *Annu. Rep. Agric. Exp. Stn. Fiji Sugar Corp.*, 20, 1979.
130. **Hunsigi, G., Dwarakinath, N., and Channaiah, C.**, Effect of some chemicals on ripening of sugarcane, *29th Annu. Conv. Deccan Sugar Technol. Assoc.*, 1, 1979.
131. **Patil, S. S., Bendigeri, A. V., Hapase, D. G., and Jadhav, A. P.**, Evaluation of the effect of cane ripeners on sugarcane maturity, *Maharashtra Sugar*, 5(2), 41, 1979.
132. **Mason, G. F.**, Chemical ripening of variety B-41227 in Trinidad, *Proc. Int. Soc. Sugar Cane Technol.*, 17(1), 663, 1980.
133. **Tianco, A. P. and Gonzales, M. Y.**, Effect of glyphosate ripener on growth response and sugar yield of sugarcane, *Proc. Int. Soc. Sugar Cane Technol.*, 17(1), 694, 1980.
134. **Dill, G. and Martin, F. A.**, The effect of N,N-bis(phosphonomethyl)glycine on some physiological components of sugar cane yield, *Proc. Am. Soc. Sugar Cane Technol.*, 37, 1977.
135. **McCatty, T.**, A review of sucrose enhancer trials in Jamaica in 1974 to 1978, *Proc. Int. Soc. Sugar Cane Technol.*, 17(1), 630, 1980.
136. **Morales, M. C. and Angulo, E. A.**, Effects of ripeners on sugar quality in cultivar H50-2036, *Proc. Int. Soc. Sugar Cane Technol.*, 17(1), 618, 1980.

137. **Julien, M. H. R., Soopramanien, G. C., and d'Espagna, M. A. L.,** Environment, flowering, rainfall and dosage rate as factors affecting response to ripener Polaris, *Proc. Int. Soc. Sugar Cane Technol.*, 17(1), 604, 1980.
138. **Yates, R. A.,** A review of recent work on chemical ripening of sugarcane, *Proc. Int. Soc. Sugar Cane Technol.*, 14, 1121, 1971.
139. **Hilton, H. W., Osgood, R. V., and Maretzki, A.,** Some aspects of MON-8000 as a sugarcane ripener to replace Polaris, *Proc. Int. Soc. Sugar Cane Technol.*, 17(1), 652, 1980.
140. **Selleck, G. W., Frost, K. R., Billman, R. C., and Brown, D. A.,** Sucrose enhancement in field scale sugarcane trials with Polaris in Florida, Hawaii, and Louisiana, *Proc. Int. Soc. Sugar Cane Technol.*, 15, 938, 1974.
141. **Rueppel, M. L.,** Treatment of sugarcane with N-(perfluoroacyl)-N-phosphonomethylglycine, U.S. Patent 4,047,926, September 13, 1977.
142. **Gourse, J. A.,** Method of increasing the yields of sugarcane with phosphonic acids, U.S. Patent 4,299,615, November 10, 1981.
143. **Evans, A. W.,** Method for increasing sugarcane yield, U.S. Patent 3,291,592, December 13, 1966.
144. **Nickell, L. G.,** Ripening of sugarcane by use of certain mono-substituted benzoic acids, U.S. Patent 3,994,712, November 30, 1976.
145. **Karabinos, J. V. and Nickell, L. G.,** Derivatives of thiophene, U.S. Patent 4,250,319, February 10, 1981.
146. **Porter, C. A.,** Method for increasing the sucrose content of growing plants, U.S. Patent 3,909,233, September 30, 1975.
147. **Takahashi, D.,** Potential sugarcane ripeners, *Rep. 35th Annu. Conf. Haw. Sugar Technol.*, 266, 1976.
148. **Teshima, A.,** Review of work with MON-8000, Embark, and Ethrel, *Rep. 38th Annu. Conf. Haw. Sugar Technol.*, 72, 1979.
149. **Dill, G. M. and Martin, F. A.,** Endogenous sucrose levels in immature internodal tissues of sugar cane as affected by plant growth regulators, *Sugar Azucar*, 76(6), 35, 1981.
150. **Yang, P. C. and Ho, F. W.,** Effects of Embark and Ethrel on sugarcane quality, yield and ratoon regrowth, *Proc. Int. Soc. Sugar Cane Technol.*, 17(1), 711, 1980.
151. **Dill, G. M. and Martin, F. A.,** The effect of ethephon on endogenous sucrose and reducing sugar levels of immature internodal tissues of sugarcane, *Proc. Plant Growth Regul. Soc. Am.*, 8, 1, 1981.
152. **Nickell, L. G. and Maretzki, A.,** Sugarcane ripening compounds — comparison of chemical, biochemical, and biological properties, *Haw. Plant. Rec.*, 58, 71, 1970.
153. **Nickell, L. G. and Tanimoto, T. T.,** Sugarcane ripening with chemicals, *Rep. Haw. Sugar Technol. 1967*, 104, 1968.
154. **Nickell, L. G. and Tanimoto, T. T.,** Ripening of sugarcane by use of quaternary amines such as chlorocholine chloride, U.S. Patent 3,493,361, February 3, 1970.
155. **Vega, N.,** Effect of Cycocel on the "ripening" of sugarcane (Spanish), *Bol. Est. Exp. Occidente (Venezuela)*, 93, 3, 1971.
156. **Carlson, A. E.,** Method for increasing sugar content of sugarcane, U.S. Patent 3,224,865, December 21, 1965.
157. **Somasundaram, A., Gopalakrishnachar, S., Srinivasamurthy, A. S., and Krishnamurthy, D. V. G.,** Studies on the effect of 2,4-D sodium salt preharvest spray on the quality of sugarcane, *Ind. J. Sugarcane Res. Dev.*, 6(3), 119, 1962.
158. **Yates, R. A.,** Artificial ripening of sugarcane with dalapon, *Trop. Agric.*, 41, 225, 1964.
159. **Gaertner, V. R. and Hamm, P. C.,** N-(2-hydroxyalkyl) derivatives of N-phosphonomethylglycine for treatment of sugarcane, U.S. Patent 4,063,922, December 20, 1977.
160. **Nickell, L. G.,** Ripening of sugarcane by use of certain quaternary ammonium halides, U.S. Patent 3,671,219, June 20, 1972.
161. **Bosshard, R. and Muller, J. C.,** Method of limiting growth of lawn grass and increasing sugar content in cane, U.S. Patent 3,898,071, August 5, 1975.
162. **Nickell, L. G.,** Chemical ripening of sugarcane using alkylarsinic acid compounds, U.S. Patent 3,992,190, November 16, 1976.
163. **Jaworski, E. G.,** Method for increasing the sucrose content of growing plants, U.S. Patent 3,981,718, September 21, 1976.
164. **Bushong, J. W., Gates, D. W., and Sullivan, T. P.,** Mefluidide (MBR 12325) — a new concept in weed control with a plant growth regulator, *Proc. Br. Crop Prot. Conf. (Weeds)*, 695, 1976.
165. **Siemer, S. R.,** Phosphorus compounds as sugarcane ripeners, U.S. Patent 4,229,203, October 21, 1980.
166. **Gaertner, V. R.,** Method for increasing the sucrose content of growing plants, U.S. Patent 4,203,756, May 20, 1980.
167. **Savory, B. and Desmoras, J.,** Method and composition to increase the sugar content of sugar cane, U.S. Patent 4,299,617, November 10, 1981.

168. **Nickell, L. G.**, Ripening of sugarcane by use of certain quaternary ammonium halides, U.S. Patent 3,660,072, May 2, 1972.
169. **Richter, S. B.**, Method of increasing the yields of sugar obtained from sugarcane, U.S. Patent 4,285,721, August 25, 1981.
170. **Koerwer, J. F.**, Pyridyloxy-phenoxyalkane carboxylic acids and derivatives as sugar enhancers for plants, U.S. Patent 4,280,832, July 28, 1981.
171. **Porter, C. A.**, Method of increasing the sucrose content of growing plants, U.S. Patent 3,826,641, July 30, 1974.
172. **Nickell, L. G.**, Vanillin as ripener for sugarcane, U.S. Patent 3,994,715, November 30, 1976.
173. **Weakley, M. L.**, Sugar production, U.S. Patent 3,860,411, January 14, 1975.
174. **Nickell, L. G. and Takahashi, D. T.**, The effects of antibiotics and other antimicrobial agents on the ripening of sugarcane, *Haw. Plant. Rec.*, 59, 15, 1975.
175. **Nickell, L. G. and Tanimoto, T. T.**, Use of isochlortetracycline in ripening sugarcane and compositions useful therein, U.S. Patent 3,505,056, April 7, 1970.
176. **Nickell, L. G. and Tanimoto, T. T.**, Method for increasing the sugar content of sugarcane, U.S. Patent 3,482,958, December 9, 1969.
177. **Vega, N.**, Effect of methyl 3,6-dichloro-o-anisoate on "ripening" (Spanish), *Bol. Est. Exp. Occidente (Venezuela)*, 93, 33, 1971.
178. **Koerwer, J. F.**, Phenoxy-phenoxyalkane carboxylic acids and derivatives as sugar enhancers for plants, U.S. Patent 4,276,080, June 30, 1981.
179. **George, E. F. and Phillips, M. R.**, The effect of 7-methylindole as a sugarcane ripener, *Br. Crop Prot. Counc. Monogr.*, 21, 211, 1978.
180. **Nickell, L. G.**, Ripening of sugarcane by use of alkyl derivatives of 3-phenoxypyridazine, U.S. Patent 3,704,111, November 28, 1972.
181. **Nickell, L. G.**, Ripening of sugarcane by use of alcohols, U.S. Patent 4,099,957, July 11, 1978.
182. **Chougule, J. D. and Patil, B. R.**, Effect of chemical ripener Asulox-40 on sugar cane quality, *Maharashtra Sugar*, 4(8), 41, 1979.
183. **Hardisty, J. A.**, The use of asulam as a sucrose enhancer in sugarcane, *Sugar News*, 56(4), 137, 1980.
184. **Hardisty, J. A.**, The use of asulam as a sucrose enhancer in sugarcane, *Proc. Int. Soc. Sugar Cane Technol.*, 17(1), 644, 1980.
185. **Levitt, G.**, Herbicidal sulfonamides, U.S. Patent 4,190,432, February 26, 1980.
186. **Levitt, G.**, Herbicidal sulfonamides, U.S. Patent 4,231,784, November 4, 1980.
187. **Levitt, G.**, Herbicidal sulfonamides, U.S. Patent 4,127,405, November 28, 1978.
188. **Nickell, L. G. and Tanimoto, T. T.**, Method of increasing sugar yield of sugarcane by treatment with endothal compounds, and compositions useful therein, U.S. Patent 3,482,959, December 9, 1969.
189. **Nickell, L. G.**, Ripening of sugarcane by use of N-valeric acid, U.S. Patent 3,870,503, March 11, 1975.
190. **Nickell, L. G.**, Use of penicillin as ripener for sugarcane, U.S. Patent 3,897,239, July 29, 1975.
191. **Copping, L. G. and Garrod, J. F.**, Method of regulating plant growth, U.S. Patent 4,139,365, February 13, 1979.
192. **Garrod, J. F. and Wells, W. H.**, The plant growth regulatory properties of a group of N-carbamoylimidazoles, *Br. Crop Prot. Counc. Monogr.*, 21, 217, 1978.
193. **Franz, J. E. and Sacher, R. M.**, Phosphinylmethyliminoacetic acid-N-oxide compounds and the sucrose increasing use thereof, U.S. Patent 4,110,100, August 29, 1978.
194. **Franz, J. E.**, Increasing sucrose content of sugarcane employing N-phenylsulfonamido-N-phosphonomethylglycine and certain derivatives thereof, U.S. Patent 3,996,040, December 7, 1976.
195. **Kupelian, R. H.**, Process for increasing sugar yield in sugarcane, U.S. Patent 3,874,872, April 1, 1975.
196. **Osgood, R. V.**, Glyphosate salts for enhancement of sugarcane ripening, *Proc. Plant Growth Regul. Working Group*, 5, 152, 1978.
197. **Osgood, R. V.**, Results of an experimental use permit program in Hawaii utilizing glyphosate (Polado) as a sugarcane ripener, *Proc. Plant Growth Regul. Working Group*, 7, 149, 1980.
198. **Nomura, N.**, Progress report on MON-8000 block tests, *Rep. 38th Annu. Conf. Haw. Sugar Technol.*, 76, 1979.
199. **Yang, P. C. and Ho, F. W.**, Chemical ripeners, *Annu. Rep. Taiwan Sugar Res. Inst.*, 10, 1978/1979.
200. **Mills, A. N.**, Results from glyphosate used as a ripener at Felixton, *Proc. S. Afr. Sugar Technol. Assoc.*, 54, 134, 1980.
201. **Clowes, M. S. J.**, Ripening activity of the glyphosate salts MON-8000 and Round-Up, *Proc. Int. Soc. Sugar Cane Technol.*, 17(1), 676, 1980.
202. **Osgood, R. V., Moore, P. H., and Ginoza, H.**, Effect of Polado on growth and dry weight partitioning of individually treated sugarcane stalks, *Annu. Rep. Exp. Stn. Haw. Sugar Plant. Assoc.*, 29, 1980.
203. **Buckman, S. J. and Pulido, M. L.**, Method for increasing yield of sugarcane, U.S. Patent 3,854,928, December 17, 1974.

204. **Pulido, M. L.,** A new sugarcane ripener, *Sugar Azucar,* 69(6), 105, 1974.
205. **Luteri, G. F.,** Sucrose ester of 2-methoxy-3,6-dichlorobenzoic acid, U.S. Patent 4,291,158, September 22, 1981.
206. **Nickell, L. G.,** Tetrahydrofuroic hydrazide for ripening sugarcane, U.S. Patent 3,992,186, November 16, 1976.
207. **Ratts, K. W.,** Method for increasing the sucrose content of growing plants, U.S. Patent 3,961,934, June 8, 1976.
208. **Pfeiffer, R. K.,** Method of increasing the sugar/sugarcane weight ratio, U.S. Patent 3,245,775, April 12, 1966.
209. **Nickell, L. G.,** Use of bacitracin as ripener for sugarcane, U.S. Patent 3,897,240, July 29, 1975.
210. **Guyot, H. M.,** Process for treating sugar producing plants to effect improved saccharose yield, U.S. Patent 3,307,932, March 7, 1967.
211. **Nickell, L. G.,** Ripening of sugarcane by use of certain alcoholic and ethoxylated compounds, U.S. Patent 3,930,840, January 6, 1976.
212. **Nickell, L. G.,** Ripening of sugarcane by use of polyethylene oxide adducts of fatty acid esters of sorbitol, U.S. Patent 3,909,238, September 30, 1975.
213. **Ahlrichs, L. E. and Porter, C. A.,** Glyphosine — a plant growth regulator for sugarcane and sugarbeets, *Proc. Br. Weed Control Conf.,* 11, 1215, 1972.
214. **Rice, E. R.,** The effect of Polaris and Polado on regrowth of two sugar cane varieties in Florida, *Sugar Azucar,* 76(6), 29, 1981.
215. **Legendre, B. L. and Martin, F. A.,** Ripening studies with glyphosine in Louisiana sugarcane, *Proc. Am. Soc. Sugar Cane Technol.,* 6, 62, 1977.
216. **Martin, F. A., Steib, R. J., and Dill, G. M.,** Ripener efficacy as affected by ratoon stunt disease, *Proc. Int. Soc. Sugar Cane Technol.,* 17(1), 614, 1980.
217. **Osgood, R. V., Moore, P. H., and Ginoza, H. S.,** Differential dry matter partitioning in sugarcane cultivars treated with glyphosate, *Proc. Plant Growth Regul. Soc. Am.,* 8, 97, 1981.
218. **Alexander, A. G. and Montalvo-Zapata, R.,** Chemical desiccation of sugarcane, *Int. Sugar J.,* 73, 261, 1971.
219. **Kaya, H.,** Cane desiccation with paraquat at the Hilo Coast Processing Company, *Proc. 35th Annu. Conf. Haw. Sugar Technol.,* 257, 1976.
220. **Arvier, A. C.,** Pre-harvest desiccation of sugarcane with paraquat in Queensland, *Exp. Agric.,* 6, 309, 1970.
221. **Samuels, G. and Beale, A.,** Reduction in sugarcane trash by desiccation with paraquat, *Agron. J.,* 68, 255, 1976.
222. **Samuels, G. and Beale, A.,** Paraquat as a preharvest desiccant for sugarcane in Puerto Rico, *J. Agric. Univ. Puerto Rico,* 60, 262, 1976.
223. **Chen, J. C. D. and Liu, P. P. D.,** Evaluation of cane leaf desiccant Gramoxone, *Sugar J.,* 27, 22, 1965.

Chapter 8

THE USE OF EXOGENOUS PLANT GROWTH REGULATORS ON CITRUS

W. C. Wilson

TABLE OF CONTENTS

I. INTRODUCTION

Citrus production is limited primarily by freezing temperatures as it will grow under a wide variety of tropical and subtropical conditions, including the extreme heat experienced in many desert areas.[1] Under subtropical conditions sweet orange (*Citrus sinensis* (L.) Osbeck), grapefruit (*C. paradisi* Macf.), mandarin (tangerine or *C. reticulata* Blanco), and lemon (*C. limon* (L.) Burm. f.) trees become dormant (or partially so) during fall and winter, and acceptable fruit maturity of most early- and mid-season cultivars occurs during this period. Spring and summer usually produce growth flushes, blooms, and fruit set, and the late orange cultivars (Valencia and its derivatives) mature during this period. In tropical regions, all citrus cultivars tend to be everbearing, i.e., they bloom, set, and mature fruit throughout the year, and fruit usually matures in a shorter period of time than in subtropical regions. Indeed, lemons, limes (*Citrus aurantifolia* Swing.), and citrons (*Citrus medica* L.) tend to be everbearing under all conditions; hence, they exhibit little or no dormancy and are very easily injured, even by light freezing conditions.

Citrus tree growth and fruit quality are sharply conditioned by climate in the growing areas.[2] Yelenosky lists five general climatic regions of the world in which citrus is grown:[3]

1. Moist marine areas, such as southern Japan, the Black Sea Coast, Adriatic Sea Coast of Yugoslavia, and parts of Turkey and New Zealand. The cold-hardy cultivars, such as satsuma mandarin, are usually grown in these colder locations, principally on trifoliate orange (*Poncirus trifoliata* (L.) Raf.) rootstock.
2. Subtropical, Mediterranean climate as typified by Spain, Israel, Italy, Turkey, parts of Australia, Lebanon, Greece, and the coastal areas of California. These areas are usually not subject to freezing conditions and are the major areas of production of lemons as well as sweet orange cultivars. The citrus fruit produced tends to have high external and internal color, few rind blemishes, and relatively thick peel. The fruits are best adapted for fresh fruit use, and commercial juice processing is usually of secondary importance.
3. Subtropical, arid areas such as the U.S. desert regions of southern California, Arizona, and possibly the Rio Grande Valley of Texas. These areas produce fruit with excellent appearance, suited primarily for fresh fruit utilization. Sizable amounts of grapefruit are often grown. These areas are occasionally subject to damaging freezes.
4. Subtropical, moist areas include Florida, parts of China, India, South Africa, Brazil, and Argentina. Florida and some of the other regions are occasionally subject to severe winter freezes. The subtropical, moist areas produce fruit with generally thin peel, fair to good internal color, but with high soluble solids and juice content. Although fruit produced in these areas is generally good for fresh fruit use, it is particularly adapted for processing. Brazil and Florida have large processing industries, primarily for production of frozen orange juice concentrate (FCOJ). Florida utilizes about 91% of its sweet orange production and 50% of its grapefruit in this manner.[4]
5. Tropical, moist areas (Hawaii, parts of Mexico, Central America, and others) include those where temperatures lower than 65°F (18.3°C) rarely occur, and rainfall is usually heavy. Fruit rarely attain good external and internal color and all cultivars tend to be everbearing, producing not one crop but a series of crops throughout the year. Fruit of all sizes and stages of maturity may occur during any month of the year, but total crop production per tree is less than in subtropical regions.[1]

Climatic conditions affect citrus leaf and fruit wax deposition and composition.[5-8] Wax accumulation often plays a role in uptake of chemicals (including growth regulators); hence, more arid conditions might reduce uptake of a growth regulator by not only drying the applied spray very rapidly, but by causing increased thickness of the epicuticular wax which can prevent uptake.[5,6]

It is well known that growth regulator activity can be greatly affected by variations in climatic conditions. Therefore, from the above discussion it should be apparent that wide variation from use of growth regulators can be expected not only between and among various cultivars, but within a single cultivar growing under several distinct climatic conditions. This chapter will bear out the prediction. Commercial use of a particular growth regulator should not be attempted in any region without thorough testing under local climatic conditions.

The purpose of this chapter is to give up-to-date research findings regarding use of growth regulators (except ethylene, per se) on all facets of citrus production and marketing. Generally, the economic uses and economics of a particular usage are not addressed except in a general way. Commerical use of plant growth regulators is still in its infancy and this type of information is not readily available from most research reports. Although controls of many plant processes, such as inducing dormancy and reducing tree regrowth following heavy pruning, are highly desirable economically, practical control methods are not yet available. However, use of preharvest drop sprays, abscission chemicals for mechanical harvesting, fruit thinning chemicals to prevent aging of fruit peel, and certain others have progressed beyond the research stage and are used regularly on a semicommercial or commercial basis. Therefore, the findings of any researcher cited here should not, of itself, imply that results are an economically accepted practice.

II. USES OF PLANT GROWTH REGULATORS ON CITRUS GROWTH AND DEVELOPMENT PROCESSES

A. Aids in Propagation and Seedling Growth

The cost of a citrus tree can be a substantial part of the initial cost of planting a citrus grove (orchard).[9] As any horticultural practice which shortens the time from seed to first fruiting reduces initial investment cost, it would be beneficial to decrease the time of germination of seed and increase both the percent germination and the growth rate of the seedling. In California it was found that soaking seed for 24 hr in water containing gibberellic acid (GA) at 1000 ppm improved germination.[10] Others have reported that germination is increased by soaking seed for various periods in water plus GA (40 ppm) and 40 ppm of naphthaleneacetic acid (NAA) and combinations including potassium nitrate (2%) and thiourea (1.5 to 2%).[11] The best treatment as measured by germination and subsequent seedling growth and development was obtained from a 12-hr soak of seeds in either GA or NAA, each at 40 ppm.[12] Sucrose (15 to 20%), indolebutyric acid (IBA), indoleacetic acid (IAA), NAA, 2,4-D (2,4-dichlorophenoxy acetic acid), 2,4,5-T (2,4,5-trichlorophenoxy acetic acid), or boric acid (10 to 30 ppm) have been tested, with boric acid particularly effective in increasing germination of mandarin seed.[13] Soaking seed for 16 to 30 hr in GA improved germination of several kinds of citrus seed and, particularly, improved germination of seed from populations which had low germination percentages.[14] However, soaking seed for 12 hr in ascorbic acid (100 ppm) gave a higher percentage of germination and survival in the field than did soaking in ethephon (Ethrel® or 2-chloroethylphosphonic acid) at 100 to 300 ppm and GA at 20 to 200 ppm.[15] Similar results have been reported by other researchers.[16]

Plant growth regulators are also reported to aid propagation. The highest number of roots was obtained from air layers of seedless lemon when the stem was treated with an aqueous solution containing a combination of IBA and NAA (1000 ppm each).[17]

Cuttings of soft or semihard wood are reported to respond well to growth regulators such as IBA.[9]

Attempts to increase the size, particularly girth, of a seedling using growth regulators so far have generally been unsuccessful. However, it was reported that GA-treated *C. madurensis* Lour. cuttings grown under temperate (vs. tropical) conditions had relatively low root weights but increased shoot elongation and total dry matter content.[18] Alar® (SADH or daminozide) reduced root, shoot, and dry matter. Under tropical conditions, CCC (Chlormequat or Cycocel) applications produced results similar to those of SADH. Fresh weight increases for seedlings treated with GA (100 ppm) plus urea (0.5%) have been reported.[19] However, it would appear that GA applications tend to increase the stem length at the expense of reduced girth.[20]

GA (200 ppm) and SADH (800 ppm) had somewhat different effects on sweet lime seedlings when applied as root treatments.[21] GA increased stem elongation while SADH initially retarded it. Each treatment resulted in seedlings with more and longer internodes, however. GA induced earlier appearance with more and shorter thorns; SADH promoted thorn elongation. With regard to assimilate translocation in 1-year-old seedlings, it was suggested that GA induced a strong apical sink, causing an enhancement of assimilate translocation to the growing apex, while SADH depressed these traits.[21]

Plant growth regulators may be useful for controlling shape of young trees of sour orange (*C. aurantium* L.), Rough lemon (*C. jambhiri* Lush.), and Cleopatra mandarin. GA (0- to 1000-ppm treatments) applied in fall or spring reduced stem height of seedlings but there were distinct cultivar differences.[22] Branching was reduced but stem girth was not affected. Other experiments showed that triiodobenzoic acid (TIBA) at 50 to 100 ppm induced wide branching angles; GA (200 to 400 ppm) both reduced branching angles and increased growth rate but caused phytotoxicity at the higher rate; and SADH (500 to 1000 ppm) had no effect on branching but reduced shoot growth rate.[23]

Plant growth regulators can reduce the time lapse between bud grafting and outgrowth of the bud. Improved bud grafting by use of IAA and IBA is reported.[24,25] Removal of wood from the bud shield in combination with the IBA treatment gave the best results.[24] Nauer et al.[26] found that applying BA (6-benzylaminopurine) and PBA [6-(benzylamino)-9-(2-tetrahydropyranyl-)9H-purine] at 2000- to 8000-ppm concentrations to navel orange buds substantially increased fall bud growth (forcing). The chemicals were applied to the buds with a cotton swab 8 days after insertion of the bud. The BA was dissolved in water and the PBA in ethanol. Soaking the buds in BA and PBA prior to bud insertion was unsuccessful. Increasing greenhouse day length to 10 hr also was beneficial.

B. Control of Vegetative Growth

Under commercial citrus nursery situations, stimulating young tree growth is desirable so that the tree can be brought into production more quickly. Although GA will stimulate growth, it does not appear to accomplish this objective satisfactorily.[24]

Attempts to use NAA to inhibit rootstock sprouting in nurseries have given mixed results.[27] Sprays were applied while the inserted bud was protected with budding tape. Good sprout control was obtained on Rough lemon and *P. trifoliata* rootstocks, but translocated NAA inhibited bud break. Treatments were partially successful on Troyer *(P. trifoliata X C. sinensis)*, but no control was obtained on Alemow *(C. macrophylla)*.

In Florida grove operations, mature trees of sweet orange, lemon, grapefruit, and tangerine cultivars often must be severely hand pruned or machine hedged and topped to prevent tree sizes from exceeding heights of 15 to 20 ft (4.6 to 6.1 m). In California, many lemons are hand or machine topped yearly to about 10 ft (3 m) to facilitate harvesting and pest control.[28] Although rootstocks to control tree size are being tested in several citrus growing areas, as yet none is commercially available. Hence, mechanical control mechanisms are necessary; regrowth by citrus is often very rapid and substantial.

Numerous growth regulators have been used in an attempt to control vegetative growth in citrus. Maleic hydrazide (MH), SADH, NAA, limonin derivatives, and a number of experimental chemicals have been tested for this purpose.[9,24,29]

Because of the desire to develop a mechanical harvesting system for lemons, considerable tree size control research has been accomplished in both Florida and California. The warm, moist tropical conditions of Florida favor Bearss lemon shoot growth which, if uncontrolled, causes the tree to resemble a bush.[30] Lateral branching is encouraged by heavy pruning, but heavy sprouting from adventitious buds results. NAA (1 to 2% ethyl ester asphalt-based solution) in water sprayed on hedged/pruned trunks up to 4 ft (1.2 m) in height successfully reduced the number and length of the shoot regrowth. Negligible fruit residues occurred following these treatments, the highest being 4.0 ppm in lemon peel oil, 9 months posttreatment, from the trees receiving the 2% NAA rate.[31] Other researchers reported that the addition of ethyl hydrogen 1-propylphosphonate (EHPP) alone, or combined with NAA, controlled sprouting but caused some leaf loss or distortion of new growth.[32] It was concluded that NAA (1%) applied as an aerosol or gel was the most effective treatment.

Similar lemon regrowth characteristics occur in California. NAA (ethyl ester formulation) applied in water prevented adventitious bud sprouting for about 1 year.[33] The water application was more effective than NAA (1%) applied in white, water-based latex paint. Krenite (ammonium ethyl carbamoylphosphonoate) applied at 0.2% in water sprays produced an acceptable degree of retardation of shoot growth for 1 year without excessive phytotoxicity to the tree or excessive peel thickness problems with the fruit.[29] EHPP also reduced regrowth but produced twisted or misshapen regrowth on the trees. Subsequent tests by Boswell et al.[34] found that a single spray containing Krenite (0.25%) plus wetting agent, applied to the tops of Lisbon lemon trees by low volume spray, significantly inhibited regrowth for over 3 years and increased yield. Again, phytotoxicity and excessive peel thickness were not observed. A second spray the following year, however, reduced yield.

Sweet orange, certain tangelos, grapefruit, and certain mandarin cultivars could also benefit from size control sprays in numerous localities throughout the world. However, their regrowth problems are generally not as severe as those of lemons.

C. Cold Hardiness

Trees of *Citrus* sp., although tropical in origin, have the ability to become cold hardened to some extent if subjected to low but not freezing temperatures in the presence of light.[3] There is no indication, however, that fruit can be cold hardened; hence, its protection is restricted to some means of artificial heating or the elimination of ice-initiating bacteria which is currently under investigation.[35]

The cold-hardened sweet orange tree is not injured after 10 hr at 26 (−3.3°C) to 24°F (−4.4°C).[2] Ice avoidance, ice tolerance, and regeneration capacities of tissues influence freeze survival. Well-developed and healthy foliage increases the cold-hardiness potential of the wood. An important observable condition associated with cold hardening is carbohydrate accumulation in leaves which, with sweet orange, accumulates most rapidly between temperatures of 41 to 59°F (5 to 15°C). Carbohydrates accumulate more slowly in stems.[36] Although reducing sugars accumulate under cold-hardening conditions, this is evidentally due to photosynthesis and not starch conversion. (This would be a good explanation for cold-hardening conditions not developing in darkness.) It is suggested that the low starch-sugar conversion restricts maximum cold hardening at low temperatures.[36]

MH will produce cold hardening of citrus,[3,24,37] but in Florida various bad side effects (delayed regrowth with malformed leaves) have prevented grower acceptance. Under northern California (Mediterranean climate) conditions, night temperatures are sufficiently low during fall that growth protection chemicals are not as important as in areas such as subtropical, moist Florida and subtropical arid Texas and Southern California where "broken winter" conditions exist. This condition is typified by irregular periods of warm or cold conditions which can prevent cold hardening or cause early breaking of any cold hardiness achieved.

Although a chemical to achieve cold hardiness of citrus trees under all conditions has not been obtained, some research progress has recently been recorded. Ancymidol (α-cyclopropyl-α-(4-methoxyphenyl)-5-pyrimidinemethanol) retarded growth of potted citrus plants

(Valencia scion on sour orange rootstock) when added at 0.10% active ingredient (a.i.) by weight to the potting media, but no increased cold hardening was observed.[38] Under Japanese growing conditions (moist, marine), NAA applied at 100 to 200 ppm in late autumn was reported to reduce damage from low temperatures.[39] NAA is reported to delay bud break of sweet orange seedlings for up to 177 days when applied at concentrations of 0.25 to 1.5%.[40] The sodium salt produced the more pronounced inhibitory effect. Translocation of NAA delayed unsprayed bud growth for up to 150 days. Carry-over inhibition effect into the second growing season was minimal. NAA is also reported to increase leaf thickness when applied to lemon trees (25 to 200 ppm).[41] The thickness increase appeared to be the result of increased cell size following the treatment.

A relatively new growth regulator, Pix® or mepiquat chloride (1,1-dimethyl piperidinium chloride), is reported to increase freezing resistance of young potted citrus trees when applied to fully extended, but not mature, shoots.[42] (Uniform regrowth had been obtained by pruning and hand defoliation of the plants.) Under Florida conditions, mepiquat applied at concentrations up to 2000 ppm to trees under normal field conditions has not prevented observable freeze damage.[43]

The cost of fossil fuels has caused most citrus growers throughout the world to minimize or abandon frost protection methods such as the use of orchard heaters. Therefore induction of cold hardiness through plant growth regulators is an important area of research which should be expanded.

III. FRUIT PRODUCTION

A. Flowering

The initiation and development of flowers involve a large number of interrelated, well-coordinated growth, senescence, and abscission processes.[44] As previously mentioned, the climate in which citrus grows affects its flowering characteristics. In tropical, moist regions where continuous flowering, fruit set, and fruit development tend to take place, flower formation usually is relatively light at any one time unless drought conditions intervene to stop growth, and this is followed by resumption of irrigation or rainfall. In subtropical regions, the advent of spring growing conditions normally produces a single bloom period which may be heavy or light depending on preceding climatic conditions and/or crop load. Some cultivars of lemons, limes, and citrons tend to be everbearing (continuously blooming) under all conditions. High temperatures reverse the flowering process and the presence of fruit can inhibit flower formation.[9]

The chemical compounds, CCC, SADH, and benzothiazole promote flowering of lemons but not of sweet oranges.[9,45] Goren and Monselise found a promotion of flower formation and fruit set on Shamouti orange when antimetabolites of nucleic acid and protein synthesis were applied in December during the flower induction period.[46] Best results were obtained from sprays (branches only) of chloramphenicol-succinate (1000 ppm), 5-fluorodeoxyuridine (5.1×10^{-3} M), and 5-bromo-3-sec-butyl-6-methyluracil (Hyvar-X herbicide) applied at 2×10^{-3} M. Salomon reported CCC applications (500, 1500, and 3000 ppm) applied by irrigation to potted lemon trees decreased tree vigor but increased flowering and fruit set with all treatments.[47] As for this date, however, this writer knows of no commercial practice to increase flower formation of citrus by use of plant growth regulators; chemicals used to reduce flowering will be discussed in Section III.C.

B. Fruit Set

Moss has suggested that fruit set in citrus appears to be controlled by the process of competitive inhibition.[9] Until the fruit is approximately 4 cm in diameter, young fruits are subject to abscission; most citrus flowers and young fruit abscise, leaving only a small percentage as the final crop to be harvested. Except for navel, most commercial orange and grapefruit cultivars in the U.S. set sufficient crops so that the need for a fruit set chemical is not necessary. However, some mandarins and mandarin-grapefruit hybrids (tangelos) benefit from growth regulators which function as fruit set chemicals. The cultivars, Orlando, Minneola, Nova, and Robinson, tend to set sparsely, but in Florida increased fruit set has been obtained through sprays of GA (10 to 75 ppm) applied to trees at full bloom.[48] Similar results were obtained in South Africa where it is recommended that all bearing blocks of Clementine tangerines be sprayed at full bloom with gibberellic acid (10 ppm).[49] Fruits retained by this method tend to be somewhat smaller and more sensitive to adverse climatic conditions than cross-pollinated fruit.[48] It has been suggested that GA causes stronger mobilization of metabolites to young ovaries and developing fruits, especially in the 3-week period after anthesis.[50] These metabolites may be essential to fruit set and development.

A rather novel method of applying GA to flowers was tested in South Africa.[51] Instead of sprays, a powder containing GA was placed in a footpath at the exit of beehives so that the bees would apply GA powder to the stigmas of Orlando tangelo flowers. The method worked well on caged test trees, but GA transfer to flowers under field conditions was negligible.

Although plant growth regulators so far have not proven to be successful as chemical fruit set agents of navel oranges in Florida and China,[48,52] their successful use in other parts of the world has been reported. Applications of GA + urea (5 to 10 ppm and 0.5%) applied prebloom and 2,4-D (20 ppm) applied with or without urea to navel oranges in May (following bloom) increased yields substantially in Egypt.[53] In India,[54] spraying sweet lime at full bloom with GA (250 to 2000 ppm), PCPA (4-chlorophenoxyacetic acid) at 25 to 100 ppm, and 2, 4, 5-T or 2,4-D (5 to 10 ppm) increased fruit set and reduced preharvest fruit drop. Also, fruit size, total soluble solids, and ascorbic acid contents were increased. In Spain,[55] GA applied to inflorescences of the cultivar Fino (seedless Clementine) mandarin at anthesis (or within 25 days post) increased fruit set and fruit weight at harvest. Applications of 2,4-D and BA produced similar increases, but not as efficiently as GA. A later report suggests that the application of GA decreased the total inflorescence leaf content of nitrogen, phosphorus, and potassium which apparently moved to the developing young fruits, perhaps aiding their retention.[56]

C. Fruit Thinning

Most commercial citrus cultivars do not require thinning, particularly if the crop is destined for processing utilization. For fresh fruit use, however, size, color, and other features are often important selling points. Some cultivars, mandarin in particular, tend to alternate bearing, thus producing heavy crops of small fruit one season, but light crops of excessively large fruit the following season. Therefore, thinning to increase size (in the "on" year) and reduce alternate bearing is often an important horticultural consideration.

The Japanese public is very conscious of fresh fruit quality and is willing to pay a premium for it. Therefore, thinning to improve fruit quality (particularly size) of satsumas is regularly practiced. A leaf-to-fruit ratio of 25:1 is desired, and the purpose of chemical thinning is to reduce the amount of hand thinning necessary to achieve this ratio.[57]

The most effective material has been NAA (200 to 300 ppm) applied 25 days following full bloom. This caused approximately a 30% increase in fruit drop and effectively thinned the crop.[58,59] It has been suggested that the mechanism of action of NAA is through increasing ethylene production by young fruits.[59] By so doing, those which are in a more senescent stage respond to the stimulation of cellulase activity, resulting in abscission.

Thinning with NAA in Japan is affected by climatic conditions, particularly temperature.[57] Excessive thinning can result if daily high temperatures are 86°F (30°C) or more following spray applications. Best results occurred when postspray temperatures were 77°F (25°C). High humidities also increased thinning, presumably by causing increased chemical uptake.

NAA was used in relatively large quantities in Japan until 1976 when its registration was not renewed.[58] Since that date, considerable research has been conducted to find other plant growth regulators which not only effectively thin satsumas, but can also be successfully registered for use by the appropriate governmental agency. One compound of this group, 2,4-DP (2,4-dichlorophenoxypropionic acid), is reported to produce fruit thinning comparable to NAA.[60] Effective chemical concentrations are about the same as those for NAA.

A new growth regulator, Figaron (IZAA or ethyl-5-chloro-H-3-indazolyl-acetate), is registered for use in Japan for thinning and quality improvement of satsumas.[57,58] This compound is applied about 50 days following anthesis. Specified concentrations range from 100 to 200 ppm. IZAA is also influenced by temperature following application, similar to NAA. IZAA causes no phytotoxicity. With satsuma, it is also reported to advance fruit color and to increase juice °Brix (sugar content). Tests with the compound in Florida, however, have generally been ineffective.[43]

The manufacturer of IZAA has stated that best results are obtained if the chemical is applied when root growth is most active and in two applications 3 weeks apart.

Ethephon has also been tested for thinning satsumas in Japan.[57] Applications of 100 ppm 30 to 35 days following anthesis effectively thinned satsumas but caused excessive defoliation. GA applied at 20 ppm during February (prior to bloom) effectively thinned satsumas by decreasing flowering.[57] The treatment decreased the number of flowers without subtending leaves, while increasing the emergence of new leaves.

In the U.S. the principal need to thin fruits has been with the cultivars Dancy tangerine and Murcott (probably an orange-mandarin hybrid commonly marketed as Honey tangerine), both of which tend to alternately bear and produce excessively heavy crops of small-sized fruit during the "on" year.[61] Various chemical treatments have been tested which include NAA (200 to 800 ppm), CPA (250 to 1000 ppm), 2,4,5-T (20 to 50 ppm), 2,4-D (25 to 100 ppm), ethephon (50 to 250 ppm), IZZA (100 to 200 ppm), spray oil (0.5 to 1%), and cycloheximide or CHI (12.5 to 2.5 ppm). Wheaton reported that NAA (200 to 500 ppm) was the most effective and consistent material tested on both cultivars, but no commercial company is interested in spending the large sums of money necessary to register it.[61] However, the auxins CPA (3-chlorophenoxy acetic acid), 2,4-D, and 2,4,5-T were also effective thinning agents, although 2,4,5-T sometimes tended to overthin and cause phytotoxicity.

In tests with ethylene releasing materials, ethephon (150 to 250 ppm) was generally effective with Dancy and caused a minimum leaf drop, but Murcott showed considerable sensitivity to it and overthinning was a problem.[61] GA (20 ppm) applied prebloom provided fair thinning of Dancy, but was ineffective on Murcott at 20 to 40 ppm. Oil (1%) thinned Dancy tangerines, but CHI overthinned in unirrigated groves and caused only fair thinning in irrigated groves. Other experiments conducted in Florida confirmed that ethephon (200 to 250 ppm) effectively thinned Dancy and increased fruit size and crop value and reduced alternate bearing.[62] Drought conditions reduced the effectiveness of thinning treatments, and applying spray applications near the end of the natural drop period in May and before the fruit became larger than 15 cm in diameter, improved thinning effectiveness. GA, alone or in combination with BA, had no effect.

Fruit thinning tests in other parts of the world with mandarin-type fruits have been similar to those mentioned above. In Queensland, Australia[63] ethephon (250 ppm) effectively thinned the mandarin Beauty of Glen Retreat when applied at the end of the natural drop period. Gallasch also reported that ethephon effectively thinned Imperial mandarin (South Australia) and, thereby, increased grower returns by $1661 (Australian)/ha compared with returns for unsprayed trees.[64] In California,[24] NAA was successfully used to thin Wilking mandarin. Similar results were reported from Israel.[65]

In some parts of the world, alternate bearing of the Valencia orange warrants correction by fruit thinning. In Australia, thinning of Valencia was accomplished by 250 ppm ethephon applied when fruit was 1.0 to 1.5 cm in diameter (about 4 weeks following bloom),[66] although crop load reduction figures varied from 14 to 29% among the crop growing districts in the cooperative test. Gallasch et al.[66] attributed the observed differences among districts primarily to various degrees of stress among the treated trees. Also suggested were varying climatic conditions at the time the sprays were applied as well as rootstock differences.[67]

Also in Australia, GA (10 to 25 ppm) has proven to be an effective means to control fruiting of Valencia on Rough lemon rootstock, but was least effective with trees on *P. trifoliata* rootstock.[67] The spray was applied in winter prior to the "on" bloom year. The effect of the GA appeared to be through increasing the number of vegetative buds vs. flower buds, and although the "on" year was not substantially reduced in some cases, the cropping during the "off" year was substantially increased. Hand thinning studies have shown that removal of only 22% of the flowers in the "on" year would greatly reduce crop differences in the 2-year cycle.[68] It has also been reported from Israel that GA applied after shoot emergence in spring is able to reduce the number of subsequent flowers.[65] Studies in California have indicated that transabscisic acid (T-ABA), but not Cis-ABA, levels in buds of the "on" trees are higher than those of "off" trees, and this delays the beginning of bud growth in the spring.[69] Prebloom applications of GA (20 ppm) effectively reduced the crop load of Valencia in Florida, but the alternate bearing problem is not of major importance in areas where the bulk of the fruit is utilized for processing purposes.[61]

NAA (500 ppm) applied to Valencia orange trees approximately 30 days following anthesis during the "on" year successfully thinned the crop and produced even cropping.[70]

In Spain trials with the Washington navel and Navelate orange using GA (25 to 200 ppm) applied in winter successfully thinned these cultivars by reducing the number of inflorescences.[71] Similar results have been reported from Israel with Shamouti oranges.[65] It was also observed that 2,4-D (12 ppm) also thinned Washington navel and Navelate by inhibiting flowering in a manner similar to that of GA.[71] It is suggested that this might be a more inexpensive method of producing this desired effect.

The auxin 2,4,5-T is probably effective for thinning purposes in many parts of the world. Unfortunately, for many years the compound contained the highly toxic substance dioxin (2,3,7,8-tetrachlorodibenzo-p-dioxin) which is believed to be a threat to the environment.[65] Since this impurity has now been largely removed by better manufacturing procedures, it is suggested that this chemical should no longer be avoided by research workers.[65] However, there are other competent researchers who believe that even this small amount of dioxin could be a health hazard.[72]

D. Preharvest Drop Control and Storage of Fruit on the Tree

One of the earliest uses of growth regulators on citrus was with 2,4-D to control preharvest fruit drop.[73] In Florida, an application of 20 ppm 2,4-D (acid equivalent) is recommended to prevent dropping of Pineapple, seedling and Temple oranges, and seedless grapefruit. Sprays should be applied dilute only and should be made in November or December.[74]

In California and other citrus-growing areas of the world, the use of 2,4-D (8 to 18 ppm) has generally been more successful than in Florida and is reported to prevent preharvest fruit drop of most cultivars.[24,73,75] An exception to this rule, however, seems to be in Cyprus where GA and 2,4-D treatments applied to Marsh grapefruit had no effect on preharvest drop or other fruit qualities.[76] A study in Israel reported that 2,4-D (18 ppm) applied to Temple oranges 6 weeks before normal harvest allowed picking to be delayed at least 1 month.[77] External and internal fruit quality was improved. The application also reduced fruit drop following an unexpected frost. In India, a mixture of 2,4-D (10 ppm) plus aureofungin (20 ppm) plus 200 ppm Planofix (NAA) was effective in reducing preharvest drop of Kinnow mandarin.[78] Tests in the Punjab showed that various mixtures of GA (50 to 100 ppm), 2,4-D (10-20 ppm), and 2,4,5-T (10-20 ppm) were effective in controlling preharvest drop of Nagpur mandarins.[79]

Because of the importance of the Japanese market to Florida and the extremely long distance this fruit must be transported, prevention of grapefruit aging is an important consideration. Also, because Florida grapefruit crop sizes continue to increase, a means of lengthening the season is important so that more orderly marketing of the crop can be accomplished.[80] Therefore, should inhibition of color change and prevention of peel aging of seedless (sparsely seeded) grapefruit be required, a combination of 2,4-D (20 ppm) and GA (10 ppm) can be used.[74,81] Similar results with export grapefruit have been reported from Australia and South Africa.[82-85] Although holding Hamlin and Valencia crops until late in the season may depress yields the following year, this reduction has not been observed with seedless (Marsh) grapefruit.[86]

E. Disease Control with Plant Growth Regulators

The citrus disease anthracnose (*Colletotrichum gloeosporioides* Penz.) is a major source of postharvest decay for Robinson tangerines in Florida.[87] Tests showed that the incidence of the disease increased as ethylene concentration in the degreening room was increased and was less for well-colored fruit than poorly colored fruit. The results of additional field tests showed that an application of ethephon (250 ppm) 5 to 7 days before harvest significantly reduced the incidence of anthracnose.[88] The mode of acion was believed to be through accumulation of ethylene in the internal portion of the fruit which induced physiological changes required for development of resistance. It was found that postharvest application of ethephon prior to degreening did not reduce the incidence of anthracnose.

In Japan[89,90] the cultivar Naruto (*C. medioglobosa* Hort. ex Tanaka) has a disorder called "yellow spot" which begins as small, yellowing areas in the rind near the stylar end of the fruit and can sometimes end with browning and fruit decay. Well-colored fruits on the outside of the tree are more susceptible than the poorly colored, inside fruit. GA treatments slightly decreased calcium and magnesium concentrations in fruit rind and also decreased the incidence of the disorder. However, ethephon treatments tended to increase it.

F. Fruit Quality

1. Rind Quality

There are a number of peel disorders which can be alleviated by applications of specific plant growth regulators.[91] It should be pointed out, however, that problems with peel are of primary interest to growers and shippers of fresh fruit just as internal qualities are of prime concern to processors.

With lemons grown in California, and probably most other Mediterranean climate areas of the world, peak production occurs during winter and spring when demand is low. Most lemons are harvested, stored (cured), then packed and shipped to market much later. However, during the summer when fresh lemon sales are strongest, the amount of fruit available is lowest. Therefore, a need to extend the lemon harvest season by prevention of premature yellowing (senescence) of the fruit is desirable. GA (5 to 40 ppm) applied in the fall is reported to be the most beneficial treatment to delay overall maturity (and yellowing).[92] These treatments also affect the second-year harvest pattern, probably from the influence of the GA on flowering, resulting in more trees producing fruit during summer when market demand is high.[24] The primary reported benefits of the GA treatment to lemons were better flexibility in harvest patterns, a longer storage life, and a reduction in the number of small, yellow lemons.[24]

There are also other materials which produce results similar to those of GA. Oil sprays, which contain potassium nitrate (5 lb/100 gal or 2.27 kg/378.5 ℓ)[28] and 2,4,5-T both caused delayed lemon fruit maturity.[24] However, the 2,4,5-T treatment was not as pronounced as with GA. Evidently, there were few phytotoxic problems with lemon unless very high levels of GA were used. Similar results were also obtained with limes.[24]

Navel oranges in the Central Valley of California have rind softening problems which develop as the fruit matures and peel color changes from green to orange.[24] These senescence changes can contribute to a number of other rind disorders such as rind staining, sticky rind, puffy peel, and water spot (also considered to be weather related).[91] GA treatments (5 to 20 ppm) applied in the early fall lessen or prevent these conditions, thus allowing a more orderly fresh fruit marketing season.

In Israel, which has a climate similar to that of California, excessive growth in peel thickness occurs with Shamouti oranges in inland areas which have a wider range of temperature than is found in the coastal areas.[92] The abnormal growth results in very rough, ugly, and unmarketable fruit, and is believed to be brought about by excessive concentrations of native growth promoters.[92] Control of the condition can be accomplished with SADH (1000 to 4000 ppm) and CCC (1500 to 2500 ppm) applied during April or May (3 to 6 weeks following bloom).[93] It is believed that the high concentration of SADH makes the commercial practice unprofitable because of chemical costs, but CCC was suggested as a commercially acceptable practice.

Creasing affects many commercial citrus cultivars throughout the world. Monselise et al.[94] state that creasing of Valencia orange is the most important single cause for rejection of this fresh fruit in Israeli packinghouses, with 26% of the discarded fruit per season being for this cause. Physiological studies showed higher pectolytic activity and a higher content of water-soluble pectins in affected fruits. These areas occur in the albedo resulting in sunken areas of weak peel. The use of GA (20 ppm) applied when fruit is 3 to 4 cm in diameter (about July, or 4 months following anthesis) caused considerable reduction in the incidence of creasing without impeding good color development. Sprays applied as late as November also reduced creasing but retarded color development. The mode of action of GA was believed to be through renewal of growth activity in the affected tissues.

In South Africa creasing of navel orange is also a major cause of cull losses from fresh fruit packinghouses and can be as high as 50% near the end of the shipping season.[95,96] It is more prevalent with Troyer citrange rootstock than with Rough lemon and is worse on the south side of a tree.[95] Recommended control measures are to apply GA (10 ppm for navel on Rough lemon and 10 to 20 ppm if on Troyer citrange) when young fruit is 30 to 50 mm in diameter (70 to 100 days after anthesis). The inclusion of nonionic surfactant (250 ppm) is also recommended.

In addition to GA, nutritional sprays containing potassium, ammonium and phosphate ions reduce fruit creasing.[94] Foliar spray applications of ammonium phosphate were more effective than potassium nitrate sprays in reducing creasing of Valencia oranges.[97] Sprays were applied during summer or approximately 3 to 4 months following anthesis.

The peel of mandarin fruit often tends to puff. In Japan, GA applied to satsuma mandarin at 100 ppm in two applications 25 and 35 days prior to harvest reduced the amount of puffy fruit, but also slowed the rate of chlorophyll degradation.[98] ABA content in peel was found to increase at a slower rate with the GA-treated fruit.

There are other reported peel disorders which can be reduced or controlled by plant growth regulators. Corky (silvery) spot of grapefruit and oranges causes losses of fresh fruit in Israel.[99] Substantial reduction of corky spot on Marsh grapefruit was accomplished with an application of 2,4-D isopropyl ester (16 to 22 ppm) applied 2 to 4 months following anthesis, or GA applied 3 months following anthesis. GA (8 ppm) applied at colorbreak is reported to cause Marsh grapefruit to be firmer late in the season and show less shrinkage in shipment.[85]

In some areas of the world fruit splitting may occur during the later growth stages of the fruit.[100] This appears to be worse in areas of high rainfall and humidity, such as in Florida, where it may cause substantial sweet orange losses in some years. The malady arises when juice sacs in the fruit begin to fill and internal fruit pressure causes the peel to rupture (split). The split itself usually includes a lengthwise cracking of the peel, or the split may involve the stylar half of the fruit. Although severity varies yearly, depending on climatic conditions, the use of plant growth regulators to reduce the problem has produced conflicting results.[101] However, in California a significant reduction in fruit splitting at the navel end of oranges was obtained by an application of 2,4-D when the fruits were small.[24]

2. Color

Two recent publications by Stewart have reviewed color research being conducted on citrus throughout the world.[102,103] His research has demonstrated that the chief carotenoid in orange and tangerine peel is β-citraurin. His conclusions regarding postharvest formation of color in fruits are that initially degreening takes place at 86°F (30°C) with 5 to 10 ppm ethylene; however, a cool coloring period is required for biosynthesis of β-citraurin.[104] This compound is temperature sensitive and is produced at temperature ranges from 77 to 59°F (25 to 15°C), a higher ethylene concentration being required with the higher temperatures. Decreasing quantities of ethylene are required to produce this color as temperatures approach 59°F (15°C). At 59°F (15°C) a significant color development will occur without additional ethylene. Below 59°F (15°C), color development dropped off rapidly. Similar findings have been reported by El-Zeftawi.[105]

Under field conditions, a similar physiology to that described probably takes place. In tropical areas, oranges and mandarins usually develop a light color break at maturity but not deep color.[1,102,103] Under subtropical conditions, particularly when fruit matures under dry conditions with cool night temperatures, brightly colored fruit consistently results. Those familiar with citrus are well acquainted with the fact that the highest quality citrus for fresh fruit use is produced in the subtropical areas. Eye appeal is generally very important in the sale of fresh fruits.

Attempts to control fruit color by use of plant growth regulators (except ethylene gas in postharvest degreening) have been restricted under Florida conditions to mandarins and their hybrids.[74] In Florida, ethephon (250 ppm) is recommended for all cultivars except Orlando tangelo which should receive only 200 ppm. An additional purpose of the sprays is to produce fruit abscission (loosening). The fruit must have achieved minimum internal quality requirements and 10 to 20% color break before this treatment can be effective.[74,106] No surfactant should be used with ethephon and it should not be combined with other materials containing surfactants; otherwise, increased leaf losses will occur.

Similar results have been obtained in Japan in tests with Ponkan, a type of mandarin.[107] A report from Korea concluded that a preharvest application of ethephon (500 ppm) increased color of satsuma fruit but resulted in excessive leaf drop.[108] Methionine (an ethylene precursor) sprayed 1 month before harvest had no affect on color, abscission, or other measurable fruit processes.

The ethylene-releasing compounds ethephon and Alsol® [2-chloroethyl-tris-(2-methoxyethoxyl)-silane] are reported to enhance rind color break and render the Washington navel orange less susceptible to certain types of handling damage;[109] however, a decrease in rind-oil rupture pressure and an increased susceptibility of the fruit to oleocellosis, a physiological rind disorder, were also observed. A later report, however, showed that an ethephon spray (480 ppm) applied after color break and 14 to 20 days preharvest reduced damage by oleocellosis.[110]

Our own observations following ethephon applications to oranges have consistently shown erratic performance and, frequently, heavy leaf losses.[43] Similar results seem to have been observed worldwide. Still, its overall advantages outweigh its disadvantages for promotion of color (and abscission) with mandarin-type fruits.[74]

Various attempts to control the leaf drop problem caused by ethephon have been made by many researchers, though little of the information seems to have been published, probably because of negative or conflicting results. At one time inclusion of zinc ion (from zinc sulfate or zinc chloride) was suggested. Japanese workers reported that inclusion of calcium acetate (5×10^{-2} *M* concentration) reduced leaf drop without affecting color development.[111] Our tests in Florida, however, showed a reduction in leaf drop, but the color and abscission advantages from use of the chemical were eliminated.[40] Similar results have been observed in California.[28] Some workers have suggested adding NAA or 2,4-D to ethephon, but our Florida tests have all given negative results from these additions.[43]

Other compounds with mode of action similar to ethephon have been tested for improving fruit color; these include Alsol®, previously mentioned, and a closely related compound CGA 15281 [2-chloroethyl methylbis (phenylmethoxy) silane]. In Florida neither compound has performed as well as ethephon; however, Alsol® is used commercially in Europe for abscission of olive fruits and causes less phytotoxicity to trees than does ethephon.[112]

Another approach to the color problem has been through use of compounds which causes the fruit to produce lycopene rather than β-citraurin.[102,103] The resultant fruit, then, tends to be red and resembles a tomato more than an orange. Best known of these compounds seems to be CPTA [2-(4-chloropehnylthio)-triethylamine hydrochloride]. H. Yokoyama and co-workers with the USDA in California have synthesized and tested a number of similar compounds.[113] Tests in Turkey with CPTA and CDEB [4-chloro-(B-(diethyl-amine)-ethyl)-benzoate] confirmed that lycopene could be produced with these compounds.[114]

Although, in time, the public could probably be educated to accept oranges which were more "red" than "orange", one of the principal problems with these chemicals is that, under field conditions, they produce an uneven or blotchy color pattern. Although reports have claimed these compounds will also increase internal fruit color, our tests have all proven negative.[43]

An additional problem might arise with fruit for processing. Although lycopene occurs naturally in some colored grapefruit ("pinks" and "reds"), the color is lost in processing so that the final product is an unappealing gray.[102] A similar situation would probably occur if intense lycopene concentrations were caused to occur in oranges, where they are not normally found.

Other than chemicals which might induce lycopene synthesis in oranges, very little has been accomplished concerning improving internal citrus fruit color. Although external peel color is not important for fruit used for processing, internal (juice) color is very important. As previously mentioned, oranges grown in the tropics usually have poor external and only fair internal color, but internal qualities such as juice content and soluble solids may be reasonably good. However, development of a plant growth regulator to improve internal color, or any of these other qualities, would be beneficial.

3. *Fruit Size*

Increased fruit size is desirable for some cultivars as small sizes often cause serious profit losses for fresh fruit packinghouses which must eliminate them in order to market fruit of a specified legal size. Mandarin-type fruits are often thinned with growth regulators for this purpose (see supra). For fruit destined for processing, however, size is usually not a consideration as pounds solids per acre (kilograms solids per hectare) are paramount, regardless of how it is obtained.

Lemons tend to produce their heaviest crops during the winter and spring when consumer demand is low (previously mentioned). Therefore, it is advantageous to produce fruit which can be picked and stored until the summer period when demand and prices are high. In California,[28] GA (10 ppm) and/or additional potash (potassium) applications are used to produce the larger, greener fruits which can be harvested and then stored up to 6 months. The small, yellow tree-ripened fruits are removed during the winter-spring period and primarily utilized for processed products.

In Australia[115] the lemon is an exception to the rule that fresh fruit purchasers prefer large fruit. When the fruit is held on the tree into the late spring it tends to grow to a size which is not acceptable in the marketplace. It also tends to be orange-yellow rather than pale yellow which is favored by buyers. A preharvest combination application of GA (10 ppm) and CCC (1000 ppm) applied in the fall delayed fruit coloring and arrested fruit growth of Australian lemons, thus extending the harvest into late spring and furnishing a larger quantity of small-sized, premium-quality fruit.[115]

4. Internal Fruit Qualities

There are a number of internal quality factors which may be improved by applications of plant growth regulators. These include color (previously mentioned), juice content, acidity, soluble solids, granulation, vitamin C content, and perhaps many more. Internal qualities appear to be more important for fruit for processing utilization; however, a perfect appearing fresh fruit will not long command a premium market price if its juice content and taste are poor.

Improvement of grapefruit flavor has been practiced for many years in Florida through the use of lead arsenate.[74] To avoid excess phytotoxicity it is recommended that it not be applied to trees less than 7 years of age. The most effective application period is 1 to 6 weeks following bloom; however, it can be applied as late as 4 months postbloom. Recommended concentrations are 4.0 to 12.5 pints (1.9 to 5.9 ℓ) of 4 lb/gal (480 g/ℓ) formulated material added to 500 gal (4732 ℓ) of water for the white flesh-colored cultivars, and 4.0 to 6.0 pints (1.9 to 2.8 ℓ) for red- and pink-fleshed cultivars. The higher rates are for increased (higher) ratio early in the shipping season. Spray coverage should be thorough.

Lead arsenate causes a reduction in total acidity of grapefruit (and, consequently, causes an increase in °Brix/acid ratio). It is also effective on oranges, but is illegal in Florida because the effect is so pronounced that most of the oranges are insipidly sweet. Research by various groups in Florida and elsewhere during past years has shown that most organic and inorganic arsenical compounds will cause acidity reduction in citrus fruits.[43] The Florida Department of Citrus has maintained a screening program for acidity reduction compounds for several years, but so far no nonarsenical compounds have been identified which appear to have commercial possibilities.[116]

In South Africa arsenical sprays have also been tested extensively and have been used commercially for many years.[117,118] South African oranges, particularly the Valencia, tend to produce fruit which has high acid content; therefore, research efforts have been aimed at reducing acidity in orange cultivars as well as grapefruit. Early findings indicated that arsenical sprays should be applied to oranges only every third year,[117] but subsequent studies showed that in the Eastern Transvaal, mature trees could be sprayed yearly with no residual effects from the arsenic.[119] The month of October (soon after bloom) was found to be the best time to apply arsenical sprays.[119]

Recent research with arsenicals in the saline waters of the Sundays River Valley has shown that the total quantity of arsenic used can be reduced by acidifying the tank mix to pH 4 with sulfuric acid.[120,121] (Phosphoric acid, however, caused considerable phytotoxicity when used.[121]) Therefore, recommended concentrations of calcium arsenate are 3/4 lb/100 Imperial gal (75 g/100 ℓ) of water for Marsh grapefruit and 1/2 lb (50 g/100 ℓ) for Valencia oranges. Lead arsenate can also be used successfully at somewhat higher concentrations.[119] Arsenic residues on fruit in both Florida and South Africa are negligible,[43,121,122] with the largest proportion found in the peel and practically none in the pulp and juice.[121,122]

In the past, inorganic arsenic has been considered to be a carcinogen and EPA clearance (or maintenance of existing clearances) or arsenicals on food crops in the U.S. has been quite difficult. However, recent nutritional research throughout the world has established arsenic as an essential nutrient for certain animal species.[123] Although arsenic has not been proven to be essential for humans, future research may indeed confirm these findings. If so, the very low residue levels noted for arsenic-treated citrus fruit, instead of presenting a health hazard, might actually be shown to be beneficial. Further, in Florida the environment might well be improved by the substitution of calcium for the heavy metal lead form of the inorganic arsenic as has been done in South Africa (see supra). Our research has also shown that organic arsenicals, particularly arsanilic acid, at comparable concentrations also reduce acidity of grapefruit,[43] but these may be a somewhat more expensive chemical to manufacture and use. Arsanilic acid currently has EPA-approved food tolerances in the U.S.[124]

Granulation of fruit (a physiological condition in which the juice sacs lose their juice and become hardened) can become a problem. The fruit may appear normal on the outside, but when cut it will have a crystalline appearance and may be of poor quality or inedible. The mandarin cultivars, particularly Dancy tangerine, seem to be most susceptible. Research in India indicated the condition with Dancy tangerine may be reduced substantially by applications of nutrient sprays containing zinc sulfate, boric acid, calcium, and potassium nitrates.[125] With the kaula mandarin, however, applications of plant growth regulators seemed to offer some promise.[126,127] NAA (300 ppm) or GA (15 ppm) applied by a single application during the August-October period produced substantial reduction in granulation. Also, ethephon (250 ppm) applied in November was beneficial.

Granulation is a major problem of navel oranges in South Africa and develops during the long overseas shipping period.[72] This is a core dryness condition, rather than the stem-end dryness, usually seen with Valencia oranges.[128,129] The malady varies in severity from season to season, but is worse when relatively warm winters precede the bloom period and this is followed by larger quantities of rain than normal during the months when the young fruit is developing. Although granulation seems to be related to a number of measurable nutritional and enzymatic quantities, these accounted for only 50% of the variation found among samples.[129] There are no reports that this condition is controllable through use of growth regulators.

Except for mandarins, granulation is not usually a problem in Florida, but fruit of the Valencia cultivar from young trees on Rough lemon rootstock is sometimes susceptible if the crop is held into the latter portion of the maturity season. In commercial practice, growers usually harvest these blocks as soon as they reach legal maturity to prevent occurrence of this condition. A similar condition will occur if fruit is frozen. To the author's knowledge, no research has been conducted to determine if plant growth regulators will lessen the severity of freeze damage to fruit.

Very little research has been conducted concerning the possibility of using plant growth regulators to increase pounds (Kg) soluble solids per acre (hectare) similar to that employed with sugarcane.[130] This would seem to be a fertile area for further research investigation. Apparently, no studies have been conducted to determine if the ascorbic acid (vitamin C) content of citrus fruit can be improved with plant growth regulators.

IV. POSTHARVEST TREATMENTS, FRUIT SHIPMENT, AND STORAGE

There are many factors to consider regarding postharvest treatments and fresh fruit storage.[131] The citrus fruit is fully ripe at harvest, it contains practically no starch reserves, and it is not a good candidate for controlled atmospheric storage. Although it is suggested that the best place to store citrus probably is on the tree, there are occasions when it must be harvested and stored, or transported very long distances in the hold of ships. Most cultivars have a long harvest period and fruit response to plant growth regulators can vary depending on the time during the season it is harvested, the time of the day it is harvested, tree condition, rootstock, location, year-to-year weather variations, etc. In assessing the performance of plant growth regulators on citrus, some or many of these varying conditions may affect activity of the chemical.

Ethephon has been used to substitute for the effects of ethylene degreening treatments for fruit. In Florida[132,133] lemons picked in early fall and degreened with 5 ppm ethylene gas are very susceptible to decay. Dipping the fruit in ethephon (500 to 1000 ppm) was found to hasten the development of marketable color by 5 to 14 days. However, there were no significant differences in the amount of decayed fruits among the treatments. Similar results appear to have been observed in Australia.[134] Also, the chemical R 33417 (2',4'-dichloro-1-cyanoethane sulfonanilide) (600 ppm) is reported to enhance the photo destruction of chlorophyll.[135] Most attempts to degreen mandarins with ethephon seem to have been more successful when applied as preharvest sprays.[87] In South Africa, however, postharvest ethephon dips of oranges are being used commercially in lieu of ethylene treatments.[72]

The problem with lime is the opposite of lemon since it is desired to retain the green fruit color which the public associates with limes. (Yellow limes are difficult to sell in most markets although internally the fruit is relatively unchanged.) Dipping Persian limes (grown in Italy) in a solution of GA (100 ppm) for 1 min allowed them to be stored up to 4 months without loss of green color if held at 45°F (7°C).[136] Other research has found that treatment of Persian and West Indian limes with N6-benzyladenine (7.5 to 30 ppm) dips (5 min to 1 hr) increased chlorophyll retention times during subsequent storage at several combinations of temperatures and humidities.[137] GA (12 to 50 ppm) dips also increased the chlorophyll retention times for the West Indian lime during subsequent storage. However, attempts to delay color change of Persian limes in Florida by dips with GA were unsuccessful.[131] It was theorized that Florida lime trees, which were in a vigorous growth condition at the time of picking, had sufficiently high levels of endogenous GA to prevent response to the exogenous applications.

In India, however, it is desirable to degreen Kagzi limes and it was found that dips of ethephon (2000 to 4000 ppm) degreened the fruit successfully in 12 days and increased total soluble solids, pH, and acidity of the fruits if held in darkness during this period.[138]

In California, as well as similar lemon growing areas, preservation of the calyx tissue ("button") on the fruit prevents stem-end rot fungus from entering through the abscission layer tissue. This can be successfully controlled by fruit dips in aqueous solutions of 2,4-D[131] or by applications of fruit waxes containing 2,4-D or 2,4,5-T.[24] Preservation of the button also improves the cosmetic appearance of the fruit, and many believe this manifests its freshness to the housewife.

Recent tests of three auxins and two ethylene biosynthesis inhibitors showed that picloram (4-amino-3,5,6-trichloropyridine-2-carboxylic acid) applied as a dip at 50 ppm was as effective as 2,4-D isopropyl ester (250 ppm) in prevention of abscission of lemon buttons.[139] Picloram, 2,4-D, and triclopyr (3,5,6-trichloropyridine-2-oxyacetic acid) prevented lemon degreening, but the ethylene biosynthesis inhibitors aminoethoxyvinylglycine (200 ppm) and aminoxyacetic acid (1848 ppm) had little effect on button abscission and no effect on degreening.

In India the Croog mandarin is harvested at three different times of the year.[140] The late, or monsoon crop, is harvested during a period of warm weather and heavy rainfall and is generally of poor quality. Improved fruit color and diminished weight loss in storage were obtained by dipping fully mature fruit for 2 min in a solution of ethephon (500 ppm) + benomyl fungicide (250 ppm) + 6% wax (emulsion).

A very good Japanese market for grapefruit has developed in recent years. Because of the long transit time in the shiphold (about 6 weeks from Florida), shipping losses can, at times, be substantial. Florida grapefruit is subject to chilling injury,[131] particularly fruit harvested in early fall which must be held at 60°F (15.5°C). However, fruit harvested by spring can be held at 50°F (10°C). Even this latter temperature is quite high when the fruits must be held for the long transit times necessary. As previously mentioned,[74,81-83] preharvest treatments with 2,4-D and GA can reduce peel senescence. Attempts to control chilling injury with plant growth regulators however, have produced conflicting results.[141] Changes in susceptibility to chilling injury were found to vary directly with the growth activities of the trees. Endogenous applications of BA, GA, and 2,4-D applied postharvest and BA and 2,4-D applied preharvest significantly altered chilling injury susceptibility, but the direction and extent of the changes were neither consistent nor predictable.

The presence of seeds in citrus fruit can occasionally cause problems. Studies in Israel found that seed numbers in fruit could be substantially reduced by application of plant growth regulators to Dancy, Temple, and Ortanique mandarin-type fruits.[142,143] Field applications of GA (20 ppm) applied at early or mid-bloom and NAA (150 ppm) applied when fruitlets were 5 to 10 mm in diameter substantially reduced the number of seeds in these fruits. No thinning occurred at these concentrations and fruit volume was increased, especially in the later treatments. Anatomical evidence indicated the mode of action was embryo abortion.

Grapefruit which are held very late into the shipping season often have sprouted seeds. Preharvest treatments with 2,4-D and GA, previously mentioned, reduced the number of sprouted seeds.[80,81]

Although controlled atmospheric storage of citrus fruits has generally been unsuccessful,[131] it is reported from Australia that long-term storage of lemons (up to 6 months) was possible when pretreated with GA and 2,4-D, then dipped in sodium o-phenylphenate and benomyl.[144] Storage conditions were 10% O_2, very low CO_2, a 10°C storage temperature, and continuous removal of the evolved ethylene.

V. GROWTH REGULATORS (ABSCISSION CHEMICALS) TO FACILITATE HARVESTING OF CITRUS FRUITS

This subject has recently been reviewed by Wilson et al.[145] Although citrus fruit can be successfully harvested without abscission chemicals, field experience has shown that chemical loosening is desirable because less tree shaking time is required, resulting in less physical abuse to the machinery and trees. However, chemicals have not allowed the construction of less powerful shakers because chemically induced fruit loosening is not always uniform, which results in about 10 to 15% adhering strongly to the tree.

Although abscission chemicals are technically classified as growth regulators, most of the currently used abscission compounds function by causing superficial peel burn followed by the fruit producing wound ethylene. The latter moves in some manner through tissue and affects the abscission zone. The only commercially available chemical that appears to function through absorption by tree and fruit, followed by conversion of the chemical into ethylene, is ethephon. In this sense it appears to function more in the classic growth regulator sense. Ethephon is the only commercially available chemical which can be used on fresh fruit. All others, because they cause superficial peel injury, must be used with fruit which will be processed rapidly.

There are a large number of physiological factors which affect abscission,[146] but temperature seems to be the most important limiting factor for early- and mid-season oranges. Although the technical aspects of predicting abscission activity are complicated,[147] field experience has shown that generally good abscission activity is achieved if daily high temperatures are 65°F (18.3°C) or greater following application of the spray. Daily high temperatures lower than this figure usually result in lessened abscission activity or none at all. Rainfall within 24 hr of a spray application often will negate its effect.

The Valencia orange is harvested in Florida from April until early July, and in California, Spain, and other Northern Hemisphere citrus-growing areas from April through October. It differs from most fruits in that fruit maturity occurs 13 to 18 months following bloom, which means that both a mature and immature crop are nearly always present when the crop is harvested. During this period, high temperatures could average close to 80°F (26.7°C) or above. Therefore, the principal limitation to use of abscission chemicals is physiological, namely Valencia undergoes a period of lessened chemical response lasting 2 to 3 weeks and usually occurring about the first of May.[148] During this period the chemical is essentially useless for loosening fruit, and as Valencia regreening occurs at this time, the terms "nonresponsive" and "regreening" have been applied to the condition. Although the condition is associated with regreening, it appears to be controlled by different mechanisms.[148]

The nonresponsive period coinciding with regreening occurs when Valencia harvest should be at a maximum. Before this period, the fruit is very responsive to abscission chemicals, but the desired Brix/acid ratios (fruit maturity) for processing oranges have not been achieved.[106] Following this period, the immature fruit on the tree usually will average 1.5 to 2.0 cm in diameter and will have achieved sufficient mass so that any mechanical shaking device will remove excessive quantities along with mature fruit.

Four chemicals are available for use in Florida,[145] and a combination of two of them has been used successfully. A summary of their uses is as follows:

A. Ethephon (Ethrel®)

This is cleared by the U.S. Environmental Protection Agency (EPA) for use in Florida on tangerines and tangerine hybrids.[74] Suggested concentrations are dilute sprays of 250 ppm for all cultivars except Orlando tangelo where the suggested rate is 200 ppm. This chemical, in addition to producing fruit loosening, enhances fruit color development.

The chief problems with ethephon, observed both in Florida and California, are its tendency to cause excessive leaf drop and its erratic fruit loosening.[145,149] This latter problem seems to have caused many Florida growers to suspend its use. Although all abscission chemicals are to some extent unpredictable in action, this chemical requires more expertise in its use, and it appears to be more subject to the vagaries of the weather than other abscissors. A surfactant should not be used with ethephon and it should not be applied as a concentrate spray.

B. Cycloheximide (Acti-Aid®)

This is cleared by EPA for use in Florida on oranges intended for processing. The chemical has generally produced good loosening of early- and mid-season oranges when applied as dilute sprays of 10 to 20 ppm. It should not be applied after the spring growth begins, otherwise, severe phytotoxicity can result. Unfortunately, its performance on the Valencia orange has been unacceptable, although usually by 6 weeks following bloom the tree (but not necessarily the immature fruit) will tolerate concentrations up to 20 ppm. Immature fruit from 1 to 6 weeks postbloom is particularly sensitive to cycloheximide injury, but this decreases somewhat after that period. Cycloheximide normally causes light to moderate rind pitting on mature fruit, thus eliminating its use for loosening fresh fruit. Use of a suitable surfactant is suggested. Some concentration of the spray appears to be possible, although best results are obtained by the use of dilute sprays.

Florida experience has shown that cycloheximide should not be used if freezing temperatures are likely to occur because it reduces the cold hardiness of the tree for an undetermined period of time. As the forecasting of freezing conditions in Florida generally cannot be made accurately except for about a 3-day period, its use, therefore, involves a certain risk during the Florida winter period.

C. 5-Chloro-3-Methyl-4-Nitro-1H-Pyrazole (Release®)

This was originally developed by Abbott Laboratories (North Chicago, Ill.), but all rights to use the compound as an abscission agent for citrus have been purchased by the Florida Department of Citrus, which is also financing toxicology studies for full EPA clearance. Currently, Release® is available in Florida for use under experimental permit on oranges destined for processing. The chemical was the first which showed the ability to loosen mature Valencia oranges while causing virtually no injury to bloom, young fruit, or foliage when used as recommended. Label recommendations (for dilute sprays) are 75 to 125 ppm for early- and mid-season oranges and 175 to 250 ppm for late-season or Valencia oranges. Concentrations above 250 ppm, however, have been reported to cause some yield reduction of Valencia oranges. Best results are obtained by use of dilute sprays and use of a suitable surfactant is recommended.

Release® causes a superficial peel injury which most often appears on the fruit as a distinct ring burn at the blossom end. Peel injury tends to be most severe near the beginning of the fruit season (December) on early- and mid-season oranges. Valencia oranges are not as subject to the injury.

D. Glyoxal Dioxime (Pik-Off®)

This is a chemical very similar in mode of action to Release® and is available for use under experimental permit by EPA. Although it does not usually produce as good abscission as some other chemicals, it is satisfactory for many uses. From a cost and residue standpoint, it is in an advantageous position and is compatible with the Valencia cultivar. It should not be used with surfactants. Dioxime should not be used after immature Valencia fruit have attained a diameter of 2.5 cm (1 in.).

Dioxime is restricted by label as to total quantity which can be applied per hectare. However, results of a series of field tests using dilute sprays showed that increased volumes of sprayed chemical per hectare substantially improved abscission performance, particularly on larger trees which required more liters per hectare (gallons per acre) than the label allowed.

E. Chemical Combinations

Two-way combinations of Release® and cycloheximide applied with surfactant as dilute sprays have given better fruit loosening than either chemical used alone. Optimum concentrations for early- and mid-season oranges have been Release® (50 to 100 ppm) and cycloheximide (1 to 5 ppm). Field experience has shown that the amount of cycloheximide should be kept as low as practicable, preferably below 2.5 ppm, because of excessive leaf drop which has been observed occasionally. Release®-cycloheximide combinations may lower cold hardiness of the tree in the same manner as cycloheximide used alone. Three-way combinations of Release®, cycloheximide, and Sweep® (chlorothalonil) are no longer being tested. Suitable surfactants should be used with combinations.

McCarty et al.[149] also tested Acti-Aid®, Release®, Pik-Off®, and Sweep® and the ethylene-releasing chemicals ethephon, Alsol®, and a similar compound, CGA-15281, under California conditions. Their conclusions were similar to those observed in Florida. The former materials caused light to moderate peel burn which was not acceptable for fresh fruit use. The ethylene-releasing compounds caused some leaf losses, and the fruit showed a slight acceleration of aging and developed "tacky rind" 10 to 12 days after harvest. GA (10 ppm) applications applied 3 months prior to harvest, or mixed with the abscission spray, had no affect on these conditions.

Effects of temperature were very similar to those observed in Florida.[140,149] When temperatures were 50°F (10°C) or less little fruit loosening occurred with navel oranges; spray application followed by rising temperatures increased abscission effectiveness, while falling temperatures reduced effectiveness.

In Australia the harvesting of citrus fruits has become an acute problem in the past few years and at least two very active abscission-mechanical harvesting research programs have been formed in Victoria and New South Wales. Both appear to be very excellent efforts which have taken the results from Florida programs and adapted them to the different climatic and growing conditions prevalent in those areas. In terms of climate, their conditions tend to be more closely related to the Mediterranean climate regions than to the subtropical, humid conditions of Florida. Their principal cultivar of interest is the Valencia because the navel orange, although grown extensively in Australia (and to a much lesser extent in Florida), is processed to a very limited extent throughout the world. The tasteless precursor of limonin is found in raw navel oranges, but upon processing limonin is formed which produces a bitter taste in the processed products.[150] The Valencia orange and early- and mid-season oranges grown in Florida contain limonin, but at relatively low levels.

Abscission tests in Australia have generally reported removal force reductions very similar to those noted in Florida.[151-153] In addition to the four previously mentioned compounds tested in Florida, the compounds ACR (dikegulac sodium) and N-252 (dithiin tetraoxide) were field tested at concentrations of 1000 to 6000 ppm and 50 to 200 ppm, respectively.[153] Tests were made on navel and lemon cultivars in addition to Valencia. ACR, like ethephon, produced abscission without wound ethylene symptoms on the rind. Lack of response in the regreening period appears to present a problem in Australia as well as Florida.

Florida and recently Australia have developed sophisticated harvesting machinery.[145,154,155] In both countries, complete mechanical systems are currently being tested.

VI. USE OF SURFACTANTS AND ADDITIVES

There are a large variety of these materials available throughout the world and some appear to be better for certain specific uses than others. Experience with abscission agents has shown that some of these are improved by addition of surfactants, but with others the surfactant causes excessive leaf drop or other phytotoxic effects.[145] Williams and Edgerton have also observed similar results with certain pome fruits.[156] About the best that can be said at this time regarding use of surfactants is that performance tests should be conducted with those which are readily available locally. Our own experience has shown there is much performance variation among the various surfactants, some even depressing abscission activity.[43] Surfactants can be aids to chemical activity; however, no surfactant has ever been able to substitute for or improve an otherwise deficient growth regulator. The entire subject of surfactants and additives appears to be more ''art'' than ''science'', hence, it is difficult to make a profound statement regarding their use.

VII. PLANT GROWTH REGULATOR REFERENCES FOR CITRUS

Many articles are cited at the end of this chapter, but space does not permit citation of the several thousands of articles, journals, and books relating to this topic. However, the author would like to mention the following book chapters and review articles which are pertinent. In addition, each of these has an extensive bibliography which the reader may wish to examine. The author recommends Coggins and Hield,[24] Monselise,[65,92] and Moss[9] for general coverage of the entire subject. In addition, the role of endogenous growth substances, internal factors and exogenous control is well treated in articles by Monselise and Goren,[157] Monselise,[158] and Goldschmidt.[159] An interesting model of the abscission process has been constructed by Biggs and Kossuth.[146]

ACKNOWLEDGMENTS

The author acknowledges the helpful reviews by the following: Drs. T. A. Wheaton and W. Grierson, IFAS, Agricultural Research and Education Center, Lake Alfred; Dr. S. F. Nagy, Florida Department of Citrus, Lake Alfred; Dr. I. M. Gilfillan, Outspan Citrus Centre, Nelspruit, South Africa; and Mr. R. M. Burns, California Cooperative Extension Service (Retired), Ventura, Calif.

REFERENCES

1. **Reuther, W.,** Climate and citrus behavior, in *The Citrus Industry,* Vol. 3, Reuther, W., Ed., Div. Agric. Serv., University of California, Berkeley, 1973, 280.
2. **Grierson, W. and Ting, S. V.,** Quality standards for citrus fruits, juice and beverages, *Proc. Int. Soc. Citricult.,* 21, 1978.
3. **Yelenosky, G.,** The potential of citrus to survive freezes, *Proc. Int. Soc. Citricult.,* 1, 199, 1977.
4. Florida Department of Agriculture and Consumer Services and U.S. Department of Agriculture Economics, Statistics and Cooperative Service, Fla. Agric. Stat., Fla. Crop Livestock Rep. Ser., 1979.
5. **Albrigo, L. G.,** Some parameters influencing development of surface wax on citrus fruits, in *Proc. 1st. Int. Citrus Congr.,* 107, 1977.
6. **Albrigo, L. G.,** Ultrastructure of cuticular surfaces and stomata of developing leaves and fruit of "Valencia" orange, *J. Am. Soc. Hortic. Sci.,* 97, 761, 1972.
7. **Albrigo, L. G.,** Variation in surface wax on oranges from selected groves in relation to fruit moisture loss, *Proc. Fla. State Hortic. Soc.,* 85, 262, 1972.
8. **Freeman, B., Albrigo, L. G., and Biggs, R. H.,** Ultrastructure and chemistry of cuticular waxes of developing citrus leaves and fruits, *J. Am. Soc. Hortic. Sci.,* 104, 801, 1979.
9. **Moss, G. I.,** The use of growth regulators in citrus culture, in *Citrus,* CIBA-GEIGY Agrochemicals Tech. Monogr. No. 4, 1975, 61.
10. **Burns, R. M. and Coggins, C. W.,** Sweet orange germination and growth aided by water and gibberellin seed soak, *Calif. Agric.,* 23, 18, 1969.
11. **Choudhari, B. K. and Chakrawar, V. R.,** Effect of some chemicals on the germination of Kagzi lime (*Citrus aurantifolia* Swingle) seeds, *J. Maharashtra Agric. Univ.,* 5, 173, 1980.
12. **Choudhari, B. K. and Chakrawar, V. R., II,** Effect of seed treatment with certain growth regulators on the shoot and root development of Kagzi lime (*Citrus aurantifolia* Swingle), *J. Maharashtra Agric. Univ.,* 6, 19, 1981.
13. **Shukla, K. S., Misra, R. L., Kaul, M. K., and Prasad, A.,** Studies on pollen germination of certain fruits, *Haryana J. Hortic. Sci.,* 7, 162, 1978.
14. **Abdalla, K. M., Wakeel, A. T. El, and Masiry, H. H. El,** Effect of gibberellic acid on seed germination of some citrus rootstocks, *Res. Bull. Faculty Agric. Ain Shams Univ.,* No. 944, Zagazig University, Cairo, Egypt, 1978.
15. **Misra, R. S. and Verma, V. K.,** Studies on the seed germination of Kinnow orange in the Central Himalayas, *Prog. Hortic.,* 12, 79, 1980.
16. **Singh, H. K., Shankar, G., and Makhija, M.,** A study on citrus seed germination as affected by some chemicals, *Haryana J. Hortic. Sci.,* 8, 194, 1979.
17. **Patil, S. B. and Chakrawar, V. R.,** Vegetative propagation of seedless lemon by air-layering, *Punjab Hortic. J.,* 19, 119, 1979.
18. **Lenz, F. and Karnatz, A.,** The effect of GA_3, Alar and CCC on citrus cuttings, *Acta Hortic.,* 49, 147, 1975.
19. **Mougheith, M. G., Hassaballa, I. A., and Rawash, M. A.,** Effect of gibberellic acid and urea sprays on seedling growth of some citrus species, *Res. Bull. Faculty Agric. Ain Shams Univ.,* No. 1077, Zagazig University, Moshtohor, Egypt, 1979.
20. **Abdalla, K. M., El-Wakeel, A. T., and El-Masiry, H. H.,** Seed-bed treatments with GA in citrus. I. Morphologic response of different rootstocks, *Res. Bull. Faculty Agric. Ain Shams Univ.,* No. 937, Zagazig University, Zagazig, Egypt, 1979.
21. **Ben-Gad, D. Y., Altman, A., and Monselise, S. P.,** The effects of root-applied GA_3 and SADH on the vegetative development of sweet lime seedlings, their assimilate distribution and starch content, *Isr. J. Bot.,* 27, 40, 1978.
22. **Abdalla, K. M., Wakeel, A. T. El, and Masiry, H. H. El,** Nursery treatments with GA in citrus. I. Morphologic response of different rootstocks, *Res. Bull. Faculty Agric. Ain Shams Univ.,* No. 936, Zagazic University, Cairo, Egypt, 1978.
23. **Hassaballa, I. A.,** Growth and branching angle response of "Benzahair" lime seedlings to growth regulator sprays, *Res. Bull. Faculty Agric. Ain Shams Univ.,* No. 1076, Zagazig University, Moshtohor, Egypt, 1979.
24. **Coggins, C. W., Jr. and Heild, H. Z.,** Plant growth regulators, in *The Citrus Industry,* Vol. 1, Reuther, W., Webber, H. J., and Batchelor, L. D., Eds., Div. Agric. Serv., University of California, Berkeley, 1968, 371.
25. **Maiti, R. B., Singh, S. M., and Singh, I. J.,** Effects of types of scion buds and plant regulators on the success of bud grafting in grapefruit *(Citrus paradisi), Indian J. Hortic.,* 16, 149, 1959.
26. **Nauer, E. M., Boswell, S. B., and Holmes, R. C.,** Chemical treatments, greenhouse temperature, and supplemental day length affect forcing and growth of newly budded orange trees, *HortScience,* 14, 229, 1979.

27. **Nauer, E. M. and Boswell, S. B.,** NAA sprays suppress sprouting of newly budded citrus nursery trees, *HortScience*, 13, 166, 1978.
28. **Burns, R. M.,** personal communication, 1982.
29. **Boswell, S. B., Burns, R. M., and Hield, H. Z.,** Inhibition effects of localized growth regulator sprays on mature lemon trees, *HortScience*, 11, 115, 1976.
30. **Lundberg, E. C. and Smith, T. S.,** A possible sprout inhibitor for Florida citrus, *Proc. Fla. State Hortic. Soc.*, 87, 20, 1974.
31. **Moye, H. A. and Willson, A. E.,** Residues of naphthaleneacetic acid (NAA) in "Bearss" lemons and their processed products following spraying of tree trunks, *Proc. Fla. State Hortic. Soc.*, 90, 272, 1977.
32. **Phillips, R. L. and Tucker, D. P. H.,** Chemical inhibition of sprouting of pruned lemon trees, *HortScience*, 9, 199, 1974.
33. **Boswell, S. B., McCarty, C. D., Ede, L. L., and Chesson, J. H.,** Control of lemon sprouts for mechanical harvesting, *Citrograph*, 60, 405, 1975.
34. **Boswell, S. B., Embleton, T. W., Jones, W. W., and Summers, L. L.,** Inhibition of regrowth and yield of mature lemon trees with ammonium ethyl carbamoylphosphonate, *HortScience*, 16, 41, 1981.
35. **Lindow, S. E., Arny, P. C., Upper, C. D., and Barchet, W. R.,** The role of bacterial ice nuclei in frost injury to sensitive plants, in *Plant Cold Hardiness and Freezing Stress; Mechanisms and Crop Implications*, Li, P. H. and Sakai, A., Eds., Academic Press, New York, 1978, 249.
36. **Yelenosky, G. and Guy, C. L.,** Carbohydrate accumulation in leaves and stems of "Valencia" orange at progressively colder temperatures, *Bot. Gaz.*, 138, 13, 1977.
37. **Hendershott, C. H.,** The influence of maleic hydrazide on citrus trees and fruits, *Proc. Am. Soc. Hortic. Sci.*, 80, 241, 1962.
38. **Yelenosky, G.,** Sensitivity of budded young orange trees and grapefruit seedlings to Ancymidol in soil mix, *HortScience*, 14, 600, 1979.
39. **Konakahara, M.,** Studies on the mechanism of frost damage and protection in citrus plantations, *Spec. Bull. Shizuoka Prefectural Citrus Exp. Stn. No. 3*, 1, 1975.
40. **Nauer, E. M., Boswell, S. B., and Holmes, R. C.,** Persistence of NAA-induced growth inhibition in sweet orange seedlings, *HortScience*, 14, 525, 1979.
41. **Chattopadhyay, P. K. and Ghosh, S. P.,** Studies on the internal structure of lemon leaves as influenced by plant growth regulators, *Curr. Sci.*, 49, 60, 1980.
42. **Peynado, A., Gausman, H. W., and Rittig, F. R.,** Effects of mepiquat chloride on citrus freezing resistance, *Proc. Plant Growth Reg. Soc. Am.*, 8, 240, 1981.
43. **Wilson, W. C.,** unpublished data, 1982.
44. **Goldschmidt, E. E.,** Abscisic acid in citrus flower organs as related to floral development and function, *Plant Cell Physiol.*, 21, 193, 1980.
45. **Nir, I., Goren, R., and Leshem, B.,** Effects of water stress, gibberellic acid and 2-chloroethyltrimethylammoniumchloride (CCC) on flower differentiation in "Eureka" lemon trees, *J. Am. Soc. Hortic. Sci.*, 97, 695, 1972.
46. **Goren, R. and Monselise, S. P.,** Promotion of flower formation and fruit set in *Citrus* by antimetabolites of nucleic-acid and protein synthesis, *Planta (Berl.)*, 88, 364, 1970.
47. **Salomon, E.,** Effect of CCC on growth distribution and fruiting in citrus, *Acta Hortic.*, 114, 11, 1980.
48. **Krezdorn, A. H. and Jernberg, D. C.,** Field evaluation of growth regulators for fruit set, *Proc. Int. Soc. Citricult.*, 2, 660, 1977.
49. **de Lange, J. H., du Plessis, S. F., Vincent, A. P., du Preez, M. B., Holmden, E. A., and Rabe, E.,** Studies on Clementine yield, fruit size and mineral composition of leaves, *Subtropica*, 3, 7, 1982.
50. **Powell, A. A. and Krezdorn, A. H.,** Influence of fruit-setting treatment on translocation of ^{14}C-metabolites in citrus during flowering and fruiting, *J. Am. Soc. Hortic. Sci.*, 102, 709, 1977.
51. **South Africa, Citrus and Subtropical Fruit Research Institute,** GA_3 trials with honey bees, *Inf. Bull.*, 70, 11, 1978.
52. **Dayuan, Wang,** Effect of BA and GA on fruit drop of citrus, *HortScience*, 16, 657, 1981.
53. **Shawki, I., El-Tomi, A., Nasr, A.,** Effect of urea, gibberellic acid and 2,4-D sprays on navel oranges, *Egypt. J. Hortic.*, 5, 115, 1978.
54. **Kumar, R., Singh, J. P., and Gupta, O. P.,** Effect of growth regulators on fruit set, fruit drop and quality of sweet lime (*Citrus limettioides* Tanaka), *Haryana J. Hortic. Sci.*, 4, 123, 1975.
55. **García-Martínez, J. L. and García-Papí, M. A.,** The influence of gibberellic acid, 2,4-dichlorophenoxyacetic acid and 6-benzylaminopurine on fruit-set of Clementine mandarin, *Scientia Hortic.*, 10, 285, 1979.
56. **García-Martínez, J. L. and García-Papí, M. A.,** Influence of gibberellic acid on early fruit development, diffusible growth substances and content of macronutrients in seedless Clementine mandarin, *Scientia Hortic.*, 11, 337, 1979.
57. **Iwahori, S.,** Use of growth regulators in control of cropping of mandarin varieties, *Proc. Int. Soc. Citricult.*, 263, 1978.

58. **Hirose, K., Iwagaki, I., and Suzuki, K.**, IZAA (5-chloroindazol-8-acetic acid ethyl ester) as a new thinnig agent of satsuma mandarin (*C. unshiu* Marc.), *Proc. Int. Soc. Citricult.*, 270, 1978.
59. **Iwahori, S. and Oohata, J. T.**, Chemical thinning of satsuma mandarin (*Citrus unshiu* Marc.) fruit by 1-naphthalene acetic acid; role of ethylene and cellulose, *Sci. Hortic.*, 4, 167, 1976.
60. **Cooke, A. R.**, personal communication, 1982.
61. **Wheaton, T. A.**, Fruit thinning of Florida mandarins using plant growth regulators, *Proc. Int. Soc. Citricult.*, in press.
62. **Jahn, O. L.**, Effects of ethephon, gibberellin, and BA on fruiting of "Dancy" tangerines, *J. Am. Soc. Hortic. Sci.*, 106, 597, 1981.
63. **Chapman, J. C.**, Ethephon for fruit thinning of Imperial and Beauty of Glen Retreat mandarins in the Central Burnett district, Queensland, *Aust. J. Exp. Agric. Anim. Husb.*, 20, 508, 1980.
64. **Gallasch, P. T.**, Thinning Imperial mandarins with ethephon increased fruit size and grower returns, *Proc. Int. Soc. Citricult.*, 276, 1978.
65. **Monselise, S. P.**, The use of growth regulators in citriculture; a review, in *Sci. Hortic.*, 11, 151, 1979.
66. **Gallasch, P. T., Bevington, K. B., Hocking, D., and Moss, G. I.**, Ethephon thinned heavy crops of Valencia oranges in three widely spaced districts of Australia, *Proc. Int. Soc. Citricult.*, 273, 1978.
67. **Moss, G. I. and Bevington, K. B.**, The influence of rootstock on the response of Valencia orange trees to applied growth-regulators, *Proc. Int. Soc. Citricult.*, 260, 1978.
68. **Moss, G.**, Balancing Valencia orange yields and quality, *Rural Res.*, No. 94, 22, 1977.
69. **Jones, W. W., Coggins, C. W., Jr., and Embleton, T. W.**, Growth regulators and alternate bearing, *Proc. Int. Soc. Citricult.*, 657, 1977.
70. **Gallasch, P. T.**, Control of alternate cropping of Valencia orange with ethephon and naphthalene acetic acid, *Aust. J. Exp. Agric. Anim. Husb.*, 152, 1978.
71. **Guardiola, J. L., Agustí, M., and García-Marí, F.**, Gibberellic acid and flower bud development in sweet orange, *Proc. Int. Soc. Citricult.*, 2, 696, 1977.
72. **Gilfillan, I. M.**, personal communication, 1982.
73. **Stewart, W. S. and Hield, H. Z.**, Effects of 2,4-dichlorophenoxyacetic acid and 2,4,5-trichlorophenoxyacetic acid on fruit drop, fruit production and leaf drop of lemon trees, *Proc. Am. Soc. Hortic. Sci.*, 55, 163, 1950.
74. **Knapp, J. L.**, Florida spray and dust schedule, *Fla. Coop. Ext. Ser.*, Circular 393-H (revised annually), 1982.
75. **Sarooshi, R. A. and Stannard, M. C.**, Effect of timing, concentration and stock/scion combination on chemical control of pre-harvest drop of navel oranges, *Aust. J. Exp. Agric. Anim. Husb.*, 15, 429, 1975.
76. **Kokkalos, T. I.**, Effect of 2,4-dichlorophenoxyacetic acid and gibberellic acid on grapefruit, *Hortic. Res.*, 1, 1981.
77. **Zur, A. and Goren, R.**, Reducing preharvest drop of "Temple" orange fruits by 2,4-D — role of cellulase in the calyx abscission zone, *Sci. Hortic.*, 7, 237, 1977.
78. **Chundawat, B. C., Gupta, O. P., and Arora, R. K.**, Studies on fruit drop in Kinnow, a mandarin hybrid cultivar, *Haryana J. Hortic. Sci.*, 4, 11, 1975.
79. **Kedar, V. P. and Gopalkrishna, N.**, Control of fruit drop in Nagpur mandarin with plant growth substances and urea, *Punjab Hortic. J.*, 16, 101, 1976.
80. **Krezdorn, A. H.**, Florida seeks to extend grapefruit season, *Calif. Citrogr.*, 62, 85, 1977.
81. **Dinar, H. M. A., Krezdorn, A. H., and Rose, A. J.**, Extending the grapefruit harvest season with growth regulators, *Proc. Fla. State Hortic. Soc.*, 89, 4, 1976.
82. **El-Zeftawi, B. M.**, Regulating pre-harvest fruit drop and the duration of the harvest season of grapefruit with 2,4-D and GA, *J. Hortic. Sci.*, 55, 211, 1980.
83. **Gilfillan, I. M., Koekemoer, W., and Stevenson, J.**, Bigger grapefruit profits from later sales, *Citrus Grower Sub-Trop. Fruit J.*, 458, 6, 1972.
84. **Gilfillan, I. M., Koekemoer, W., and Stevenson, J.**, Extension of the grapefruit harvest season with gibberellic acid, *Proc. Int. Soc. Citricult.*, 3, 335, 1973.
85. **Gilfillan, I. M. and Stevenson, J. A.**, Changes in shape, firmness and internal quality of export grapefruit between packhouse and sale point, *Citrus Grower Sub-Trop. Fruit J.*, 507, 5, 1976.
86. **Ramirez, J. M., Krezdorn, A. H., and Rose, A. J.**, Influence of date of harvest on yields of "Hamlin" and "Valencia" oranges and "Marsh" grapefruit, *Proc. Fla. State Hortic. Soc.*, 90, 61, 1977.
87. **Brown, G. E. and Barmore, C. R.**, The effect of ethylene, fruit color, and fungicides on susceptibility of "Robinson" tangerines to anthracnose, *Proc. Fla. State Hortic. Soc.*, 89, 198, 1976.
88. **Barmore, C. R. and Brown, G. E.**, Preharvest ethephon application reduces anthracnose of Robinson tangerines, *Plant Dis. Rep.*, 62, 541, 1978.
89. **Ichii, T. and Hamada, K.**, Studies of "rind yellow spot", a physiological disorder of Naruto (*Citrus medioglobosa* Hort. ex Tanaka), *J. Jpn. Soc. Hortic. Sci.*, 46, 442, 1978.
90. **Ichii, T. and Hamada, K.**, Formation of n-hexanal from rind tissues as related to aging and disorder of rind of Naruto (*Citrus medioglobosa* Hort. ex Tanaka), *J. Jpn. Soc. Hortic. Sci.*, 49, 65, 1980.

91. **Grierson, W.,** Physiological disorders of citrus fruits, *Proc. Int. Soc. Citricult.*, No. 622, in press.
92. **Monselise, S. P.,** Understanding of plant processes as a basis for successful growth regulation in citrus, in *Proc. Int. Soc. Citricult.*, 1, 250, 1978.
93. **Erner, Y., Goren, R., and Monselise, S. P.,** Reduction of peel roughness of "Shamouti" orange with growth regulators, *J. Am. Soc. Hortic. Sci.*, 101, 513, 1976.
94. **Monselise, S. P., Weiser, M., Shafir, N., Goren, R., and Goldschmidt, E. E.,** Creasing of orange peel — physiology and control, *J. Hortic. Sci.*, 51, 341, 1976.
95. **Gilfillan, I. M., Stevenson, J. A., and Wahl, J. P.,** Control of creasing in navels with gibberellic acid, *Proc. Int. Soc. Citricult.*, in press.
96. **Gilfillan, I. M., Stevenson, J. A., Holmden, E., Ferreira, C. J., and Lee, A.,** Gibberellic acid for reducing creasing in navels in the Eastern Cape, *Citrus Subtrop. Fruit J.*, 605, 11, 1980.
97. **Bar-Akiva, A.,** Effect of foliar application of nutrients on creasing of "Valencia" oranges, *HortScience*, 10, 69, 1975.
98. **Kuraoka, T., Iwasaki, K., and Ishii, T.,** Effects of GA_3 on puffing and levels of GA-like substances and ABA in the peel of satsuma mandarin (*Citrus* unshu Marc.), *J. Am. Soc. Hortic. Sci.*, 102, 651, 1977.
99. **Goren, R., Monselise, S. P., and Ben Moshe, A.,** Control of corky (silvery) spots of grapefruit by growth regulators, *HortScience*, 11, 421, 1976.
100. **Stewart, I.,** personal communication, 1982.
101. **Wheaton, T. A.,** unpublished data, 1982.
102. **Stewart, I.,** Citrus color — a review, *Proc. Int. Soc. Citricult.*, 1, 308, 1977.
103. **Stewart, I.,** Color as related to quality of citrus, in *Citrus Nutrition and Quality*, ACS Symp. Ser. 143, Nagy, S. and Attaway, J. A., Eds., American Chemical Society, Washington, D.C., 1980, 129.
104. **Wheaton, T. A. and Stewart, I.,** Optimum temperature and ethylene concentrations for postharvest development of carotenoid pigments in citrus, *J. Am. Soc. Hortic. Sci.*, 98, 337, 1973.
105. **El-Zeftawi, B. M.,** Chemical and temperature control of rind pigment of citrus fruits, *Proc. Int. Soc. Citricult.*, 33, 1978.
106. **Wardowski, W., Soule, J., Grierson, W., and Westbrook, G.,** Florida citrus quality tests, *Fla. Coop. Ext. Serv. Bull.*, No. 188, University of Florida, Gainesville, 1979.
107. **Iwahori, S., Tominaga, S., and Oohata, J. T.,** Degreening of ponkan (*Citrus reticulata* Blanco) fruit by ethephon (2-chloroethylphosphonic acid), *Bull. Faculty Agric. Kagoshima Univ.*, 27, 7, 1977.
108. **Oh, S.-D., Kim, Y.-Y., Hong, S.-B., and Chung, S.-K.,** Effect of pre-harvest application of methionine and ethephon on colour and quality of satsuma fruits (*Citrus unshiu* Marc.), *J. Korean Soc. Hortic. Sci.*, 19, 103, 1978.
109. **Levy, Y., Greenberg, J., and Ben-Anat, S.,** Effect of ethylene-releasing compounds on oleocellosis in "Washington" navel oranges, *Sci. Hortic.*, 11, 61, 1979.
110. **Erner, Y.,** Reduction of oleocellosis damage in Shamouti orange peel with ethephon preharvest spray, *HortScience*, 57, 129, 1982.
111. **Iwahori, S. and Oohata, J. T.,** Alleviative effects of calcium acetate on defoliation and fruit drop induced by 2-chloroethylphosphonic acid in citrus, *Sci. Hortic.*, 12, 265, 1980.
112. CIBA-GEIGY Corp., personal communication, 1980.
113. **Yokoyama, H., Hsu, W., Poling, S. M., Hayman, E., and De Benedict, C.,** Bioregulators and citrus fruit color, *Proc. Int. Soc. Citricult.*, 3, 717, 1977.
114. **Valadon, L. R. G. and Mummery, R. S.,** Effects of two triethylamines on the carotenogenesis of Turkish lemons and oranges, *Z. Pflanzenphysiol.*, 90, 11, 1978.
115. **El-Zeftawi, B. M.,** Effects of gibberellic acid and cycocel on colouring and sizing of lemon, *Sci. Hortic.*, 12, 177, 1980.
116. **Wilson, W. C., Kenny, D. S., and Holm, R. E.,** The Florida Department of Citrus cooperative chemical screening programs for citrus, *Proc. Int. Soc. Citricult.*, 2, 692, 1977.
117. **Crous, P. A.,** The reduction of acidity in Valencias by use of arsenical sprays, *Citrus Grower*, 94, 1, 1941.
118. **Mynhardt, C. Z.,** Arsenical sprays, *Citrus Grower*, 273, 1, 1956.
119. **Basson, W. J.,** Arsenical sprays reduce acids in citrus juice, *Farming S. Afr.*, 52, July, 1959.
120. **Gilfillan, I. M. and Stevenson, J. A.,** Acidification of calcium arsenate sprays in non-saline areas, *Citrus Subtrop. Fruit J.*, 13, March, 1976.
121. **Gilfillan, I. M., Wahl, P., Holmden, E., Reay, N., and Stevenson, J. A.,** Acidification of calcium arsenate sprays, *Citrus Subtrop. Fruit J.*, 5, March, 1976.
122. **Kesterson, J. W., Braddock, R. J., Koo, R. C. J., and Reese, R. L.,** Arsenic and lead content of expressed Florida grapefruit oil as related to cultural sprays, *HortScience*, 10, 65, 1975.
123. **Metz, W.,** The essential trace elements, *Science*, 213, 1332, 1981.
124. *The Merck Index*, 9th ed., Windholz, M., Ed., Merck & Co., Rahway, N.J., 1976, 106.
125. **Singh, R. and Singh, R.,** Effect of nutrient sprays on granulation and fruit quality of "Dancy" tangerine mandarin, *Sci. Hortic.*, 14, 235, 1981.

126. **Singh, R. and Singh, R.,** Effect of GA_3, Planofix (NAA) and Ethrel on granulation and fruit quality in "Kaula" mandarin, *Sci. Hortic.*, 14, 315, 1981.
127. **Chakrawar, V. R. and Singh, R.,** Studies on citrus granulation. IV. Effect of some growth regulators on granulation and quality of citrus fruits, *Haryana J. Hortic. Sci.*, 7, 130, 1978.
128. **Van Noort, G.,** Dryness in navel fruit, *Proc. 1st. Int. Citrus Symp.*, 3, 1333, 1969.
129. **Gilfillan, I. M. and Stevenson, J. A.,** Postharvest development of granulation in South African export oranges, *Proc. Int. Soc. Citricult.*, 1, 299, 1977.
130. **Nickell, L. G.,** Controlling biological behavior of plants with synthetic plant growth regulating chemicals, in *Plant Growth Substances,* Mandava, N. B., Ed., American Chemical Society, Washington, D.C., 1979, chap. 10.
131. **Grierson, W. and Hatton, T. T.,** Factors involved in storage of citrus fruits: a new evaluation, *Proc. Int. Soc. Citricult.*, 1, 227, 1977.
132. **Wardowski, W. F., Barmore, C. R., Smith, T. S., and DuBois, C. W.,** Curing of Florida lemons with ethephon, *Proc. Fla. State Hortic. Soc.*, 87, 216, 1974.
133. **Jahn, O. L.,** Degreening of Florida lemons, *Proc. Fla. State Hortic. Soc.*, 87, 218, 1974.
134. **Rippon, L. E.,** Colouring of lemons with ethephon, *Rural Newsl. (Aust.),* 68, 38, 1978.
135. **Hall, A. E. and Coggins, C. W., Jr.,** Chlorophyll destruction in lemon fruit by light and 1,4-dichloro-1-cyanoethanesulphonanilide, *Physiol. Plant.*, 44, 221, 1978.
136. **Bleinroth, E. W., Hansen, H. A., Ferreira, V. L. P., and Angelucci, E.,** Storage of Tahiti limes and Sicilian lemons at low temperature and with GA, *Coletanea Inst. Tecnol. Alimentos,* 7, 343, 1976.
137. **Blunden, G., Jones, E. M., Passam, H. C., and Metcals, E.,** Increases in chlorophyll retention times of limes after postharvest immersion in N_6-benzyladenine and gibberellic acid, *Trop. Agric.*, 56, 311, 1979.
138. **Rana, R. S. and Chauhan, K. S.,** The effect of post-harvest Ethrel treatment on degreening of fruits of "Kagzi" lime, *Punjab Hortic. J.*, 16, 30, 1976.
139. **Einset, J. W., Lyon, J. L., and Johnson, P.,** Chemical control of abscission and degreening in stored lemons, *J. Am. Soc. Hortic. Sci.*, 106, 531, 1981.
140. **Ramana, K. V. R., Setty, G. R., Moorthy, N. V. N., Saroja, S., and Nanjundaswamy, A. M.,** Effect of ethephon, benomyl, thiabendazole and wax on colour and shelf-life of Coorg mandarins (*Citrus reticulata* Blanco), *Trop. Sci.*, 21, 265, 1979.
141. **Ismail, M. A. and Grierson, W.,** Seasonal susceptibility of grapefruit to chilling injury as modified by certain growth regulators, *HortScience,* 12, 118, 1977.
142. **Feinstein, B., Monselise, S. P., and Goren, R.,** Studies on the reduction of seed number in mandarins, *HortScience,* 10, 385, 1975.
143. **Lewin, I. J. and Monselise, S. P.,** Further studies on the reduction of seeds in mandarins by NAA sprays, *Sci. Hortic.*, 4, 229, 1976.
144. **Wild, B. L., McGlasson, W. B., and Lee, T. H.,** Long term storage of lemon fruit, *Aust. Citrus News,* 53, 10, 1977.
145. **Wilson, W. C., Coppock, G. E., and Attaway, J. A.,** Growth regulators facilitate harvesting of oranges, *Proc. Int. Soc. Citricult.*, in press.
146. **Biggs, R. H. and Kossuth, S. V.,** Physiological model — a citrus harvest aid for "Valencia" fruits, *Acta Hortic.*, 120, 71, 1981.
147. **Wilson, W. C.,** Use of growth chamber data to predict field performance of abscission sprays for oranges, *Proc. Plant Growth Regul. Soc. Am.*, 7(Abstr.), 41, 1980.
148. **Wheaton, T. A., Wilson, W. C., and Holm, R. E.,** Abscission response and color changes of "Valencia" oranges, *J. Am. Soc. Hortic. Sci.*, 102, 580, 1977.
149. **McCarty, C. D., Boswell, S. B., Burns, R. M., and Atkin, D. R.,** Chemical fruit abscission trials: progress, but not practical yet, *Calif. Citrogr.*, 67, 9, 1981.
150. **Ting, S. V., Fisher, J. F., and Nagy, S. F.,** personal communication, 1982.
151. **Hutton, R. J.,** Abscission chemicals for late "Valencia" oranges in the Murrumbidgee irrigation areas of New South Wales, *Proc. Int. Soc. Citricult.*, 257, 1978.
152. **Hutton, R. J.,** Development of a mechanical harvesting system for citrus fruit, *Proc. Int. Soc. Citricult.*, in press.
153. **El-Zeftawi, B. M.,** Effects of various abscission chemicals on fruit loosening of citrus, *Proc. Int. Soc. Citricult.*, 255, 1978.
154. **Brown, G. A. and Hutton, R. J.,** Machines for harvesting citrus fruit, *Proc. Int. Soc. Citricult.*, 99, 1978.
155. **Wedd, S. and Hutton, R. J.,** Progress in mechanical harvesting of citrus, *Agric. Gaz. New South Wales,* 90, 1979.
156. **Williams, M. W. and Edgerton, L. J.,** Fruit Thinning of Apples and Pears with Chemicals, U.S. Department of Agriculture Inf. Bull. No. 289, 1981.

157. **Monselise, S. P. and Goren, R.,** The role of internal factors and exogenous control in flowering, peel growth and abscission in citrus, *HortScience*, 13, 134, 1978.
158. **Monselise, S. P.,** Citrus fruit development: endogenous systems and external regulation, *Proc. Int. Soc. Citricult.*, 2, 664, 1977.
159. **Goldschmidt, E. E.,** Endogenous growth substances of citrus tissues, *HortScience*, 11, 95, 1976.

Chapter 9

COTTON

George W. Cathey

TABLE OF CONTENTS

I. INTRODUCTION

Cotton (*Gossypium* spp.) is basically a perennial woody shrub that evolved in tropical, dry areas of the world, but is now widely grown as a herbaceous annual under both semiarid and humid conditions. This striking example of plant adaptation has been accomplished primarily through breeding and selection. Despite these adaptive changes, however, cotton still exhibits many attributes of its tropical origin. For example, it grows best in warm temperatures and high light intensity, is somewhat drought tolerant, tends to store starch in the stem and roots, and will often continue or resume growth late in the season. Alterations of some of these factors would be desirable.

Since most plant growth and development processes are regulated by natural plant hormones, many of these processes may be manipulated either by altering the plant hormone level or by changing the capacity of the plant to respond to its natural hormones. Plant growth regulator chemicals are synthetic hormones that have the capacity to do this. Indeed, other chapters of this book demonstrate that every phase of plant growth, from seed germination to crop maturity, can be modified by applications of selected exogenous growth regulators.

With the exception of harvest-aid chemicals, the use of growth regulators in cotton culture is relatively new. Yet it seems logical that such chemicals could be used just as effectively to manipulate performance of cotton as it has with many other agronomic and horticultural crops. Thus, plant growth regulator chemicals could become another tool in the cotton producer's reserve for ensuring efficient production.

Research with plant growth regulators on cotton has increased significantly during the past few years, with the major emphasis being directed primarily to the areas of (1) improved seed germination; (2) early flower production and increased early fruit retention; (3) improved quality and yield; (4) control of excessive vegetative growth; (5) early termination of reproductive and vegetative growth; and (6) improved harvest-aid systems. The primary aim of this chapter is to discuss those products that are currently in commerical use as growth regulators on cotton as well as those that are in the final stages of exerpimental testing. Occasional references will be directed to experimental compounds that show promise.

II. GERMINATION AND SEEDLING VIGOR

Cotton requires 150 or more days from planting to harvest. So, to use the entire growing season effectively and for maturation to occur before development is checked by autumn cold, the seeds are often planted relatively early in the spring, before weather conditions are favorable for optimum germination and seedling growth. Because of its tropical origin, the germinating cotton seed and seedlings suffer from many stresses associated with cool early-season temperate zone environments. Cold soil slows germination and delays or prevents seedling emergence; with continued low temperatures, seedling growth of those that do emerge is very slow. Cold weather also provides a favorable environment for diseases that damage the seed and seedlings. In addition, other stresses are sometimes created by cultural practices such as improper soil incorporation of herbicides, fungicides, insecticides, or fertilizers. There is a real need to find ways to increase the tolerance of planting seed to these various stresses. Some of this improvement may be accomplished through breeding efforts, but the use of chemicals for this purpose may ultimately be more feasible. Any treatment that would improve the performance of planting seed would be valuable.

Several efforts have been made during the past several years to mitigate the effects of adverse environmental conditions on seed germination and seedling growth. Most of these efforts have been directed toward seed treatments as protectants against disease organisms, soil treatments with fungicides, herbicides, systemic insecticides, and soil treatments that alter the physical character of the soil. Many of these treatments provide improved performance of planting seed by eliminating some of the early-season stresses within the germination zone. Some, however, may create additional stresses that have deleterious effects on either germination or seedling growth. For example, the herbicide trifluralin (α,α,α-trifluoro-2,6-dimitro-*N*,*N*-dipropyl-*p*-toluidine), when improperly incorporated in the soil, may inhibit lateral root development, increase seedling diseases, and reduce fertilizer uptake.[1,2] Systemic fungicide and insecticide treatments may reduce both germination and seedling survival.[3]

More recently, synthetic growth regulator chemicals have been tested as seed and seedling treatments to cause further improvements in germination and growth. Some of these have shown some advantages, but none is presently in widespread commercial use. The deleterious effects caused by trifluralin have been reduced by soil incorporation of cottonseed oil, oleic acid, or D-α-tocopherol.[2] IAA (indole-3-acetic acid) and kinetin may also help overcome trifluralin damage.[4] Cole and Wheeler[5] obtained satisfactory stands of cotton at suboptimum temperatures by preconditioning seed before planting with either water alone, gibberellic acid (GA), or AMP (adenosine-3′-5′-cyclic monophosphate). Chilling injury was reduced and both germination and seedling growth improved by each treatment. Thomas and Christiansen[6] obtained similar results by hydrating seed in water for 6 hr at 31°C, followed by a 24-hr chill treatment at 5°C in water. They reported increased yields of plantings from poor quality seed treated in this manner. Christiansen and Ashworth[7] used an antitranspirant spray to prevent chilling injury to seedling cotton. Water loss by the seedlings was reduced by approximately 40%, and both seedling death rate and cold stress inhibition of subsequent growth were reduced. Gausman et al.[8] reported a reduction in wind-blown sand damage to both the leaves and stems of cotton seedlings sprayed with mepiquat chloride (1,1-dimethyl-piperidinium chloride). The treated plants had more turgid leaves and stems with fewer lenticels than did the untreated plants. Ergle[9] reported that gibberellins induced more rapid emergence and taller seedlings, but caused no effect on seedling survival or agronomic performance. In a similar study, Bird and Ergle[10] found that cotton cultivars differed in their response to GA_3 and suggested that the cultivars may vary in their levels or natural gibberellins.

Seed treatments with ABA abscisic acid (ABA), Cycocel (2-chloroethyltrimethylammonium chloride), IAA, NAA (1-naphthaleneacetic acid), or 2,4-D (2,4-dichlorophenoxy acetic acid) have provided little benefit to either seed germination or seedling growth.[11-13] In fact, IAA, NAA, and ABA were all shown to inhibit germination[11,12] and, in some cases, reduce stands and delay maturity.[11] Halloin[12] did find, however, that ethephon [(2-chloroethyl)phosphonic acid], GA_3, and kinetin all partially overcame the inhibitory effect of ABA. More recently, Reddy and Probhaker[14] reported that Cycocel prevented root rot in seedlings.

III. VEGETATIVE DEVELOPMENT

In many areas where cotton is produced, the cultural practices employed and the environmental conditions that exist are often conducive to sustaining growth and development of large plants with an abundance of fruiting positions. In addition, if either poor insect control or inclement weather promotes early loss of fruiting forms by shedding, the poorly fruited plants grow even taller, bear dense foliage, and tend to lodge. This rank condition is not only conducive to the rotting of bolls as they mature, but makes it difficult to obtain desirable chemical defoliation in preparation for mechanical harvesting. Insect control also becomes more difficult because of the increased attractiveness of lush vegetation to most damaging insects, and the difficulty of obtaining adequate penetration of the insecticide through the plant canopy. Therefore, a real need exists for the prevention of excessive vegetative growth in situations that are conducive to such growth.

A phase of cotton research that provides one approach to this problem centers on the efforts to reduce plant development to a size that will alleviate the several problems associated with rank growth. Early attempts to alter plant size in Mississippi were first made by various mechanical treatments.[15] None of these ever proved entirely satisfactory, however, and efforts were then directed toward chemical treatments to accomplish the same objective. The plant growth retardant Cycocel (also known as CCC and chlormequat) has been used in many areas of the world to significantly reduce plant height. However, significant reductions in seed cotton yields are frequently reported when this chemical is used. In addition, fiber and seed quality may be adversely affected. Thomas[16] was able to significantly reduce plant height with spray applications of CCC to greenhouse cotton; seed cotton yield, however, was reduced by about the same percentage as was the plant height. He reported only a minor reduction in flowering, but a significant reduction in boll set during the second through fifth week of flowering. The limitation in productivity became evident 2 to 4 weeks after treatment, and persisted for 2 to 5 weeks depending upon the CCC concentration. In a more recent study Thomas[17] was able to reduce plant height by 40 cm without a significant reduction in yield. De Silva[18] reported that cotton in Uganda responded to CCC with a reduction in both plant height and yield. A significant loss in fruit from treated plants occurred during the sixth and seventh week of flowering. Results from this study were similar to those obtained in Arizona[19] and in Mississippi.[20,21]

Marani et al.[22] applied CCC at 50 and 100 g/ha and CMH (N-dimethyl-N-β-chlorethyl-hydrazonium chloride) at 480 and 720 g/ha. Both chemicals significantly decreased growth rate. When applied at the beginning of flowering, neither CCC at 50 g/ha nor CMH at either rate decreased lint yield or quality. Application of CCC at 100 g/ha did decrease boll retention and yield. These results were improvements over earlier efforts cited by the authors. Singh[23] applied CCC to four cultivars of cotton in India where excessive vegetative growth was a problem. Application of CCC 70 to 80 days after planting retarded growth and increased the number of bolls per plant, boll weight, and lint yield. Sprays of 40 to 160 ppm CCC increased yields 15 to 45%. Singh[24] also found evidence that CCC promoted drought resistance in cotton. TIBA (2,3,5-triiodobenzoic acid) was used by Thomas[25] to reduce plant size and dry weight, but the chemical also significantly reduced seed cotton yield. Similar treatments with CCC in the same experiment reduced plant size without affecting yield.

The experimental growth regulator chemicals BAS 0660W (dimethyl-morpholium chloride) and BAS 0640W [dimethyl-N-(β-chlorethyl)hydronium chloride] caused significant reductions in plant height, and increases in early flowering, early maturity, and lint percent.[26] These two chemicals were forerunners of one of the most recent plant growth regulators to become commercially available to the cotton grower, i.e., 1,1-dimethyl-piperidinium chloride (also known as mepiquat chloride and Pix). The product has been under investigation in cotton since 1975 and the results from these efforts have been at variance on occasion, particularly when treatment effect on seed cotton yield was measured. There has been general agreement, however, with respect to time and rate of application and chemical effects on plant size and appearance. Willard[27] reported that the generally accepted practice has been to apply 50 g/ha when the plants begin to flower and, under some circumstances, apply a second treatment of 25 g/ha approximately 3 weeks later. Approximately 1 week after treatment with mepiquat chloride the leaves appear dark green in color[28-30] and the growth rate becomes noticeably reduced.[30-32] Under conditions of luxuriant moisture and fertility, a 20 to 30% reduction in plant height can be expected from applications of mepiquat chloride.[32] Similar reductions in lateral branch length also occur. Walter et al.[30] found a 22% reduction in canopy width of plots treated with mepiquat chloride. Reduction in plant size was a result of reduced internode length and not a result of reduced fruiting branch numbers or the number of fruiting positions per plant. The number of bolls harvested per plant was reduced, however, but individual boll weight was increased so that no difference in total yield of seed cotton occurred. Conversely, Feaster et al.[31] found smaller bolls on mepiquat chloride-treated plants. Flower production was increased, but seed cotton per flower was reduced due to a combination of smaller bolls and reduced boll set. Erratic yield responses to mepiquat chloride were also reported by Briggs[33] in Arizona. He concluded that environmental factors have a major role in determining the final yield response to this chemical.

Mepiquat chloride-treated leaves are thicker and have a reduced surface area.[28,30] The chlorphyll a/b ratio is increased,[34] and CO_2 uptake is decreased for about 30 days after treatment and then increases as the plants become larger.[34,35] The change is apparently due to biomass dilution of the chemical. Low concentrations of mepiquat chloride were shown to stimulate CO_2 uptake.[35] Gausman et al.[28] described mepiquat chloride-treated leaves as having an increased total thickness with no difference in epidermal thickness. The palisade cells were longer and thinner, and the spongy parenchyma layer was thicker. There were fewer but larger mesophyll air spaces, so that the total air space was greater.

That root growth and water relations are affected by mepiquat chloride applications has been well documented.[36-39] Cappy and Cothren[38] used fiber optic methods to observe the effect of mepiquat chloride treatments on root growth of cotton plants. They reported that the rate of root density increase was significantly less on treated plants than on the untreated controls. The greatest difference occurred during the early flower stage of plant development. Plant water status was more favorable under three moisture levels after treatment with mepiquat chloride,[37] and treated plants had a higher leaf water potential as stress increased.[39] Pot weights in greenhouse experiments indicate significantly less total water use by treated plants, but field experiments showed no difference in water savings on a seasonal basis.[39]

IV. REPRODUCTIVE DEVELOPMENT

Cotton normally begins to flower and set fruit about 65 days after emergence and continues until environmental factors stop growth. Flower production is relatively low at first, but gradually increases for 4 or 5 weeks and then declines rather abruptly as the plants become loaded with developing bolls. There is a regulatory mechanism within the plant that causes an unusually large percentage of the fruiting forms (flower buds and young bolls) to shed. The plant frequently matures bolls from fewer than half the flowers produced. The exact nature of this mechanism is not understood, but it could conceivably be either nutritional as suggested by Eaton[40] or hormonal as proposed by Horowitz.[41] More likely, however, an interplay of both mechanisms is involved.

The use of synthetic growth regulator chemicals to improve yield and quality has been studied by numerous investigators. In addition to the traditional efforts of using chemicals to increase yield through increased boll set, other approaches, such as the chemical alteration of plant processes, are being investigated. Attempts to increase fruitfulness with growth regulator chemicals have been less successful than other areas of cotton plant manipulation. Major beneficial effects on initiation and retention of cotton fruit were reported by Phillips et al.[42] when field plants in Arkansas were sprayed with chlordimeform [N-(chloro-O-tolyl)-N,N-dimethylformamide]. However, no differences in flowering rate, boll production, or yield were detected in similar studies conducted in an insect-free environment in Mississippi.[43] Conflicting results have also been obtained from the use of acephate (O,S-dimethyl acetylphosphoramidothioate) to alter plant growth and development.[44] Multiple applications of acephate to field plants in 1977 caused an increase in flowering rate, boll production and size, and yield. Similar treatments in 1978 resulted in no effect on any of the parameters measured. Acephate was shown to move from treated leaves to all parts of the cotton plant, with preferential translocation to areas of rapid growth.[45] The fruiting forms appeared to be the main sink.

When insecticides are used as test chemicals for plant growth regulation, it is often difficult to distinguish between the plant response to the chemical per se, and the response to relief from insect damage. There is ample evidence, however, that some insecticide chemicals can have physiological effects on flowering and fruiting. Hacskaylo and Scales[46] reported that flower formation, boll set, and plant growth were retarded and plant maturity hastened when cotton grown under insect-free conditions was sprayed with dieldrin (1,2,3,4,10-10-hexachloro-6,7-epoxy-1,4,4a,5,6,7,8,8a-octahydro-endo-exo-1,4,5,8-dimethanonaphthalene) and DDT [1,1,1-trichloro-2,2-bis(4-chlorophenyl)ethane]. Conversely, plants sprayed with guthion [O,O-dimethyl S-(4-oxo-1,2,3-benzotriazin-3(4H-yl)methyl) phosphorodithioate] produced more flowers and more bolls and had a longer maturation period than the controls.

Brown et al.[47,48] used multiple applications of toxaphene-DDT, calcium arsenate, and methyl parathion [O,O-dimethyl O-(p-nitrophenyl)phosphorotrithioate] to determine insecticide effects on growth and development of field cotton. Treatments with toxaphene-DDT increased boll production in both 1960 and 1961. A concurrent reduction in boll size the first year offset the boll number advantage and yield was not affected. Yield was increased the second year in the toxaphene-DDT plots. Treatments with calcium arsenate reduced both boll production and yield, but hastened maturity each year. Roark et al.[49] reported no effect on plant growth and development from treatments with toxaphene, DDT, or toxaphene-DDT mixtures. Methyl parathion apparently has no effect on boll numbers or seed cotton yields, but tends to increase average node number of the first fruiting branch, delay flower production, increase average boll period, and delay maturity.[48-50]

Walhood[13] applied gibberellic acid (GA_3) directly to cotton flowers and young fruit in California and found a significant increase in boll set. Attempts to increase yields by spray applications to entire plants or entire fields were only moderately effective. Slightly higher yields occurred in some cases, but the results were erratic and showed no relationship between GA_3 rates and yield increases. Boll numbers were increased when individual plants were treated, but since boll size was reduced, only a slight yield increase occurred. Walhood[51] reported earlier that application of GA_3 to terminals of individual plants that had stopped their growth caused a resumption in growth. So either delayed senescence or prolonged growth may account for the occasional yield increases. Lane[52] reported that the application of GA_3 to cotton in Texas caused the plants to be taller and yield slightly less than the untreated controls. Subbiah and Mariakulandia[53] found that flower bud and boll abscission were reduced on GA_3-treated plants, but yields were not affected.

Work by Varma[54] suggests that the cotton fruit retention/abscission ratio depends on the balance between nitrogen and the endogenous growth regulators GA and ABA within the plant tissue. He reported the GA and ABA levels to be low in retained fruit as compared to abscising ones, and nitrogen levels to be high in retained bolls and low in those abscising. Work by Rogers[55] also suggests that the retention/abscission ratio of cotton fruit depends more on the balance between ABA and other hormone levels than on the absolute amount of ABA present. No significant effects on yield or quality were found during 2 years of testing GA_3 in Mississippi.[56] Plants were slightly taller and delayed in maturity. Boll production was increased slightly, but the bolls were smaller and no yield increase occurred.

Several reports from India indicate that IAA and NAA may decrease boll abscission and increase cotton yields. Negi and Singh[57] reported an increase in boll numbers and an 8 to 10% increase in yield at first harvest when plants were sprayed with 10 ppm NAA. A slight decrease in yield occurred when a 20-ppm concentration was used. Murty et al.[58] sprayed cotton with 30 ppm NAA at flower initiation and again at peak flowering and obtained a decrease in boll shedding and increased yields. Varma[59,60] found that the abscission promoting effects of abscisic acid were completely counteracted when NAA was applied to either flower buds or young bolls.

Low rates of TIBA (2,3,5-triiodobenzoic acid) applied to plants 2 weeks before first bloom caused a slight increase in boll set and slightly earlier termination, but had no significant effect on yield.[25] Rates higher than 50 ppm reduced both boll set and yield of greenhouse cotton. Freytag and Coleman[61] reported yield increases of 16% from the use of TIBA on field-grown cotton in Texas. The TIBA treatments increased boll size and the number of bolls per plant. The position of the first fruiting branch was also lowered. They suggested that TIBA inhibited IAA transport and decreased endogenous ethylene concentrations.

Cytokinins delay or prevent senescence and promote the ability of organs to compete for metabolites.[62] Rodgers[63] made comparative analyses of retained and naturally abscising cotton fruits and found that abscission was negatively correlated with the concentration of cytokinins. Logically, then, spray applications of this chemical to cotton plants should reduce flower bud and young boll shed and promote fruit growth. Varma,[59,60] however, found that cytokinin treatments promoted boll abscission except when application was made directly to the abscission zone. Numerous formulations containing cytokinins are marketed as plant growth stimulants for a wide range of crops. Although several of these have been tested for yield enhancement in cotton, the author is unaware of any reports in the literature of significant yield increases from their use. Cothren and Cotterman[64] tested one such product for 2 years in Arkansas and found trends toward increased yields in the cytokinin treated plots, but the increases were not significantly different from the untreated controls either year. Similar results were obtained by Cathey[65] in Mississippi after 3 years of testing. Cothren and Cotterman[64] reported significant decreases in transpiration and nitrogen loss from treated leaves which suggest that cytokinins may alter the metabolism of cotton plants in favor of increased yields.

Yield and quality of cotton might also be affected by alterations in fiber properties. For example, elongation and secondary wall thickening of cotton fibers overlap in time,[66] so if the duration of either process is extended, the total yield or the quality of the fiber might be significantly improved. Studies of some effects of growth regulators on fiber properties have been made in recent years. Bhatt et al.[67] reported that low concentrations (20 mg/ℓ) of Cycocel gave coarser fiber without affecting other fiber characters, whereas higher concentrations (100 to 200 mg/ℓ) increased length and fineness, but decreased strength, maturity, and yield. Cycocel was also shown to cause a significant number of abnormal bolls to be produced,[68] and the fiber from these bolls was coarser, stronger, and more mature. Bhatt et al.[67] also found that IAA improved length and fineness of fiber, and that low concentrations of NAA increased fiber fineness but had the reverse effect at higher concentrations. Gibberellic acid was shown to significantly increase fiber length in one variety of cotton in India;[69] in a separate study, however, GA_3 had no effect on fiber length or any other fiber property of another Indian variety.[70]

In addition to modifying plant growth and development per se, plant growth regulator chemicals may have effects that indirectly influence production. Some of these may be of practical value, while others may not. For example, TD-1123 (potassium 3,4-dichloro-5-isothiazole carboxylate) was shown to cause male sterility in cotton flowers without significantly reducing female fertility.[71] Treatment rates, concentrations, and timing of applications have not been completely worked out, but if chemical male sterility could be achieved throughout the growing period, the production of hybrid cotton seed might be feasible. Ervin et al.[72] reported that the growth retardants CCC and mepiquat chloride mitigated symptom expressions of *Verticillium* wilt of cotton and increased the yield of plants grown on wilt-infested land. Alterations in plant structure by applications of mepiquat chloride may alter the microclimate within the plant canopy sufficiently to cause significant reductions in boll rots, thereby improving both yield and quality. Snow et al.[73] found this to be the case in Louisiana during years of abundant moisture. Boll rot damage in Mississippi was also reduced by mepiquat chloride treatments in 1981.[74] The short, compact, open canopy plants resulting from mepiquat chloride treatments might conceivably permit more efficient insect control through superior penetration and plant coverage by insecticide applications. Conversely, the darker green color of mepiquat chloride-treated plants may be more attractive to damaging insects. This is an area of research that needs further investigation.

Carns[75] expressed doubt that boll set and yield of cotton can be increased directly by the application of synthetic growth regulators. He suggested that instead of this approach, research should be directed towards chemical alteration of plant processes such as increased photosynthesis rate, altered photosynthate export from leaves, and altered respiration rates. Guinn et al.[76] presented evidence to show that photosynthesis limits the yield of cotton, even when the plants are grown under optimum environmental conditions. They enriched the atmosphere with CO_2 and increased the photosynthesis rate by 31 to 65% and boll production by 65%. Freytag et al.[77] investigated the effects of soil-injected ethylene on yields of cotton in Texas. Significant yield increases occurred during each year of a 2-year study when treatments (1.59 and 3.14 kg/ha) were imposed at the eighth node stage of plant development. None of the quality attributes was affected. Experiments with two experimental growth regulator chemicals (BTS-44 and DPX-6634) that had shown indications of altering photorespiration in cotton were conducted during 1980 and 1981 in Mississippi.[78] Both chemicals exhibited trends toward increased boll production and yield, but the yields were not significantly different from the controls either year.

V. CROP TERMINATION

A relatively new concept in cotton production in many areas is timely termination of growth. Objectives have been twofold in nature: namely, early harvest with its inherent preservation of quality, and reduction in populations of damaging insects. Cotton is inherently a perennial which is cultured as an annual; in many areas where high yields are produced, the plants frequently continue vegetative and reproductive growth until late in the season or, at times, even until frost. This late-season growth contributes little or nothing to yield, but diverts energy from maturing bolls, delays harvest, increases boll rot, and provides food for diapausing (overwintering) insects. New growth in terminals attracts insects; the abundance of food causes increased damage to late-maturing bolls, as well as increasing overwintering insect populations that would otherwise remain at lower levels.

Plants with an abundance of late-season growth are also less responsive to most defoliant chemicals and mechanical harvest is less efficient. Delayed harvest allows deterioration of lint and seed of the bolls that have already opened in the lower portion of the plant; as the fall season progresses, inclement weather conditions become more prevalent which may delay harvest even further. In addition, late-season cool weather will generally adversely affect defoliation and harvest efficiency. Late-season growth can be limited to some extent by managerial practices such as selection of short-season "determinate" varieties, regulation of fertility and water, and early-season insect control. None of these has proven entirely satisfactory, however. The search continues for an appropriate chemical growth regulator and management practice that will cause a rapid removal of late-season fruiting forms and terminate further plant growth without interfering with the maturation of bolls already present, or causing an excessive yield reduction.

Most of the crop termination investigations for cotton have been conducted by Kittock et al.[79] in Arizona and Thomas et al.[21] in Mississippi. The primary objective of research in Arizona has been suppression of the pink bollworm (*Pectinophora gossypiella* Saunders), while that in Mississippi has been directed toward harvest efficiency. More recent work in Mississippi, however, was concerned with suppression of late-season populations of insects, including boll weevil (*Anthonomus grandis* Boheman), bollworm (*Heliothis zea* Boddie), and tobacco budworm (*H. virescens* F.).[80] While 30 or more chemicals have been subjected to preliminary evaluations as cotton plant growth terminators, those most extensively tested include CCC, chlorflurenol (methyl 2-chloro-9-hydroxyfluorenol-9-carboxylate), 2,4-D, TD-

1123, BAS-0660, DPX-1840 [3,3a-dihydro-2-(p-methoxyphenyl)-8*H*-pyrazolo[5,1-a]isoindol-8-one], ethephon, dicamba (3,6-dichloro-o-anisic acid), silvex [2-(2,4,5-trichlorophenoyl)propionic acid], and glyphosate [*N*-(phosphonomethyl)-glycine]. Most of these have been evaluated separately, in various combinations, and in sequential applications; many have been eliminated from further testing because of their demonstrated adverse effect on one or more parameters of efficient production. Most have been (arbitrarily) classified as: (1) slow acting and persistent or (2) fast acting and nonpersistent.[81] The most desirable results have been obtained with combination treatments of chemicals from both groups, either in a single application or in sequential applications.[19,20,80-85] CCC and chlorflurenol were the most effective of the slow acting persistent group, while 2,4-D and TD-1123 proved most satisfactory as fast acting nonpersistent chemicals. 2,4-D, however, exhibited some deleterious effects that might cause it to be unacceptable to growers. For example, it caused the leaves to be nonresponsive to defoliant chemicals[84] and it translocated to the seed causing poor germination and seedling growth. Some of the other chemicals were effective crop terminators, but they were either inconsistent in their effect or produced undesirable effects on yield and quality.[84-86] There is general agreement among investigators as to the most efficient use rate for each chemical and the most desirable time of treatment application. CCC and chlorflurenol are each applied at 0.56 kg/ha, TD-1123 at 0.56 to 1.12 kg/ha, and 2,4-D at 0.028 to 0.041 kg/ha. Satisfactory results have been obtained in most cases when treatments were applied during late August or early September.

The graph in Figure 1 shows the cotton production pattern, by weeks of flowering, that occurs in the Mississippi Delta. This is a composite graph made from the data of several years and shows that portion of the total yield potential that is normally affected by terminator chemicals. Most terminators have some adverse effect on bolls that are less than 18 to 20 days old, or those that set during the 3 weeks before treatment. Bolls that set after August 21 in cotton with a production pattern such as the one in the graph would be affected to some degree by treatments made on September 11. The flower buds and very young bolls would be eliminated and the older bolls affected in varying degrees depending on their age.

Thomas and Hacskaylo[87] found that DPX-1840 was readily absorbed and translocated to stem tips of greenhouse cotton where it eventually retarded growth without serious effects on boll development. Additional studies with field plants revealed that this chemical satisfactorily curtailed late growth and fruit set without material reductions in yield when applied in mid to late August.[20] DPX-1840 was also found to be more effective than CCC in this respect.[88] Similar results were obtained when various combinations of chlorflurenol and TD-1123 were applied in late August and early September.[80] Leaves, flower buds, small bolls, and insect populations were significantly reduced, with only minimal yield reductions.

In addition to reduced late-season insect populations, significantly earlier harvests may be obtained from some crop termination treatments.[78,89] Wolfenbarger and Davis[89] applied chlorflurenol and 2,4-D to cotton 35 days after first flower and reported 90 to 95% open bolls 30 days later. Yields were also increased because of reduced boll damage. These results indicate that crop termination treatments might make it possible to harvest before October 1. This would be extremely advantageous to cotton growers in the mid-South.

Bariola et al.[90] and Kittock et al.[83,85] reported highly significant correlations between pink bollworm larvae numbers and immature boll numbers that remained in the field at harvest. Some of their crop terminator treatments completely eliminated the immature bolls and reduced pink bollworm larvae by as much as 97%. Hopkins and Moore[91] used low rates of the defoliant chemical thidiazuron (N-phenyl-N′-1,2,3-thiadiazue-5-ylurea) to reduce feeding sites and insect populations without adversely affecting either yield or quality. The use of defoliant chemicals, at any use rate, for this purpose is highly risky, however. Too often, excessive defoliation occurs and a significant loss in yield and quality results.[92] The herbicide glyphosate [N-(phosphonomethyl)glycine] might also suffice as a crop terminator on cotton.

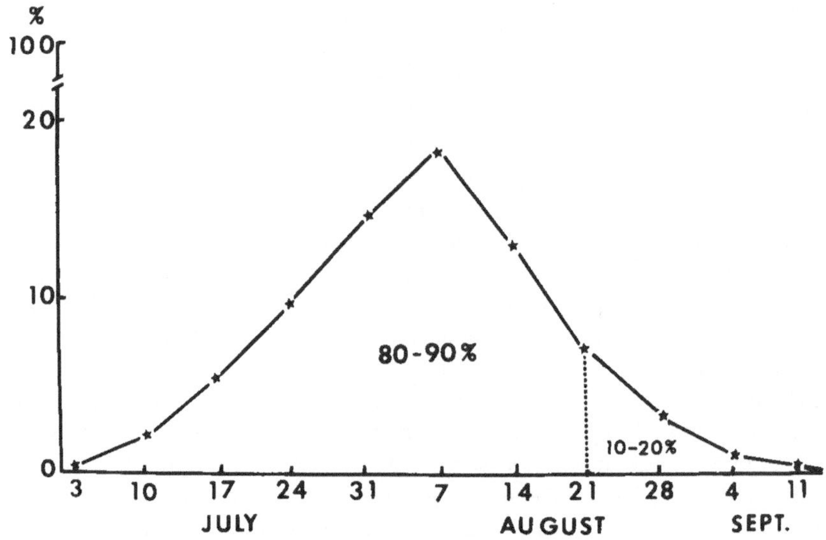

FIGURE 1. Cotton production pattern for the Mississippi Delta.

It was shown to inhibit regrowth development for as long as 7 weeks after application.[87] At some use rates, however, deleterious effects on seed were noted, so further testing of this chemical is warranted.

VI. HARVEST AIDS

The use of chemicals to condition cotton plants for harvest received its greatest impetus with the advent of mechanical harvesters during World War II. The practice has since increased until over three quarters of the 5 to 6 million hectares grown in the U.S. annually is now treated with harvest-aid chemicals of some kind.[93] It is considered one of the major beltwide practices, and has become an important step in the production of high quality fiber. In addition, since incentives for high yields have resulted in cultural practices that tend to force plants into late-season vegetative and reproductive growth, the benefits to be derived from the use of harvest-aid chemicals have increased because of the difficulty of effectively conditioning late-season plants for mechanical harvest.

The practice, however, is not always profitable. Research has indicated that yields may be lowered, and the quality of both fiber and seed reduced, when these chemicals are improperly used.[92-98] Many factors determine the profit or loss that may accrue from the use of harvest aids; it is sometimes difficult to determine if the benefits derived exceed the costs involved. Some advantages may not be measured by monetary standards. Factors difficult to measure include those associated with early harvest, ease of machine operation, difference in machine wear, insect pests, boll rots, seed and fiber damage, and regrowth. Chemicals in general used as harvest-aids on cotton may be classified as defoliants, desiccants, regrowth inhibitors, or boll openers; a limited amount of overlap of the categories occurs with some chemicals. Most however, are used to perform a specific function.

After cotton defoliation was accidentally discovered in the late 1930s, research indicated that it was feasible to remove leaves from a cotton crop before it was harvested.[99] Interest intensified with the advent of mechanical harvesters and during the next several years, several thousand chemicals were screened in the search for efficient defoliants. From this extensive search, only about 20 chemicals were ever recommended by public service agencies; of the 20, fewer than half are now extensively used. The search continues; a few promising chemicals are being tested at present under experimental label, and some of these may be available to the public within a few years. It appears unlikely that chemically induced defoliation results from any specific physiological action by the defoliant chemicals, since there is such variance in their chemical structures and in the injuries produced. The physiological basis for leaf abscission is beyond the scope of this chapter; rather, the reader is referred to the excellent review by Carns[100] for more information on the subject.

Formulations of sodium and magnesium chlorate were among the earliest products used extensively as cotton defoliants, and both are still in widespread use across the belt. They are relatively inexpensive products and are very effective, especially when leaves are fully matured; they have little effect on immature leaves or on regrowth vegetation. Sodium chlorate is highly flammable and is usually formulated with a fire retardant such as sodium borate or urea; magnesium chlorate is hygroscopic and thereby offers its own fire-retardant properties. Both chemicals are labeled commercially under a variety of trade names that are equally effective when applied at equivalent active ingredient rates.

The two organophosphorus compounds (S,S,S-tributyl phosphorotrithioate and tributyl phosphorotrithioite) are highly efficient defoliants and frequently cause leaf fall before excessive drying occurs. They often remove immature leaves and are relatively effective in removing regrowth vegetation. Cacodylic acid (hydroxydimethylarsine oxide) is formulated as sodium cacodylate and is used extensively as a cotton defoliant in the western U.S. where leaves are consistently tougher than those further east. Limited testing of the product in Mississippi indicates a tendency towards excessive desiccation of leaves in the terminal portion of the plant.[93]

Two of the more recently developed defoliant chemicals are dimethipin (2,3-dihydro-5,6-dimethyl-1,4-dithiin-1,1,4,4-tetraoxide) and thidiazuron. Both products have undergone extensive testing across the cotton belt under experimental labels, and are expected to be available for grower use during the 1982 season.[101,102] The two chemicals are about equal in their effect on cotton leaf abscission, and the effect is comparable with that caused by either the chlorates or the organophosphates. Both chemicals, however, are superior to either the chlorates or the phosphates in the inhibition of regrowth development.[91] Once these two products become available the farmer will have a wider choice of products to use in the harvest-aid program.

Numerous adjuvant compounds have been used by growers to improve the effectiveness of defoliant chemicals. They are formulated to accomplish such things as alter droplet size; reduce drift; improve coverage; retain chemical in contact with the leaf longer; and aid penetration. Their effectiveness has not been well established in all cotton growing areas or in all plant growing environments. Improvements have been reported by researchers in Arizona,[103] but results in Mississippi have been inconsistent, even when used in adverse environments.[93] Therefore, establishing the need for such additives is difficult. More recently, cotton growers have begun to add chemicals with either abscission or desiccation-inducing properties to the defoliant mixture. The value of this practice is also controversial.

The addition of an endothall [7-oxabicyclo(2.2.1)heptane-2,3-dicarboxylic acid] formulation usually causes slightly earlier leaf drop, but by the seventh day after treatment results are similar to those from the defoliant treatment alone. There is a possibility however, of a synergistic effect when endothall is used in combination with defoliants. Recent work by Sterett et al.[103] showed such an effect of endothall and ethephon. Sixty to ninety grams per hectare of paraquat (1,1′-dimethyl-4,4′-bipyridinium ion) is frequently added to defoliant solutions applied late in the season after adverse environmental conditions develop. Paraquat aids in abscission and/or desiccation of regrowth vegetation and juvenile leaves in plant terminals.[104] Morgan and Durham[105] used gibberellic acid to enhance ethylene-induced abscission, and suggested that GA_3 might improve the performance of several of the defoliant chemicals.

Attempts are frequently made to induce senescence and partial defoliation in plants before they are mature enough for complete defoliation. This technique is used to improve air circulation by removal of mature leaves and, by increasing senescence in the immature leaves, to cause them to be more responsive to a subsequent defoliant treatment. This practice usually involves the addition of small quantities of a defoliant chemical to a late insecticide application, or the application of a low rate of the defoliant alone 2 or 3 weeks before defoliation is planned. This is not a widespread practice, however, and is not generally recommended because of inconsistent results. Too often, excessive defoliation occurs, and a loss in yield and quality results.[92]

Plants have been successfully conditioned for increased response to defoliant chemicals, however, by use of the experimental growth regulator TD-1123.[106] Defoliation was increased by as much as 25%, without an adverse effect on yield or quality, when this chemical was applied 10 days before the regular defoliant treatment. A similar effect by this chemical was noted by Arle.[107] Sequential treatments of TD-1123 and defoliant have an apparent synergistic effect on several physiological events that occur following defoliant applications.[108] For example, most parameters that are affected by a defoliant chemical are not altered by TD-1123, but become more pronounced and occur earlier in sequentially treated leaves.

Desiccation is a substitute for defoliation and is used for rapid drying of leaves and other plant parts preparatory to harvest. The practice is standard on about 1.5 million ha in the high plains of Texas, Oklahoma, and New Mexico, where stripper harvest methods are used. Complete removal or drying of leaves by defoliation, desiccation, or frost is essential for satisfactory results with stripper harvesting. Complete leaf removal with defoliants is often difficult in some areas, so desiccant chemicals are used extensively to allow harvesting before frost. Desiccant chemicals are also used occasionally in other parts of the U.S. if conditioning is needed late in the season when defoliants act too slowly or are ineffective, or when regrowth is abundant. Although some desiccation may occur when excessive rates of defoliant chemicals are used, the two chemicals that are used primarily for cotton desiccation, arsenic acid and paraquat, are much more effective. Desiccants can be used nearer to scheduled harvest dates than can defoliants. This allows additional time for maturation of late bolls. Because growth stops almost immediately after desiccants are applied, they should not be used until 85% or more of the bolls are open.[95-98]

Regrowth vegetation may, under certain conditions, appear at intervals during the growing season, but it most frequently appears towards the end of the season after the majority of the bolls are mature. In addition, terminal and axillary buds of actively growing plants are often activated after chemical defoliation, and new growth develops before the crop can be harvested. Regrowth vegetation is of major importance to the cotton production industry in many areas. The low responsiveness of juvenile leaves to most defoliants and the difficulty encountered in obtaining adequate spray coverage of this type of vegetation reduce harvest-aid efficiency. If vegetation grows profusely, the added expense of a second application frequently is necessary before harvest. If allowed to remain on the plant, the new succulent vegetation may interfere with picking efficiency and may increase green stain in lint. The regrowth problem can be minimized in some years by proper management of fertilizer and moisture, but complete control is difficult because rainfall and temperature factors cannot be controlled.

During the 1950s amitrole (3-amino-S-triazole) and maleic hydrazide (1,2-dihydro-3,6-pyridazinedione) were used, with some success, to inhibit regrowth development that followed defoliation.[109] In addition to regrowth inhibition, amitrole also exhibited abscission-inducing properties and enhanced the effectiveness of defoliant chemicals when added to the defoliant solution. Because of residue in the seed, however, these chemicals are no longer marketed for use on cotton. More recently glyphosate was shown to suppress regrowth development for as long as 7 weeks when used either alone or in combination with a defoliant.[86] Glyphosate also enhanced the effectiveness of defoliant chemicals. Deleterious effects on seeds of immature bolls were found, however, and further testing of this chemical is required before it can be recommended for use on cotton. Thus, no chemicals are presently marketed in the U.S. for the express purpose of inhibiting regrowth development. The experimental defoliant chemicals dimethipin and thidiazuron discussed above have exhibited strong tendencies toward suppression of regrowth vegetation that frequently develops after chemical defoliation.[91,101,102] So the use of these products, either alone or in combination with other defoliants, might help to alleviate the late-season regrowth problem. The problem might also be minimized if chemical crop termination should become a standard practice.

The most recent harvest-aid practice in cotton production is the use of chemicals to accelerate boll dehiscence. The indeterminate character of cotton causes bolls to be set over a 4- to 6-week period, and because of environmental conditions, the boll opening period may last for as long as 10 weeks. In most cases this dictates two or more harvests, and frequently the late-set bolls fail to open before freezing or other inclement weather conditions develop. In addition, quality of early-set bolls deteriorates as the time between dehiscence and harvest is extended. Accelerated boll dehiscence might alleviate some of these problems by allowing earlier harvest thereby preserving quality; increasing the percentage of total yield gathered at first harvest or possibly eliminating the need for a second harvest; or reducing the number of late bolls lost to freezing temperatures. It appears logical that these goals might be accomplished without yield or quality reduction unless bolls are forced open too far in advance of full maturation or natural dehiscence. Leffler[110] reported that bolls that developed under favorable conditions did not open until 2 weeks after they had reached maximum dry weights. Other workers have shown that most of the difference in time from anthesis to dehiscence between early- and late-season bolls can be attributed to delayed dehiscence after the late-set bolls have reached full maturity.[111]

Morgan et al.[112] presented evidence to show that control of boll dehiscence is similar to that of abscission and suggested that it could be manipulated in a similar fashion. The process of dehiscence and boll flaring has been shown to be a mechanical process related to boll maturity,[113,114] vascular structure of the elements of the carpel,[114,115] and dehydration of the entire boll.[114-116] Lipe and Morgan,[117] however, suggested that ethylene was also involved. They reported that cotton fruiting forms produce large amounts of ethylene from the day of anthesis until the time of young boll shedding and again just before boll dehiscence. A similar pattern of ethylene production by cotton bolls was observed by Shen et al.[118] In addition, they found that the ethylene released by bolls was decreased with decreasted temperatures. This may account in part for the delayed dehiscence of apparently mature bolls in cool weather. Delayed boll dehiscence under conditions of optimum moisture and nutrition may also be related to reduced ethylene production. Guinn[119,120] studied the ethylene production pattern of young cotton bolls and found that both water and nutritional stresses stimulated ethylene production by these organs.

Ethylene levels in plant tissue can be increased through the use of the plant growth regulator chemical ethephon. The chemical is converted to ethylene inside of plant tissue and has been used on cotton plants to cause leaf abscission and boll dehiscence. Leaf abscission, however, is generally not satisfactory unless the maximum recommended rate is used, the plants are fully mature, and the temperature is relatively high.[121] Conversely, satisfactory boll dehiscence results have been obtained when ethephon was used at varying rates under a variety of conditions.[122-128] Results most frequently reported include increased boll opening rate, increased percentage of crop gathered at first harvest, and increased yield. Three years of testing with ethephon in Mississippi resulted in a 16% increase in yield at first harvest, no effect on total yield, and a slight reduction in quality of fiber and seed from bolls undehisced at treatment time.[123-125] Similar results were reported by Dunster et al.[121] for tests conducted in the western U.S., and by workers in India.[126,128] Singh and Kumar[126] found the material to be more effective on late- rather than early-maturing varieties. Shen et al.[118] observed that ethephon applications to cotton leaves led to increased ethylene release in bolls, whereas application to bolls failed to increase production in leaves. They also reported two peaks of ethylene production in bolls following treatment with ethephon; the first occurred 2 days after treatment and the second just before boll dehiscence. Only one peak occurred in the control and it was 8 days later than that found on the treated plant. This work seems to imply that boll dehiscence will be more pronounced if ethephon treatments are made before the plants are defoliated. Other research, however, has shown no difference in effect when the chemical is applied before defoliation, with the defoliant chemical, or after defoliation.[123,124]

If the process of boll dehiscence is related to dehydration and desiccation and if the water supply to mature bolls has been deleted by the formation of an abscission zone as suggested by Morris[115] and Simpson and Marsh,[114] it appears logical that a desiccant chemical applied to these bolls would accelerate the process. This would be especially beneficial when the environment is not conducive to rapid drying. Moisture loss from cotton bolls during the final stages of maturation was shown to be more rapid in the interior or seed cotton fraction than in the carpel.[129] Arsenic acid treatments to these bolls caused accelerated moisture loss from both fractions, but the loss was more pronounced in the carpel than in the interior or seed cotton fraction.

Desiccant chemicals on cotton have been studied primarily for preparation for chemical harvest, i.e., desiccation of leaves and/or defoliation; however, accelerated boll opening was also observed.[130] Kirby and Stelzer[131] presented data to show an 8 to 10% increase in boll opening when paraquat was used in a cotton harvest-aid test. In other harvest-aid studies paraquat was reported to cause 90% defoliation within 2 weeks of treatment and to shorten the boll opening period by 2 weeks with no adverse effect on yield or quality.[132,133] More recently the desiccant type harvest-aid chemicals have been used for the express purpose of accelerating boll opening and their effectiveness compared with that of ethephon. Singh and Kumar[126] treated upland cotton in India with ethephon, paraquat, and cacodylic acid, and reported no significant difference between chemicals for either boll opening or yield effects. All were effective in opening undehisced bolls and increasing the yield of late maturing varieties, but were less effective on early maturing varieties. None of the chemicals had an adverse effect on fiber or seed quality. Other research indicates that ethephon is more effective as a boll opener, and causes fewer deleterious effects on yield and quality, than either of the desiccant type chemicals.[123-125,129] All of these chemicals, including ethephon, tend to have an adverse effect on one or more components of bolls that are unopened at treatment time; boll size and lint micronaire are affected most frequently. If treatments are applied when 60% or more of the bolls is open, however, adverse effects are minimized. Treatments with arsenic acid or paraquat sometimes cause yield reductions in plots harvested with spindle picker harvesters. This may be related to the severity of desiccation. Bolls treated with desiccant chemicals often are only partially opened and fail to flare sufficiently for all cotton to be extracted by the spindle pickers.

VII. SUMMARY

Growth regulator chemicals have been used in cotton production since the advent of mechanical harvesters; most of the research and usage has been concerned with harvest-aid programs. Only in recent years has serious consideration been given to the use of synthetic chemicals to modify other aspects of cotton growth and development. Specific goals of this relatively new effort include (1) improved germination and seedling vigor; (2) improved balance between reproductive and vegetative growth; (3) early flowering and increased boll set; (4) improved yields and quality; (5) suppression of late-season growth and early maturity;

and (6) improved harvest-aid systems. A review of the literature and assessment of unpublished reports indicate that many of these goals can be realized without serious effects on productivity. The degree of success apparently is determined by choice and concentrations of chemicals, timing of application, and environmental conditions subsequent to treatment. An additional requirement for success, however, is that the capacity of the plant to respond to growth regulators is not limited by inadequate managerial control of factors such as fertility, moisture, and pests that may also alter plant development.

Although research has indicated that selected growth regulators can enhance seed germination and improve seedling vigor, the work is as yet only academic. With the exception of seed and soil treatments for protection against pests, synthetic chemicals have not been used extensively as growth regulators per set at this early stage of plant development. Rather, most work has been directed toward alterations of the natural plant processes during the reproductive and maturation stages of growth.

Most of the growth regulator chemicals used on cotton are inhibitors rather than promoters of growth, and most modify plant development rather than basic productivity. Attempts to increase the fruitfulness of the cotton plant directly with synthetic growth regulators have, in general, given negative results. However, a considerable amount of success has been attained with growth regulators in areas such as vegetative growth suppression, forced abscission of late-season fruit, accelerated boll dehiscence, increased earliness, and improved harvest-aid systems. Thus, indirect yield benefits may result from reduced losses and/or increased harvest efficiency.

As profit margins narrow and management of factors limiting yields becomes more proficient, producers are likely to seek additional means of improving production efficiency. Chemical growth regulators represent a logical next step, so there is apt to be an increased interest in these products as another practice to ensure efficient production.

REFERENCES

1. **Cathey, G. W. and Sabbe, W. E.,** Effects of trifluralin on fertilizer phosphorus uptake patterns by cotton and soybean seedlings, *Agron. J.*, 64, 254, 1972.
2. **Christiansen, M. N. and Hilton, J. L.,** Prevention of trifluralin effect on cotton with soil supplied lipids, *Crop Sci.*, 14, 489, 1974.
3. **Ranney, C. D.,** Multiple cottonseed treatments: effect on germination, seedling growth, and survival, *Crop Sci.*, 12, 346, 1972.
4. **Hassawy, G. S. and Hamilton, K. C.,** Effects of IAA, kinetin, and trifluralin on cotton seedlings, *Weed Sci.*, 19, 265, 1971.
5. **Cole, D. F. and Wheeler, J. E.,** Effect of pregermination treatments on germination and growth of cotton seed at suboptimal temperatures, *Crop Sci.*, 14, 451, 1974.
6. **Thomas, R. O. and Christiansen, M. N.,** Seed hydration-chilling treatment effects on germination and subsequent growth and fruiting of cotton, *Crop Sci.*, 11, 454, 1971.
7. **Christiansen, M. N. and Ashworth, E. N.,** Prevention of chilling injury to seedling cotton with anti-transpirants, *Crop Sci.*, 18, 907, 1978.
8. **Gausman, H. W., Heilman, M. H., Fryrear, D. W., and Rittig, F. R.,** Delay of senescence and wind tunnel sand damage of cotton plants by mepiquat chloride, in *Proc. Beltwide Cotton Prod. Res. Conf.*, National Cotton Council, Memphis, 1981, 46.
9. **Ergle, D. R.,** Compositional factors associated with the growth response of young cotton plants to gibberellic acid, *Plant Physiol.*, 33, 344, 1958.
10. **Bird, L. S. and Ergle, D. R.,** Seedling growth differences of several cotton varieties and the influence of gibberellin, *Agron. J.*, 53, 171, 1961.
11. **Coats, G. E.,** Effect of growth regulators on germinating cotton, *Miss. Agric. Exp. Stn. Bull.*, 752, 8, 1967.

12. **Halloin, J. M.,** Inhibition of cotton seed germination by abscisic acid, in *Proc. Beltwide Cotton Prod. Res. Conf.*, National Cotton Council, Memphis, 1979, 279.
13. **Walhood, V. T.,** Effect of gibberellins on yield and growth of cotton, in *Proc. Beltwide Cotton Prod. Res. Conf.*, National Cotton Council, Memphis, 1958, 27.
14. **Reddy, C. S. and Probhaker, A. S.,** Preliminary studies on the effect of seed soaking with cycocel on the germination and early growth of groundnut and cotton, *Food Farming Agric.*, 10, 330, 1979.
15. **Thomas, R. O.,** unpublished data, 1963.
16. **Thomas, R. O.,** Effects of application and concentration of 2-chloroethyl trimethylammonium chloride on plant size and fruiting response of cotton, *Crop Sci.*, 4, 403, 1964.
17. **Thomas, R. O.,** Field comparisons of selected growth retardants, in *Proc. Beltwide Cotton Prod. Res. Conf.*, National Cotton Council, Memphis, 1972, 49.
18. **De Silva, W. H.,** Some effects of the growth retardant chemical CCC on cotton in Uganda, *Cotton Grower Rev.*, 48, 131, 1971.
19. **Kittock, D. L., Arle, H. F., and Bariola, L. A.,** Current status of chemical termination of cotton fruiting, in *Proc. Beltwide Cotton Prod. Res. Conf.*, National Cotton Council, Memphis, 1974, 55.
20. **Thomas, R. O. and Hacskaylo, J.,** Cotton fruiting and yield responses to growth retardants used for suppressing late-season plant development, in *Proc. Beltwide Cotton Prod. Res. Conf.*, National Cotton Council, Memphis, 1974, 61.
21. **Thomas, R. O.,** Cotton flowering and fruiting response to application timing of chemical growth retardants, *Crop Sci.*, 15, 87, 1975.
22. **Marani, A., Zur, M., Eshel, A., Zimmerman, H., Carmeli, R., and Karaduaid, B.,** Effect of time and rate of application of two growth retardants on growth, flowering, and yield of upland cotton, *Crop Sci.*, 13, 429, 1973.
23. **Singh, S.,** Revolution in cotton yield with CCC, *Indian Farming*, 20, 5, 1970.
24. **Singh, S.,** Studies on the effects of soil moisture stress on the yield of cotton, *Indian J. Plant Physiol.*, 18, 359, 1975.
25. **Thomas, R. O.,** Effects of two growth retardants on flowering and boll production of greenhouse cotton plants, in *Proc. Beltwide Cotton Prod. Res. Conf.*, National Cotton Council, Memphis, 1967, 222.
26. **Follin, J. C.,** Remarkable action of two growth regulators — BAS 0660W and BAS 0640W — on the cotton plant, *Coton Fibres Trop.*, 29, 449, 1973.
27. **Willard, J. I.,** Pix plant regulator research: past and future, in *Proc. Plant Growth Regul. Working Group Conf.*, Sullivan, E. F., Treas., Great Western Sugar Co., Longmont, 1979, 116.
28. **Gausman, H. W., Stabenow, J., Rittig, F. R., Escobar, D. E., and Garza, M. V.,** Mepiquat chloride effects on cotton leaf anatomy, in *Proc. Plant Growth Regul. Working Group Conf.*, Sullivan, E. F., Treas., Great Western Sugar Co., Longmont, 1980, 8.
29. **Gausman, H. W., Rittig, F. R., Namkin, L. N., Rodriguez, R. R., Escobar, D. E., and Garza, M. V.,** Effects of 1,1 dimethyl-piperdium chloride on cotton (*Gossypium hirsutum* L.) leaf chlorphyll, in *Proc. Plant Growth Regul. Working Group Conf.*, Sullivan, E. F., Treas., Great Western Sugar Co., Longmont, 1978, 137.
30. **Walter, H., Gausman, H. W., Rittig, F. R., Namken, L. N., Escobar, D. E., and Rodriguez, R. R.,** Effect of mepiquat chloride on cotton plant leaf and canopy structure and dry weights of its components, in *Proc. Beltwide Cotton Prod. Res. Conf.*, National Cotton Council, Memphis, 1980, 32.
31. **Feaster, C. V., Briggs, R. E., and Turcotte, E. L.,** Pima cultivar responses to Pix, in *Proc. Beltwide Cotton Prod. Res. Conf.*, National Cotton Council, Memphis, 1980, 81.
32. **Heilman, M. D.,** Interactions of nitrogen with Pix on the growth and yield of cotton, in *Proc. Beltwide Cotton Prod. Res. Conf.*, National Cotton Council, Memphis, 1981, 47.
33. **Briggs, R. E.,** Varietal responses to Pix treated cotton in Arizona, in *Proc. Beltwide Cotton Prod. Res. Conf.*, National Cotton Council, Memphis, 1981, 47.
34. **Gausman, H. W., Stein, E., Rittig, F. R., and Leamer, R. W.,** Leaf Co_2 uptake and chlorphyll ratios of Pix-treated cotton, in *Proc. Plant Growth Regul. Working Group Conf.*, Sullivan, E. F., Treas., Great Western Sugar Co., Longmont, 1979, 117.
35. **Gausman, H. W., Rittig, F. R., Escobar, D. E., and Rodriguez, R. R,,** Effect of mepiquat chloride (Pix) on CO_2 uptake of cotton leaves, in *Proc. Plant Growth Regul. Working Group Conf.*, Sullivan, E. F., Treas., Great Western Sugar Co., Longmont, 1980, 1.
36. **Urwiler, M. J. and Cothren, J. I.,** Cotton responses to "Pix": field and laboratory studies, in *Proc. Plant Growth Regul. Working Group Conf.*, Sullivan, E. F., Treas., Great Western Sugar Co., Longmont, 1979, 110.
37. **Stuart, B. L., Wendt, C. W., and Abernathy, J. R.,** The effects of Pix (1,1-dimethyl-piperidinium chloride) on the water relations of cotton in *Proc. Beltwide Cotton Prod. Res. Conf.*, National Cotton Council, Memphis, 1981, 56.

38. **Cappy, J. J. and Cothren, J. I.**, Root growth and development of Pix-treated cotton plants, in *Proc. Plant Growth Regul. Working Group Conf.*, Sullivan, E. F., Treas., Great Western Sugar Co., Longmont, 1980, 22.
39. **Stuart, B. L., Wendt, C. W., and Abernathy, J. R.**, The influence of mepiquat chloride on the plant water status of cotton, in *Proc. Plant Growth Regul. Working Group Conf.*, Sullivan, E. F., Treas,, Great Western Sugar Co., Longmont, 1980, 172.
40. **Eaton, F. M.**, Physiology of the cotton plant, in *Ann. Rev. Plant Physiol.*, 6, 299, 1955.
41. **Horowitz, H.**, Influence des conditions du milieu sur la formation et la chute des organes floraux chez le cottonnier, *Coton Fibres Trop.*, 17, 23, 1962.
42. **Phillips, J. R., Herzog, G. A., and Nicholson, W. F.**, Effect of chlordimeform on fruiting characteristics and yield of cotton, *Ark. Farm. Res.*, 27, 4, 1977.
43. **Cathey, G. W.**, unpublished data, 1978.
44. **Cathey, G. W.**, unpublished data, 1979.
45. **Cathey, G. W., Ross, B. W., and Harvey, A. J.**, Absorption, translocation and accumulation of ^{14}C-acephate in cotton plants, in *Proc. Plant Growth Regul. Working Group Conf.*, Sullivan, E. F., Treas., Great Western Sugar Co., Longmont, 1981, 165.
46. **Hacskaylo, J. and Scales, A. L.**, Some effects of guthion alone and in combination with DDT and of a dieldrin-DDT mixture on growth and fruiting of the cotton plant, *J. Econ. Ent.*, 52, 396, 1959.
47. **Brown, L. C., Lincoln, C., Frans, R. F., and Waddle, B. A.**, Some effects of toxaphene-DDT and calcium arsenate on growth and development of cotton, *J. Econ. Ent.*, 54, 309, 1961.
48. **Brown, L. C., Cathey, G. W., and Lincoln, C.**, Growth and development of cotton as affected by toxaphene-DDT, methyl parathion, and calcium arsenate, *J. Econ. Ent.*, 55, 298, 1962.
49. **Roark, B., Pfrimmer, T. R., and Merkle, M. E.**, Effects of some formulations of methyl parathion, toxaphene, and DDT on the cotton plant, *Crop Sci.*, 3, 338, 1963.
50. **Beasley, C. A.**, Increase in average node of first fruiting branch in two varieties of cotton as induced by methyl parathion, in *Proc. Beltwide Cotton Prod. Res. Conf.*, National Cotton Council, Memphis, 1969, 99.
51. **Walhood, V. T.**, The effect of gibberellins on boll retention and cut-out in cotton, in *Proc. Beltwide Cotton Prod. Res. Conf.*, National Cotton Council, Memphis, 1957, 24.
52. **Lane, H. C.**, Response of cotton on the Texas high plains to foliar treatments of gibberellc acid, in *Proc. Beltwide Cotton Prod. Res. Conf.*, National Cotton Council, Memphis, 1958, 26.
53. **Subbiah, K. K. and Mariakulandia, A.**, Application of gibberellic acid and naphthalene acetic acid in preventing bud and boll shedding in Cambodia cotton, *Madras Agric. J.*, 59, 350, 1972.
54. **Varma, S. K.**, Variations in the endogenous growth regulators and nitrogen content in retained and abscising bolls of cotton, *Indian J. Plant Physiol.*, 21, 18, 1978.
55. **Rogers, J. P.**, Cotton fruit development and abscission: changes in endogenous abscisic acid-like substances, *S. Afr. J. Sci.*, 76, 471, 1980.
56. **Cathey, G. W.**, unpublished data, 1981.
57. **Negi, L. S. and Singh, A.**, A preliminary study on the effect of some hormones on yield of cotton, *Indian Cotton Grower Rev.*, 10, 153, 1956.
58. **Murty, P. S., Ragu, D. N., and Rao, G. V.**, Effect of plant growth regulators on flower and boll drop in cotton, in *Food Farming Agric.*, p. 9, 1976.
59. **Varma, S. K.**, Reversal of abscisic acid promoted abscission of flower buds and bolls of cotton (*Gossypium hirsutum* L.) with other regulators, *Indian J. Exp. Biol.*, 14, 309, 1976a.
60. **Varma, S. K.**, Role of abscisic acid in the phenomena of abscission of flower buds and bolls of cotton (*Gossypium hirsutum* L.) and its reversal with other regulators, *Biol. Plant.*, 18, 421, 1976b.
61. **Freytag, A. H. and Coleman, E. A.**, Effect of multiple applications of 2,3,5-triiodobenzoic acid (TIBA) on yield of stormproof and nonstormproof cotton, *Agron. J.*, 65, 610, 1973.
62. **Letham, D. S.**, Chemistry and physiology of kinetin-like compounds, *Ann. Rev. Plant Physiol.*, 18, 349, 1967.
63. **Rodgers, J. P.**, Cotton fruit development and abscission: fluctuations in the levels of cytokinins, *J. Hortic. Sci.*, 56, 99, 1981.
64. **Cothren, J. T. and Cotterman, C. D.**, Evaluation of Cytozyme Crop+ as a foliar application to enhance cotton yields, *Ark. Farm Res,*, 29, 2, 1980.
65. **Cathey, G. W.**, unpublished data, 1981.
66. **Schurbert, A. M., Benedict, C. R., Berlin, J. D., and Kohel, R. J.**, Cotton fiber development: kinetics of cell elongation and secondary wall thickening, *Crop Sci.*, 13, 704, 1973.
67. **Bhatt, J. G., Raman, C. V., Sankaranaraganan, T. G., and Iver, S. K.**, Changes in lint character of cotton varieties by growth regulators, *Cotton Grower Rev.*, 49, 160, 1972.
68. **Bhatt, J. G. and Ramanujam, T.**, Abnormal bolls in cycocel-treated cotton plants: their fiber characters and yield, *Cotton Grower Rev.*, 49, 354, 1972.

69. **Bhatt, J. G. and Ramanujam, T.,** Some responses of a short-branch cotton variety to gibberellin, *Cotton Grower Rev.*, 48, 136, 1971.
70. **Sitaram, M. S. and Abraham, E. S.,** Note on effect of gibberellic acid on quality of Laxmi cotton, *Cotton Grower Rev.*, 50, 150, 1973.
71. **Olvey, J. M., Fisher, W. D., and Patterson, L. L.,** TD-1123: a selective male gametocide, in *Proc. Beltwide Cotton Prod. Res. Conf.*, National Cotton Council, Memphis, 1981, 84.
72. **Erwin, D. C., Tasi, S. D., and Kahn, R. A.,** Growth retardants mitigate verticillium wilt and influence yield of cotton, *Phytopathology*, 69, 283, 1979.
73. **Snow, J. P., Crawford, S. H., Berggren, G. T., and Marshall, J. G.,** Growth regulator tested for cotton boll rot control, *La. Agric.*, 24, 1981, 3.
74. **Minton, E. B.,** personal communcation, 1981.
75. **Carns, H.,** Hormonal influence on adscission and flowering, in *Proc. Beltwide Cotton Prod. Res. Conf.*, National Cotton Council, Memphis, 1979, 276.
76. **Guinn, G., Hesketh, J. D., Fry, K. E., Mauney, J. R., and Radin, J. W.,** Evidence that photosynthesis limits yield of cotton, in *Proc. Beltwide Cotton Prod. Res. Conf.*, National Cotton Council, Memphis, 1976, 60.
77. **Freytag, A. H., Wendt, C. W., and Lira, E. P.,** Effects of soil-injected ethylene on yields of cotton and soybeans, *Agron. J.*, 64, 524, 1972.
Group Conf., Sullivan, E. F., Treas., Great Western Sugar Co., Longmont, 1980, 197.
79. **Kittock, D. L., Arle, H. F., Henneberry, T. J., and Bariola, L. A.,** Chemical termination of late-season cotton fruiting in Arizona and California, 1972 to 1976, U.S. Department of Agriculture Publ. ARS-W52, Washington, D.C., 1978.
80. **Thomas, R. O., Cleveland, T. C., and Cathey, G. W.,** Chemical plant growth suppressants for reducing late-season cotton bollworm-budworm feeding sites, *Crop Sci.*, 19, 861, 1979.
81. **Kittock, D. L. and Arle, H. F.,** Termination of late season cotton fruiting with plant growth regulators, *Crop Sci.*, 17, 320, 1977.
82. **Bariola, L. A., Kittock, D. L., Arle, H. F., Vail, P. V., and Henneberry, T. J.,** Controlling pink bollworms: effects of chemical termination of cotton fruiting on populations of diapausing larvae, *J. Econ. Entomol.*, 69, 633, 1976.
83. **Kittock, D. L., Arle, H. F., and Bariola, L. A.,** Chemical termination of cotton fruiting in Arizona in 1974, in *Proc. Beltwide Cotton Prod. Res. Conf.*, National Cotton Council, Memphis, 1975, 71.
84. **Kittock, D. L., Henneberry, T. J., and Bariola, L. A.,** Chemical termination for insect control in cotton: past, present, and future, in *Proc. Beltwide Cotton Prod. Res. Conf.*, National Cotton Council, Memphis, 1979, 62.
85. **Kittock, D. L., Mauney, J. R., Arle, H. F., and Bariola, L. A.,** Termination of late season cotton fruiting with growth regulators as an insect-control technique, *J. Environ. Qual.*, 2, 405, 1973,
86. **Cathey, G. W. and Barry, H. R.,** Evaluation of glyphosate as a harvest-aid chemical on cotton, *Agron. J.*, 69, 11, 1977.
87. **Thomas, R. O. and Hacskaylo, J.,** Cotton response to DPX-1840 growth retardant following root and foliar uptake, in *Proc. Beltwide Cotton Prod. Res. Conf.*, National Cotton Council, Memphis, 1973, 36.
88. **Thomas, R. O.,** unpublished data, 1972.
89. **Wolfenbarger, D. A. and Davis, J. W.,** Termination of cotton plants with chemicals and the effect on populations of boll weevils and tobacco budworms, in *Proc. Beltwide Cotton Prod. Res. Conf.*, National Cotton Council, Memphis, 1976, 46.
90. **Bariola, L. A., Henneberry, T. J., and Kittock, D. L.,** Status of chemical termination of cotton plant fruiting as a means for controlling the pink boll worm and boll weevil, in *Proc. Beltwide Cotton Prod. Res. Conf,*, National Cotton Council, Memphis, 1979, 130.
91. **Hopkins, A. R. and Moore, R. F.,** Thidiazuron: effect of application on boll weevil and bollworm population densities, leaf abscission, and growth of the cotton plant, *J. Econ. Entomol.*, 73, 768, 1980.
92. **Thomas, R. O.,** Some effects on boll maturity associated with bottom defoliation timing and with partial crown defoliation, in *Proc. Cotton Defoliation Physiol. Conf.*, National Cotton Council, Memphis, 1965, 92.
93. **Cathey, G. W.,** Harvest-aid chemicals and practices for cotton, *Outlook Agr.*, 10, 191, 1979.
94. **Crowe, G. B. and Carns, H. R.,** The Economics of Cotton Defoliation, Miss. Agric. Exp. Stn., Bull. No. 552, State College, 1957, 26.
95. **Albert, W. B.,** Additional studies on the effects of recommended herbicide and defoliation practices upon cotton seed germination, in *Proc. Cotton Defoliation Physiol. Conf.*, National Cotton Council, Memphis, 1964, 17.
96. **Brown, L. C. and Hyer, A. H.,** Chemical defoliation of cotton. V. Effects of premature treatments on boll components, fiber properties, germination, and yield of cotton, *Agron. J.*, 48, 50, 1956.
97. **McMeans, J. L., Walhood, V. T., and Carter, L. M.,** Effects of green-pick, defoliation and desiccation practices on quality and yield of cotton, *Gossypium hirsutum, Agron, J.*, 58, 91, 1966.

98. **Walhood, V. T.,** Effects of defoliants and desiccants on some boll properties, in *Proc. Cotton Defoliation Physiol. Conf.*, National Cotton Council, Memphis, 1954, 67.
99. **Tharp, W. H., Neek, W. E,, Jones, D. L., Haddon, C. B., Parroitt, I. M., Talley, P. J., and Smith, A. L.,** *Cotton Defoliation,* Prog. Rep., National Cotton Council, Memphis, 1949, 20.
100. **Carns, H. R.,** Abscission and its control, in *Ann. Rev. Plant Physiol.*, 19, 295, 1966.
101. **Taylor, W. K.,** DROPP: thidiazuron experimental cotton defoliant, in *Proc. Beltwide Cotton Prod.-Mech. Conf.*, National Cotton Council, Memphis, 1981, 70.
102. **Ames, R. B.,** Status of HARVADE 5F as a cotton defoliant, in *Proc. Beltwide Cotton Prod.-Mech. Conf.*, National Cotton Council, Memphis, 1981, 71.
103. **Sterett, J. P., Leather, G. R., and Tozer, W. E.,** Synergistic interaction between endothall and ethephon in abscission, *Plant Physiol.*, 51, S-29, 1973.
104. **Kirby, B. W. and Stelzer, L. R,,** Paraquat as a harvest-aid chemical: effect on lint and seed quality, in *Proc. Beltwide Cotton Prod. Res. Conf.*, National Cotton Council, Memphis, 1968, 68.
105. **Morgan, P. W. and Durham, J. I.,** Ethylene-induced leaf abscission is promoted by gibberellic acid, *Plant Physiol.*, 55, 308, 1975.
106. **Cathey, G. W.,** Evaluation of potassium 3,4-dichloroisothiazole-5-carboxylate as a harvest-aid chemical on cotton, *Crop Sci.*, 18, 301, 1978.
107. **Arle, H. F.,** Conditioning cotton for defoliation, in *Proc. Beltwide Cotton Prod. Res. Conf.*, National Cotton Council, Memphis, 1976, 49.
108. **Cathey, G. W., Elmore, C. D., and McMichael, B. L.,** Some physiological responses of cotton leaves to foliar applications of potassium 3,4-dichloroisothiazole-5-carboxylate and S,S,S-tributyl phosphorotrithioate, *Physiol. Plant,* 51, 140, 1981.
109. **Brown, L. C. and Hyer, A. H.,** Chemical defoliation of cotton. III. A study of seed and fiber from cotton plants treated with amino triazole, *Agron. J.*, 46, 580, 1954.
110. **Leffler, H. R.,** Development of cotton fruit. I. Accumulation and distribution of dry matter, *Crop Sci.*, 68, 855, 1976.
111. **Walhood, V. T. and Counts, B.,** Boll, fiber and seed properties of early and late season bolls, in *Proc. Beltwide Cotton Prod. Res. Conf.*, National Cotton Council, Memphis, 1955, 51.
112. **Morgan, P. W., Byer, E. M., Lipe, J. A., and McAfee, J. A.,** Ethylene, a regulator of cotton boll shed and boll opening, in *Proc. Beltwide Cotton Prod. Res. Conf.*, National Cotton Council, Memphis, 1971, 42.
113. **Baranov, P. A. and Maltzev, M.,** Boll dehiscence, in *The Structure and Development of the Cotton Plant,* Ogis-Isogis, Moscow, 1937, 77.
114. **Simpson, M. E. and Marsh, P. B.,** Vascular anatomy of cotton carpels as revealed by digestion in ruminal fluid, *Crop Sci.*, 17, 819, 1977.
115. **Morris, D. A.,** Capsule dehiscence in *Gossypium, Cotton Grower Rev.*, 41, 167, 1964.
116. **Reid, R. K. and Pinckard, J. A.,** Changes in the cellular structure of the cotton peduncle related to water transport and boll opening, *Crop Sci.*, 21, 717, 1981.
117. **Lipe, J. A. and Morgan, P. W.,** Ethylene: role in fruit abscission and dehiscence processes, *Plant Physiol.*, 50, 765, 1972b.
118. **Shen, Y. Q., Fang, B. C., and Sheng, M. Z.,** Approach to the physiological function of ethephon in ripening of cotton bolls, *Acta Botanica Sin.*, 22, 236, 1980.
119. **Guinn, G.,** Water deficit and ethylene evaluation by young cotton bolls, *Plant Physiol.*, 57, 403, 1976.
120. **Guinn, G.,** Nutritional stress and ethylene evaluation by young cotton bolls, *Crop Sci.*, 16, 89, 1976.
121. **Dunster, K. W., Dunlap, R. L., and Gonzales, F. J.,** Influence of ethrel plant regulator on boll opening and defoliation of western cotton, in *Proc. Plant Growth Regul. Working Group Conf.*, Sullivan, E. F., Treas., Great Western Sugar Co., Longmont, 1980, 15.
122. **Cothren, J. T.,** Boll opening responses of cotton to ethrel and GAF-7767141, in *Proc. Plant Growth Regul. Working Group Conf.*, Sullivan, E. F., Treas., Great Western Sugar Co., Longmont, 1980, 83.
123. **Cathey, G. W. and Luckett, K.,** Using growth regulator chemicals to enhance cotton harvest, in *Proc. Plant Growth Regul. Working Group Conf.*, Sullivan, E. F., Treas., Great Western Sugar Co., Longmont, 1980, 7.
124. **Cathey, G. W. and Luckett, K.,** Some effects of growth regulator chemicals on cotton earliness, yield, and quality, in *Proc. Beltwide Cotton Prod. Res. Conf.*, National Cotton Council, Memphis, 1980, 35.
125. **Cathey, G. W., Luckett, K. E., and Rayburn, S. T.,** Accelerated cotton boll dehiscence with growth regulator and desiccant chemicals, *Field Crop Res.*, 5, 113, 1982.
126. **Singh, G. and Kumar, S.,** Effect of some defoliants on boll opening and yield of cotton, *Indian J. Agric. Sci.*, 48, 632, 1978.
127. **Maksymowicz, W., Williams, J. M., and Pritchard, D. W.,** Enhancement of harvest efficiency and yields of cotton by GAF's GAFGRO, in *Proc. Plant Growth Regul. Working Group Conf.*, Sullivan, E. F., Treas., Great Western Sugar Co., Longmont, 1981, 94.

128. **Singh, O. S. and Singh, O.,** Studies on eco-physiology of boll dehiscence in cotton, *Indian J. Ecol.*, 2, 43, 1975.
129. **Cathey, G. W.,** Acceleration of boll dehiscence with desiccant chemicals, *Agron. J.*, 71, 505, 1979.
130. **Thomas, R. O.,** Performance of commercial and experimental defoliants and desiccants at Stoneville, Mississippi, in *Proc. Beltwide Cotton Prod. Res. Conf.*, National Cotton Council, Memphis, 1958, 12.
131. **Kirby, B. W. and Stelzer, L. R,,** Paraquat as a harvest-aid chemical: effect on lint and seed quality, in *Proc. Beltwide Cotton Prod. Res. Conf.*, National Cotton Council, Memphis, 1968, 68.
132. **Dippenar, M. C.,** Chemical defoliation of cotton, *Crop Prod.*, 4, 47, 1975.
133. **Kirdikere, C. B. and Koraddi, U. N.,** Effect of defoliants on cotton, *Curr. Res.*, 4, 61, 1975.

Chapter 10

PLANT GROWTH REGULATING CHEMICALS — CEREAL GRAINS

J. Jung and W. Rademacher

TABLE OF CONTENTS

I. INTRODUCTION

Cereal plants are of outstanding importance for the nutrition of the population of the world. Wheat, rice, maize, barley, oats, and other cereal grains account for approximately 65 and 50% of the world's total production of edible dry matter and protein, respectively.[1,2] In order to meet permanently increasing demands it appears inevitable that productivity of these species must be enhanced as much as possible.

Besides genetic and climatic factors, the growth and yield of a given crop are mainly determined by the amount of nutrients available. This means that under practical conditions growth steering can be achieved to a certain extent by the specific degree of fertilization. In fact, optimal plant nutrition is one of the major prerequisites to obtain for instance wheat yields in the range of 10 t/ha or even more.

However, application of fertilizers is only a coarse method and processes such as grain formation can only be influenced indirectly. Often its beneficial effects are even accompanied by adverse phenomena, such as increased susceptibility to harmful biotic and abiotic factors. Plant growth regulators — besides fungicides, herbicides, and insecticides — offer another agrochemical tool for making plants use nutrients more efficiently and for exploiting their genetic and physiological potentials on a higher level.

The following part of this chapter will give a survey of the importance plant growth regulators have at present and may have in future for the production of cereal grains. However, emphasis will be laid on plant growth retardants for the control of lodging, since currently this is the only major application for plant growth regulators in cereals. Actually, out of some 60 different substances with growth regulating activity that are commercially available,[3] only three — chlormequat chloride, mepiquat chloride, and ethephon — are used in cereal production. Future roles for plant growth regulators to make plants withstand adverse environments and increase yield formation will be the other topics. For further information the reader is referred to treatises by Wittwer,[4] Scott,[5] Bangerth,[6] Jeffcoat,[7] Nickell,[3] and Hawkins and Jeffcoat.[8]

II. AIMS FOR PLANT GROWTH REGULATING COMPOUNDS

Plant growth regulators — naturally occurring phytohormones and synthetic compounds — are able to influence and modify growth and differentiation in plants at high levels of activity without having a nutritive character. Many processes throughout the life cycle of a cereal plant can be seen as targets for compounds of this type. By promoting, inhibiting, or modifying such processes, results might be gained which directly or indirectly lead to increased yields. Table 1 gives a summary of these conceptions seen from two different points of view.

Besides showing activity in relation to the points mentioned in Table 1 there are, however, more requirements a growth regulator must satisfy:

1. The effect of an active compound must be reliable and predictable in order to give precise prescriptions and recommendations for specific crops under specific environmental conditions.
2. In order to exclude risks for man and the environment, no problems due to toxic side effects must arise from the plant growth regulator or its metabolites. Compounds with very high activity and specificity would be ideal.
3. The cost-benefit ratio has to be at a favorable level; this is not only relevant for the user of a plant growth regulator, but is equally important for the company producing it. An estimated sum of at least $25 million (U.S. currency) has to be spent for the development and introduction of a new active compound.[9] Consequently, to be eco-

Table 1
AIMS FOR PLANT GROWTH REGULATORS

Target processes	Possible results
Sequence and duration of developmental steps	Yield increase by:
Germination	Earlier sowing date
	Enhanced establishment
Anatomical and morphological structure: shape of shoot and root	Enhanced resistance to environmental stresses
Stress behavior, e.g., towards:	
Drought	Increased formation of yield-determining organs (harvest index)
Low temperature	
Diseases	
Utilization of:	Increased photoproductivity
Solar irradiation	
Water	
Nutrients	
Flowering	Further advantages by:
Sink formation	Increased quality
Transport and partitioning of assimilates	Less labor costs
Senescence	Promoted mechanical harvesting
Dormancy	Positive effects on postharvest behavior and shelf life

nomically viable, a new growth regulator must have a specific use for at least one major world crop.

III. CONTROL OF LODGING BY PLANT GROWTH RETARDANTS

Lodging is one of the principal problems in intensive cereal cultivation with high N application rates, especially under European climatic conditions. Apart from losses in yield which usually lie between 20 and 40%, the quality of grains is adversely affected and mechanical harvesting is considerably hampered. With the introduction of chlormequat chloride (CCC), especially suited for stem stabilizing in wheat (see Figure 1), an important contribution has thus been made. Since its commercialization in 1964, CCC has become the most widely used plant growth regulator. Today application of this compound in combination with increased nitrogen dressings is a standard technique in many wheat growing areas of Europe. In the Federal Republic of Germany approximately 60% of the wheat acreages is treated with this growth retardant.[10]

Practically all wheat varieties so far tested react to CCC, although to different degrees. In order to achieve an optimum effect, the rate of application has to be adjusted to the particular variety. Additionally, the growth stage and local cultivation conditions must be taken into consideration. When CCC is used properly, it definitely offers possibilities for optimizing the nitrogen yield curve in wheat production as shown by the example in Figure 2.

In addition to the so-called "physiological" form of lodging, increasing attention has recently been paid to the parasitic variant which is mainly caused by the *Cercosporella herpotrichoides* fungus and which is primarily attributable to the close sequence of certain cereal species in the rotation. Even before lodging damage occurs, there can be an adverse influence on the transport of nutrients and water in the cereal plant as the result of tissue damage at the base of the stalk. Additional treatment with systemic fungicides has therefore also been introduced on a considerable scale as a further flanking measure to improve the action of nutrients, especially N. This procedure can counteract the effect of one-sided cereal crop rotations on the spread of "foot" diseases, and the level of yield achieved with

FIGURE 1. Effect of chlormequat chloride on the lodging behavior of wheat. The strip beginning in the right section has been treated with CCC.

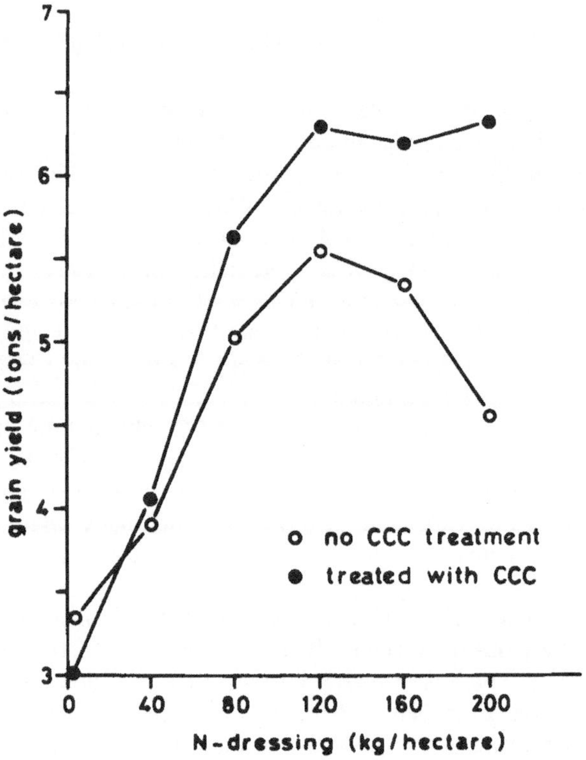

FIGURE 2. Effect of chlormequat chloride on yield of winter wheat at different stages of N fertilization. (From yield trials according to Dilz, K., *Stikstof,* 51, 174, 1966.)

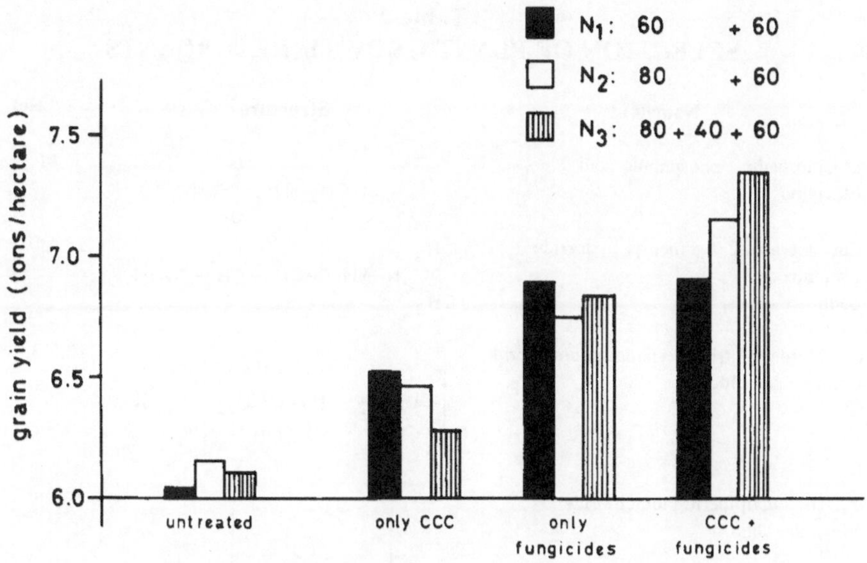

FIGURE 3. Increased yields from the use of chlormequat chloride and systemic fungicides with three different rates of N dressing in winter wheat (mean values of five trials each).

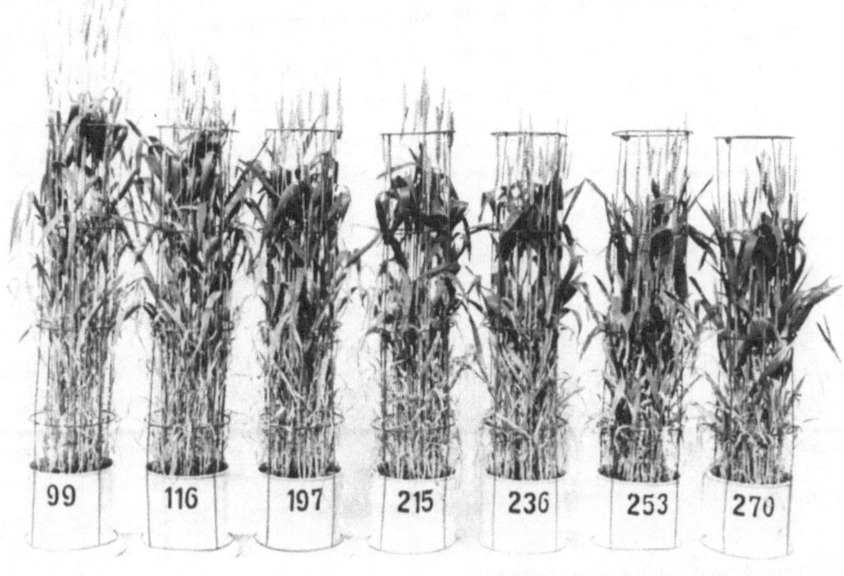

FIGURE 4. The response of "Opal" wheat plants to concentrations of 0 to 3000 ppm of chlormequat chloride (foliage spray).

appropriate nitrogen dressings and CCC treatment can be consolidated and raised further. (Figure 3).

CCC was first described by Tolbert in 1960 as being active as a growth retardant on wheat and several other species.[12,13] The length of wheat plants is reduced while stem diameter and robustness are increased, thus stabilizing treated plants against rain and wind (Figure 4). The significance of these findings for wheat production, especially under European conditions, was soon recognized and CCC became commercially available in 1964.[14,15] The

Table 2
SELECTION OF PLANT GROWTH RETARDANTS

Name(s)	Structure	Ref.
(2-Chloroethyl) phosphonic acid Ethephon	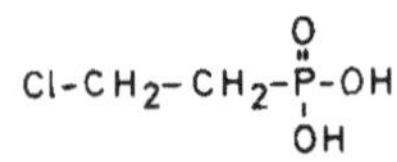	3, 48
Succinic acid 2,2-dimethyl hydrazide Daminozide Sadh	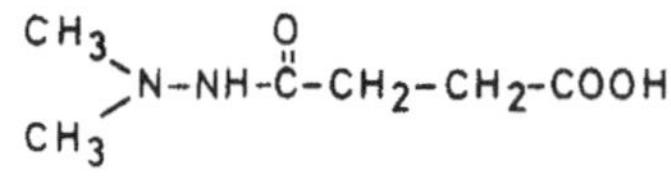	29, 30
(2-Chloroethyl)trimethylammoniumchloride Chlormequat chloride CCC	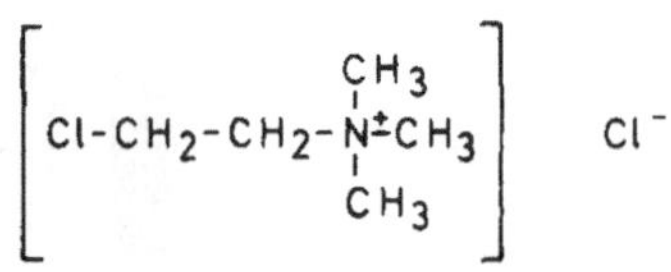	12, 13, 29, 30
1,1-Dimethylpiperidiniumchloride Mepiquat chloride DPC	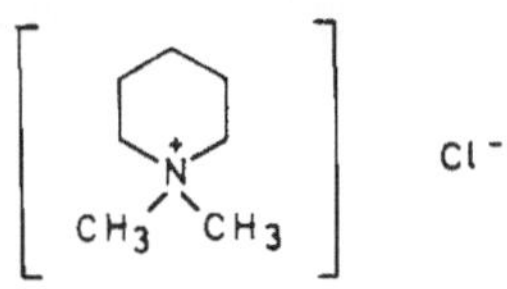	19, 29
(2-Isopropyl-5-methyl-4-trimethylammoniumchloride)phenyl-1-piperidiniumcarboxylate	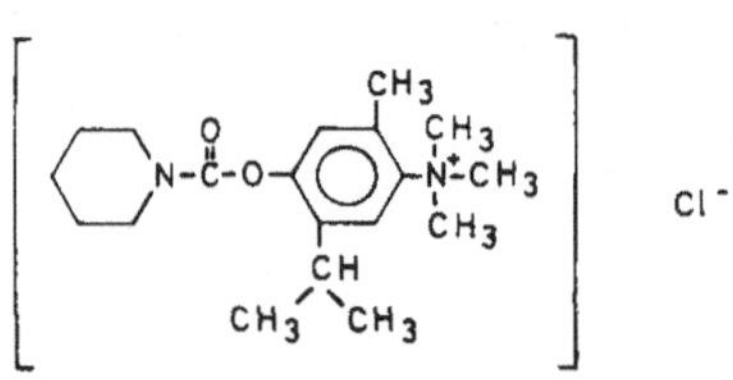	29, 30
AMO-1618 2,4-Dichlorobenzyltributylphosphoniumchloride	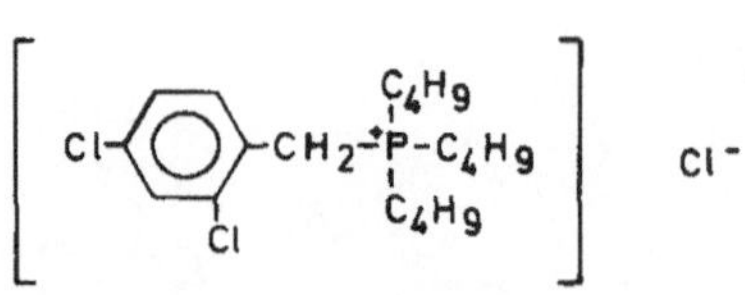	29, 30
Chlorphonium 3-[1,2,4-Triazolyl-(1)]-1-(4-chlorophenyl)-4,4-dimethyl-pentan-1-on LAB 117 682	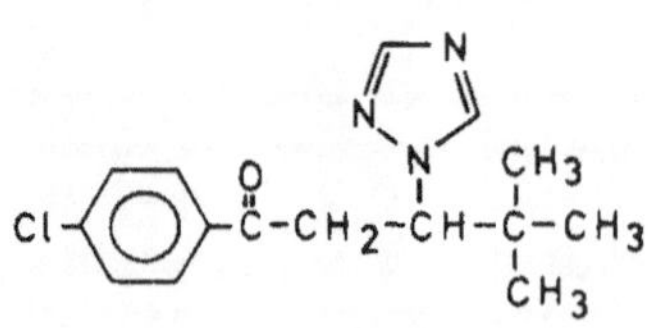	27
α-Cyclopropyl-α(4-methoxyphenyl)-5-pyrimidine methanol Ancymidol	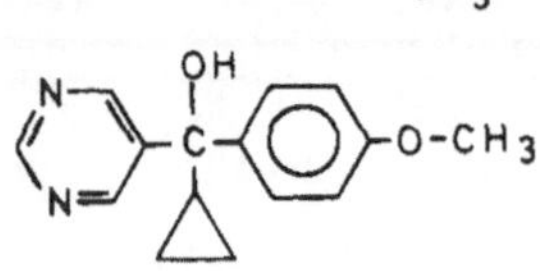	29
5-(4-Chlorophenyl)-3,4,5,9,10-pentaazatetracyclo-5,4,$10^{2,6}$, $O^{8,11}$-dodeca-3,9-diene LAB 102 883 BAS 106.. W	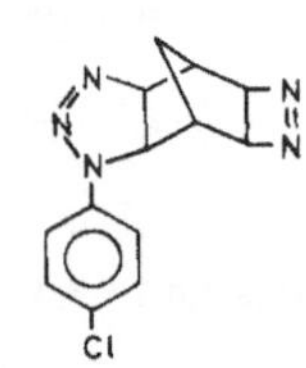	27

response of rye and oats to CCC treatment is relatively weak, and barley, maize, and rice are almost insensitive to this retardant. Therefore a search for more active compounds in the particular cereals has been made. Especially in winter barley and winter rye, mepiquat chloride (DPC) in combination with ethephon or ethephon on its own showed good activity as stem stabilizers and were introduced onto the market in 1979.[10,16-22] In addition to CCC,

Table 3
ACTIVITIES OF 9 GROWTH RETARDING COMPOUNDS IN WHEAT, BARLEY, RYE, MAIZE, AND RICE

Molar Concentrations Required for a 50% Reduction of Shoot Growth[a]

Name of compound	Wheat	Barley	Rye	Maize	Rice
Ethephon	ND	4.6×10^{-3}	ND	ND	$\geqslant 10^{-2}$
Daminozide	7.3×10^{-3}	1.1×10^{-2}	1.1×10^{-2}	$\geqslant 10^{-2}$	2.4×10^{-3}
CCC	3.5×10^{-2}	1.5×10^{-2}	3.1×10^{-2}	$\geqslant 10^{-2}$	1.5×10^{-2}
DPC	6.3×10^{-2}	1.1×10^{-2}	1.6×10^{-2}	1.5×10^{-2}	6.5×10^{-3}
AMO-1618	1.0×10^{-2}	5.2×10^{-3}	4.6×10^{-3}	$\geqslant 10^{-2}$	1.8×10^{-3}
Chlorphonium	4.5×10^{-4}	7.1×10^{-4}	3.0×10^{-4}	ND	1.0×10^{-4}
LAB 117 682	1.7×10^{-4}	1.8×10^{-4}	1.4×10^{-4}	2.0×10^{-5}	9.4×10^{-6}
Ancymidol	3.0×10^{-5}	4.0×10^{-5}	1.3×10^{-5}	1.7×10^{-6}	2.3×10^{-7}
BAS 106.. W	2.8×10^{-5}	1.8×10^{-5}	1.8×10^{-5}	7.9×10^{-8}	1.2×10^{-7}

Note: ND, not determined.

[a] Hydroponic culture; retardants applied via the medium.

DPC, and ethephon, several other types of chemicals show growth retarding activity and continuous efforts are being made to find and develop new and better suited substances. Table 2 gives a selection of some well-known and recently developed retardants.

Besides searching new active compounds, the combination of different growth retardants is another approach to control lodging in cereals. As mentioned above, a mixture of DPC and ethephon is used in barley. Simultaneous application of the two retardants offers some advantages as compared with the use of the components singly.[23] Similarly, other experimental combinations consisting of CCC, ancymidol, and ethephon showed promising results when applied to rye, wheat, barley, and oats.[20,24-26]

A comparison of the biological activities exerted by various retardants in wheat, barley, rye, maize, and rice is presented in Table 3. From the bioassay data the potential of compounds like ancymidol, BAS 106.. W, and the triazole as compared to the other types of retardants can clearly be seen. However, it must be pointed out that these values — obtained in an artificial system — will not necessarily reflect the usefulness of the respective compounds under practical conditions. Daminozide for instance is not used in cereal growing. Furthermore, problems due to toxicity, persistence, and economical feasibility are not rendered by these data.[27,28]

The mode of action of most plant growth retardants is generally seen in influencing the gibberellin (GA) system of the treated plant, thereby slowing both cell division and elongation.[29] A simultaneous application of an active gibberellin will neutralize growth retarding activity.[27,30] Furthermore, the stunting effect achieved by growth retardants (Figure 5) resembles closely the situation in dwarf rice and maize varieties where certain steps in GA biosynthesis are blocked genetically.[31,32]

In this relationship the situation seems quite clear for ancymidol, BAS 106.. W, and the triazole retardants. These compounds were all shown to block with high specificity the oxidative steps leading from *ent*-kaurene to *ent*-kaurenoic acid in the course of GA biosynthesis (Figure 6).[33-36]

In the gibberellin-producing fungus *Gibberella fujikuroi* and in cell-free systems of this fungus and of higher plants, CCC, chlorphonium, and AMO-1618 inhibit the formation of copalylpyrophosphate from geranylgeranylpyrophosphate. The subsequent conversion of copalylpyrophosphate to *ent*-kaurene is also reduced by these substances, but to a lower degree (Figure 6).[37,38] It must be emphasized, however, that CCC and DPC in contrast to

FIGURE 5. Effect of BAS 106.. W on the growth behavior of maize. Left to right, seed treatment with 0, 25, 50, and 100 g active ingredient per 100 kg seeds.

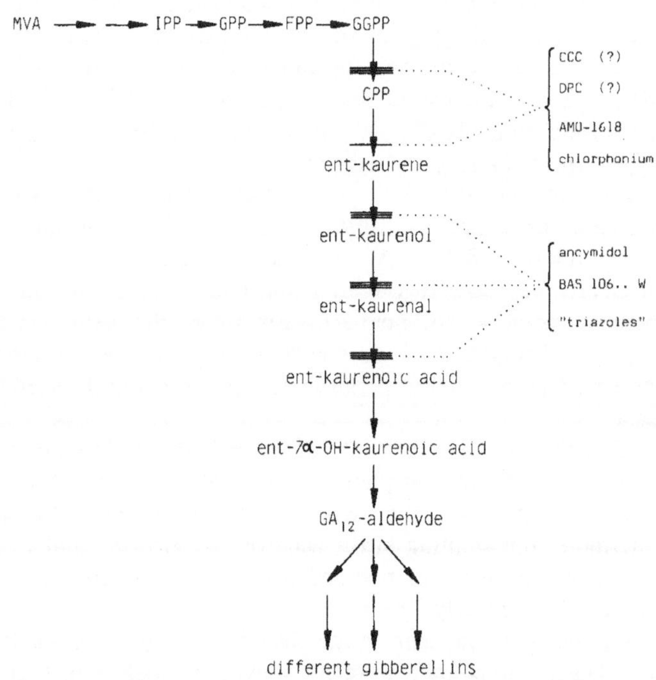

FIGURE 6. Pathway of gibberellin biosynthesis showing steps blocked by plant growth retardants. MVA, mevalonic acid; IPP, isopentenyl pyrophosphate; GPP, *trans*-geranyl pyrophosphate; FPP, *trans*-farnesyl pyrophosphate; GGPP, *trans*-geranylgeranyl pyrophosphate; CPP, copalyl pyrophosphate.

their effectiveness in the intact fungus *Gibberella fujikuroi*,[39,40] are almost inactive in enzyme preparations of different sources.[35,41-44] Although data from cell-free systems of certain plants cannot necessarily be regarded as being representative of intact other plants, it becomes clear that factors other than the inhibition of gibberellin biosynthesis are also involved in the effects caused by certain growth retardants. Thus reports from Douglas and Paleg indicate that AMO-1618 and other retardants cause a lack of phytosterols which were necessary for normal growth.[45-47]

Ethephon is known to release ethylene after application to plants. In certain species and under certain conditions, ethylene inhibits both cell division and elongation.[48,49]

Evidence is available that daminozide stimulates the rate of endogenous ethylene production.[50] Another or additional explanation for the mode of action of daminozide could be that it causes active gibberellins to metabolize more rapidly and GA transport to be greatly reduced.[51-53]

IV. RESISTANCE TO LOW TEMPERATURES

As a limiting factor, low temperatures are of importance not only for plant productivity, but also for plant distribution in large areas of the temperate and other climatic regions. Susceptibility to low temperatures can vary in different species, and different mechanisms of damaging and of natural resistance are known.[54-59]

The northward extension of high yielding wheat varieties is primarily restricted by too low winter temperatures. In many areas considerably increased yields could be expected for instance in maize if more rapid emergence and early growth could be achieved from earlier sowing, without increasing chilling injury. From these examples it can be seen that especially in marginal areas it is highly important to find ways of increasing plant resistance and tolerance to low temperatures.

Besides breeding such genotypes, attempts have been made to enhance the low temperature resistance of cereal plants by chemical treatment. The widespread use of CCC as a seed treatment in the U.S.S.R. can be attributed mostly to the fact that this increases the winter resistance of wheat seedlings. Improved overwintering of wheat treated with CCC was first shown in model trials.[60] Since then increased hardiness of wheat to cold after CCC treatment has been reported by several authors.[61-64]

Zadoncev et al.[65] emphasize particularly the lowering of the tillering node after seed treatment with CCC as the reason for improved wintering of wheat under field conditions (Figure 7). The seed treatment carried out in the U.S.S.R. in 1974 over an area of 2 million ha of winter and spring wheat underlines the practical significance of the effect of this treatment.[66]

Besides CCC, other compounds such as 2-amino-6-methyl benzoic acid and long-chained alkylene diamines have been found to protect wheat and other cereals from frost damage.[67,68] The usefulness of these substances still has to be evaluated, however.

V. RESISTANCE TO DROUGHT

In many areas water is the major factor limiting crop production. In a situation of drought, plant growth may be restricted severely when the rate of water consumption exceeds that of water uptake. Therefore artificial watering is indispensable in certain areas. It is estimated that about 5% of the U.S. wheat acreage is under irrigation.[69] Worldwide, approximately 200 million ha is irrigated.[70] Besides drainage and evaporation, much of the water naturally available or applied in irrigation disappears through transpiration. The artificial reduction of transpiration may guarantee growth — or at least survival — of crop plants under suboptimal moisture conditions. Thus the demand for irrigation water could be reduced or the irrigation intervals might be extended.

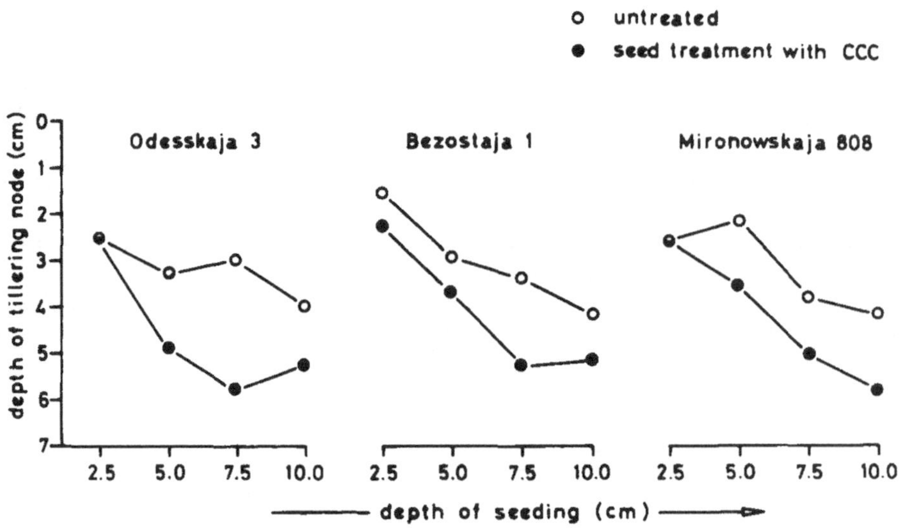

FIGURE 7. Depth of the tillering node of wheat in relation to seed treatment with CCC and the depth of seeding.[65]

FIGURE 8. *Effect of an antitranspirant treatment (left) on the wilting behavior of barley plants.*

Several evolutionary mechanisms are known by which plants became adapted to environments with limited water supply.[71] With a view to inducing such mechanisms artificially, many attempts have been made by crop physiologists to find antitranspirants that are applicable under practical conditions.[72-77] Besides testing substances which exert a purely physical effect, such as film forming or reflecting materials, emphasis has been put on growth substances which influence plant-water relations physiologically. At present most activities in this field concentrate on the regulation of stomatal movement as a short-term reversible control (see example in Figure 8). However, changes in the physiological behavior, anatomy, morphology, and developmental sequence of the whole plant should be regarded as being equally important in determining water consumption.[78]

By using antitranspirants, significant reductions in water consumption and even increases in water use efficiency have been achieved. However, in most cases this was only possible at the expense of absolute production. At the moment no commercially applicable method is available for reducing water consumption in cereals or in other crop plants on a field scale. Therefore, the purpose of the following can only be to give some insights into current trends in this area.

A. Controlling Stomatal Movement

Stomata can be seen as valves in the epidermis of the plant controlling both the uptake of CO_2 for photosynthetic production and the intensity of transpiration. In a situation of severe water stress, this inverse connection implies that a status has to be found by the plant "somewhere between starvation and desiccation." Under natural conditions the opening and closing of the stomata are regulated by the CO_2 concentration in the stomatal pore and the water status of the plant. A situation of water deficiency is mediated by a rapid increase of abscisic acid (ABA) which causes the guard cells to close, thus diminishing transpiration.[79-83] Because of its rapid metabolism, UV instability, and complicated synthesis, ABA is not used commercially as an antitranspirant. Similarly, derivatives of ABA (e.g., its methylester), ABA metabolites (e.g., phaseic acid and dihydrophaseic acid), and other naturally occurring compounds closely related to ABA (e.g., xanthoxin and vomifoliol) are of restricted usefulness. Farnesol — another sesquiterpenoid — was detected to appear as a natural antitranspirant in water-stressed *Sorghum sudanense,* but was phytotoxic when applied exogenously.[84] However, recent results indicate that this problem can be overcome under certain conditions.[85] (For further naturally occurring compounds related to ABA see Bearder.[86])

Many efforts are made to find synthetic analogs of ABA and other types of compounds with more promising attributes. At present, several substances are known to equal or even exceed the activity of abscisic acid on stomatal closure.[87,88] Table 4 gives a selection of compounds with antitranspirant activity. Certain fatty acids, phenylmercuric acetate, alkenylsuccinic acid, salicylaldoxime, and certain inhibitors of photosynthesis are further synthetic antitranspirants.[75,77,89] Most probably the activity of these substances is related to more or less specific influences on stomatal movement, too.

Anticytokinins[90] offer another approach for the control of transpiration physiologically on the stomatal level. These compounds can be expected to act as antagonists to the endogenous cytokinins which are known to promote transpiration via stomatal opening.[91,92] After application to detached wheat leaves through their cut ends, it was indeed possible to demonstrate the activity of two anticytokinins as antitranspirants in synergism to ABA (Table 5). However, spray applications to intact plants showed no such effects. Inadequate uptake through the epidermis or too rapid metabolism of the anticytokinins are considered responsible for these negative findings.[93]

B. Other Possibilities of Reducing Susceptibility to Drought

Besides regulating stomatal aperture, the role of ABA in a situation of low water supply seems to be of a more general nature.[78] In well-watered wheat plants, ABA induces symptoms typical of prolonged water stress: reduced number of leaves, less leaf surface with reduced density of stomata, and formation of trichomes.[94,95] In peas a thicker layer of wax deposits has been observed after ABA treatment.[96] Applied at suitable concentrations, ABA stimulates root growth and increases volume flux through the roots.[97-100] Furthermore, accumulation of proline, a process possibly protecting cereals and other plants from desiccation,[101] is induced by ABA.[102,103]

Table 4
SELECTION OF COMPOUNDS WITH ANTITRANSPIRANT ACTIVITY

Name(s)	Structure	Ref.
(+)-Abscisic acid	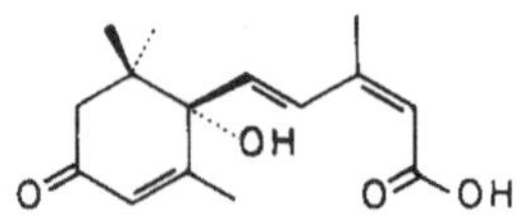	79, 86
(+)-Vomifoliol	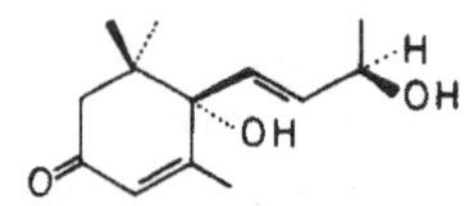	86
Xanthoxin	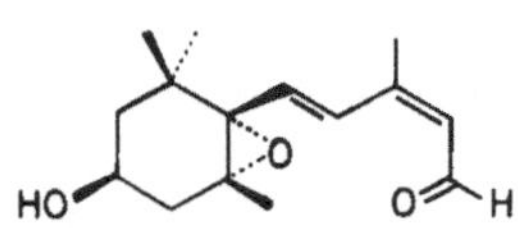	86
All-*trans*-farnesol	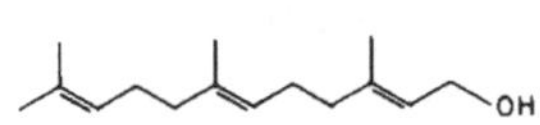	84, 86
2,4-Pentadienoic acid, 5-(4-chloro-2H-benzo-pyran-3-yl)-3-methyl	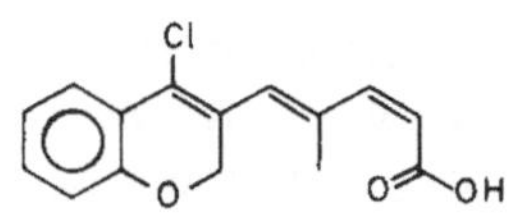	88
4-Cyclopentylamino-2-methylthio-pyrrolo (2,3-*d*)-pyrimidine ACK 1	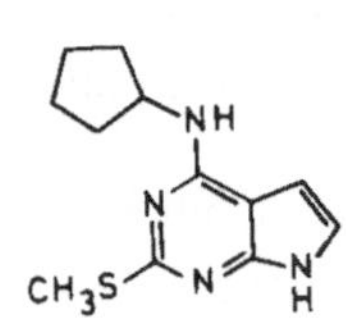	90, 93
3-Methyl-7-pentylamino-pyrazolo(4,3-*d*)-pyrimidine ACK 2	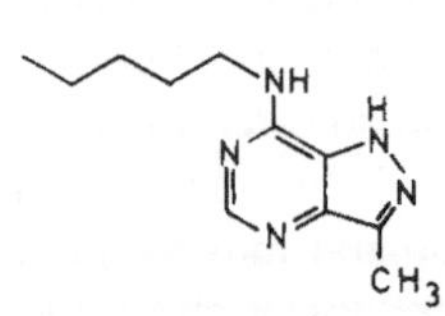	90, 93

In principle all these processes constitute targets for synthetic long-term antitranspirants. Compounds structurally related to ABA will probably be the most promising, although other growth regulators have already been shown to exert equivalent activities. Thus the beneficial effect of CCC on wheat and other plants in a situation of water stress is well documented.[104-106]

VI. REGULATION OF YIELD FORMATION

Yield formation in crop plants is a highly complex process determined by internal as well as external factors. In countries with developed agriculture, 5 to 7 t/ha of wheat are harvested on an average. Under extremely favorable conditions, top yields of twice as much can be achieved. However, Aufhammer indicates a theoretical grain yield potential of more than 20 t/ha.[107] In particular he points out that under average conditions in wheat

- Only about 20 to 30% of the tillers develops ears
- Only about 40 to 80% of the florets per spikelet is fertile
- The full dry weight of about 65 mg is achieved only by a proportion of the kernels[108]

Table 5
EFFECTS OF TWO ANTICYTOKININS (ACK) ON THE TRANSPIRATION OF WHEAT PRIMARY LEAVES[a]

Treatment	Amount of water transpired per leaf area (% of control)
ABA 10^{-6} *M*	68
ABA 10^{-7} *M*	80
ACK 1 10^{-6} *M*	75
ACK 1 10^{-7} *M*	83
ACK 2 10^{-6} *M*	72
ACK 2 10^{-7} *M*	80
ABA 10^{-7} *M* + ACK 1 10^{-7} *M*	65
ABA 10^{-7} *M* + ACK 2 10^{-7} *M*	68

[a] For chemical structures see Table 4.

Obviously the wheat (and other cereal) varieties of today adjust themselves anatomically and morphologically to a yield capacity which is only a fraction of the originally initiated potential. Photosynthetic production of carbon assimilates apparently does not play an important role as a limiting factor in cereals.[109-111] Recent review articles[112,113] rather indicate that under given environmental conditions, yield in cereals is mainly determined by processes related to sink development and sink supply such as

- The control of lateral bud growth and thus the number of tillers and ears per unit of area
- The regulation of the number of kernels per ear
- The regulation of kernel size
- The loading of assimilates at the source and deloading at the sink end of the phloem pathway (assimilate partitioning)
- Duration of grain fill

The authors emphasize the importance of endogenous plant hormones for these processes. Thus it seems quite likely that growth regulators offer a tool for influencing cereals towards increased yields.[114]

The literature contains a vast amount of contributions dealing with attempts to increase yield in various crops by means of growth regulators. In wheat, long-term experience with CCC has confirmed that yield is generally enhanced by 5 to 20%, an effect not related to increased lodging resistance.[115-117] Frequently this phenomenon is attributable to a greater number of ear-bearing tillers per unit of area;[115,118] to a larger number of grains per ear;[119,120] and evidently to an extension of the assimilation phase during kernel development.[117] In addition, the beneficial effects of CCC in situations of environmental stress, as outlined above, have to be considered, too.

Further positive results for cereals have been found, e.g., after application of cytokinins to wheat and rice,[121,122] abscisic acid to barley,[123] and dinoseb to maize.[124,125] Similarly, triacontanol, a compound reported to stimulate various aspects of vegetative growth in rice, barley, and maize, [126-128] has been found to increase grain yield of maize.[129] It is difficult to reproduce any of these results, however, and, with regard to triacontanol, Bhalla stated that "consistently inconsistent results were obtained."[130] (For additional references see Michael and Beringer[112] and Nickell.[3]) Probably this difficulty is due to the variability and complexity of the various biological processes leading to yield. Furthermore, there is not sufficient knowledge about the type of growth regulator required at specific stages of development and under certain external conditions.

VII. CONCLUSIONS

Plant growth regulators do play an important role in intensive cereal production as antilodging agents. Furthermore, applications are known to increase the plant resistance to certain environmental stresses.

Besides permanently looking for better substitutes for the plant growth regulators already in use, new types of compounds for new indications have to be found. In order to get higher and more constant yields while obeying the limits set by the area of available land and cost of energy, the number of ways to use growth regulators seems almost to be unlimited. It has to be emphasized, however, that searching for these future compounds is a complicated operation. Far more basic research is required to get a better understanding of the biological processes that limit crop productivity and thus find more suitable screening parameters. Similarly, the utilization of future growth regulators will most probably require a high degree of skill by the user.

Aims for plant growth regulators are aims for plant breeding and vice versa. All the effects that can be achieved with chemical regulators may eventually be attained by breeding as well. However, in contrast to breeding, plant growth regulators will probably offer a quicker approach to certain pragmatic solutions.

REFERENCES

1. **Evans, L. T.,** Crops and world food supply, crop evolution and the origins of crop physiology, in *Crop Physiology,* Evans, L. T., Ed., Cambridge University Press, London, 1975, 1.
2. **Harlan, J. R.,** The plants and animals that nourish man, *Sci. Am.,* 235, 89, 1976.
3. **Nickell, L. G.,** *Plant Growth Regulators — Agricultural Uses,* Springer-Verlag, New York, 1982.
4. **Wittwer, S. H.,** Phytohormones and chemical regulators in agriculture, in *Phytohormones and Related Compounds: A Comprehensive Treatise,* Vol. 2, Letham, D. S., Goodwin, P. B., and Higgins, T. J. V., Eds., Elsevier/North-Holland, Amsterdam, 1978, 599.
5. **Scott, T. K., Ed.,** *Plant Regulation and World Agriculture,* Plenum Press, New York, 1979.
6. **Bangerth, F., Ed.,** *Anwendungsmöglichkeiten von Phytohormonen und Wachstumsregulatoren in der Pflanzenproduktion,* Verlag Eugen Ulmer, Stuttgart, 1980.
7. **Jeffcoat, B., Ed.,** *Aspects and Prospects of Plant Growth Regulators,* Monogr. No. 6, British Plant Growth Regulator Group, Wessex Press, Wantage, 1981.
8. **Hawkins, A. and Jeffcoat, B., Eds.,** *Opportunities for Manipulating Cereal Productivity,* Monogr. No. 7, British Plant Growth Regulator Group, Wessex Press, Wantage, 1982.
9. **Lürssen, K.,** Economic aspects of the development of plant growth regulators, in *Aspects and Prospects of Plant Growth Regulators, Monogr. No. 6,* Jeffcoat, B., Ed., British Plant Growth Regulator Group, Wessex Press, Wantage, 1981, 241.
10. **Jung, J.,** Possibilities for optimalization of plant nutrition by new agrochemical substances — especially in cereals, in *Plant Regulation and World Agriculture,* Scott, T. K., Ed., Plenum Press, New York, 1979, 279.
11. **Dilz, K.,** Stikstofbemesting van granen, *Stikstof,* 51, 174, 1966.
12. **Tolbert, N. E.,** (2-Chloroethyl)-trimethylammoniumchloride and related compounds as plant growth substances. I. Chemical structure and bioassay, *J. Biol. Chem.,* 235, 475, 1960.
13. **Tolbert, N. E.,** (2-Chloroethyl)-trimethylammoniumchloride and related compounds as plant growth substances. II. Effect on growth of wheat, *Plant Physiol.,* 35, 380, 1960.
14. **Linser, H., Mayer, H. H., and Bodo, G.,** Über die Wirkung von Chlorcholinchlorid auf Sommerweizen, *Die Bodenkultur,* 12, 279, 1961.
15. **Jung, J. and Sturn, H.,** Wachstumsregulierende Wirkung von Chlorcholinchlorid (CCC), *Landw. Forsch.,* 17, 1, 1964.

16. **Zeeh, B., König, K. H., and Jung, J.,** Development of new plant growth regulators with biological activity related to CCC, *Kemia (Helsinki),* 9, 621, 1974.
17. **Hoffmann, G., Schulzke, D., Heyter, F., Kramer, W., and Kühnel, F.,** Camposan, ein neuer Halmstabilisator in Winterroggen, *Nachrichtenblatt Pflanzenschutz DDR,* 28, 249, 1975.
18. **Sadeghian, E. and Kühn, H.,** Der Einfluss von Ancymidol and Ethrel auf den Gibberellinsauregehalt von Getreidepflanzen, *Z. Pflanzenernährg. Bodenk.,* 139, 309, 1976.
19. **Jung, J. and Dressel, J.,** Pyridazinium- und Piperidiniumsalze als Wachstumsregulatoren, *Z. Pflanzenernahrg. Bodenk.,* 140, 375, 1977.
20. **Kühn, A., Schuster, W., and Linser, H.,** Halmverkürzung bein Winterroggen durch kombinierte Anwendung von CCC und Ethephon, *Z. Acker- u. Pflanzenbau,* 145, 22, 1977.
21. **Behrendt, S., Schott, P. E., Jung, J., Bleiholder, H., and Lang, H.,** Die Verbesserung der Standfestigkeit zu Wintergerste mit Hilfe eines Wachstumsregulators, *Landw. Forsch. Sonderheft,* 35, 277, 1978.
22. **Knittel, H., Behrendt, S., and Schott, P. E.,** Einfluss eines Wachstumsregulators auf Wachstum und Ertrag von Wintergerste bei unterschiedlichem Lagerdruck, *Z. Acker- u. Pflanzenbau,* 150, 50, 1981.
23. **Schott, P. E. and Rittig, F. R.,** New findings on the biological activity of mepiquat-chloride, in *Chemical Manipulation of Crop Growth and Development,* McLaren, J. S., Ed., Butterworths, London, 1982, 415.
24. **Kühn, H., Höfner, W., and Linser, H.,** Verstärkte Halmverkürzung bei Getreide durch Kombination von Wachstumsregulatoren (CCC, Ethephon, Ancymidol), *Landw. Forsch. Sonderheft,* 35, 271, 1979.
25. **Brückner, U. and Höfner, W.,** Verstärkte Halmverkürzung und Ertragssteigerung durch kombinierte Anwendung der Wachstumsregulatoren CCC und Ancymidol bei Sommerweizen, *Z. Acker- u. Pflanzenbau,* 149, 251, 1980.
26. **Höfner, W., Kühn, H., and Brückner, U.,** Halmverkürzung bei Hafer im Gefässversuch durch CCC und Ancymidol, *Z. Pflanzenernährg. Bodenk.,* 144, 215, 1981.
27. **Rademacher, W. and Jung, J.,** Comparative potency of various synthetic plant growth retardants on the elongation of rice seedlings, *Z. Acker- u. Pflanzenbau,* 150, 363, 1981.
28. **Rademacher, W., Jung, J., and Fritsch, H.,** Comparative potency of various synthetic plant growth retardants in different test systems, *Plant Physiol.,* 67(Suppl.), 576, 1981.
29. **Dicks, J. W.,** Mode of action of growth retardants, in *Recent Developments in the Use of Plant Growth Retardants, Monogr. No. 4,* Clifford, D. R. and Lenton, J. R., Eds., British Plant Growth Regulator Group, Wessex Press, Wantage, 1980, 1.
30. **Cathey, H. M.,** Physiology of growth retarding chemicals, *Ann. Rev. Plant Physiol.,* 15, 1964, 271.
31. **Murakami, Y.,** Dwarfing genes in rice and their relation to gibberellin biosynthesis, in *Plant Growth Substances 1970,* Carr, D. J., Ed., Springer-Verlag, New York, 1972, 166.
32. **Phinney, B. O.,** Gibberellin biosynthesis in the fungus *Gibberella fujikuroi* and in higher plants, in *Plant Growth Substances,* Mandava, N. B., Ed., American Chemical Society, Washington, D.C., 1979, 57.
33. **Coolbaugh, R. C. and Hamilton, R.,** Inhibition of *ent*-kaurene oxidation and growth by α-cyclopropyl-α-(p-methoxyphenyl)-5-pyrimidine methyl alcohol, *Plant Physiol.,* 57, 245, 1976.
34. **Coolbaugh, R. C., Hirano, S. S., and West, C.,** Studies on the specificity and site of action of ancymidol, a plant growth regulator, *Plant Physiol.,* 62, 571, 1978.
35. **Hildebrandt, E.,** Der Einfluss von ausgewählten Wachstumschemmern auf die Gibberellinbiosynthese in zellfreien Systemen, Diploma dissertation, University of Göttingen, 1982.
36. **Hildebrandt, E., Graebe, J. E., Rademacher, W., and Jung, J.,** Mode of action and biological activity of new potent plant growth retardants: BAS 106 ..W and triazole compounds, poster no. 762 presented at 11th Int. Conf. on Plant Growth Substances, Aberystwyth, Wales, 1982.
37. **Graebe, J. E. and Ropers, H.-J.,** Gibberellins, in *Phytohormones and Related Compounds: A Comprehensive Treatise,* Vol. 1. Letham, D. S., Goodwin, P. B., and Higgins, T. J. V., Eds., Elsevier/North-Holland, Amsterdam, 1978, 107.
38. **Sembdner, G., Gross, D., Liebisch, G.-W., and Schneider, G.,** Biosynthesis and metabolism of plant hormones, in *Hormonal Regulation of Development I, Molecular Aspects of Plant Hormones,* MacMillan, J., Ed., Springer-Verlag, New York, 1980, 281.
39. **Ninnemann, H., Zeevart, J. A. D., Kende, H., and Lang, A.,** The plant growth retardant CCC as inhibitor of gibberellin biosynthesis in *Fusarium moniliforme, Planta,* 61, 229, 1964.
40. **Rademacher, W. and Jung, J.,** Inhibition of gibberellin biosynthesis in *Fusarium moniliforme* and *Sphaceloma manihoticola* by different types of plant growth retardants, poster no. 154 presented at 11th Int. Conf. on Plant Growth Substances, Aberystwyth, Wales, 1982.
41. **Dennis, D. T., Upper, C. D., and West, C. A.,** An enzymic site of inhibition of gibberellin biosynthesis by AMO-1618 and other plant growth retardants, *Plant Physiol.,* 40, 948, 1965.
42. **Anderson, J. D. and Moore, T. C.,** Biosynthesis of (−)-kaurene in cell-free extracts of immature pea seeds, *Plant Physiol.,* 42, 1527, 1967.
43. **Fall, R. R. and West, C. A,,** Purification and properties of kaurene synthetase from *Fusarium moniliforme, J. Biol. Chem.,* 246, 6913, 1971.

44. **Frost, R. G. and West, C. A.**, Properties of kaurene synthetase from *Marah macrocarpus, Plant Physiol.*, 59, 22, 1977.
45. **Douglas, T. J. and Paleg, L. G.**, Inhibition of sterol biosynthesis by 2-isopropyl-4-dimethylamino-5-methylphenyl-1-piperidine carboxylate methyl chloride in tobacco and rat liver preparations, *Plant Physiol.*, 49, 417, 1972.
46. **Douglas, T. J. and Paleg, L. G.**, Plant growth retardants as inhibitors of sterol biosynthesis, *Plant Physiol.*, 54, 238, 1974,
47. **Douglas, T. J. and Paleg, L. G.**, Inhibition of sterol biosynthesis and stem elongation of toacco seedlings induced by some hypocholesterolemic agents, *J. Exp. Bot.*, 32, 59, 1981.
48. **Abeles, F. B.**, *Ethylene in Plant Biology*, Academic Press, New York, 1973.
49. **Goodwin, P. B.**, Phytohormones and growth and development of organs of the vegetative plant, in *Phytohormones and Related Compounds: A Comprehensive Treatise*, Vol. 2, Letham, D. S., Goodwin, P. B., and Higgins, T. J. V., Eds., Elsevier/North-Holland, Amsterdam, 1978, 31.
50. **Jindal, K. K., Andersen, A. S., and Dalbro, S.**, Ethylene production in alar treated apple branches, *Plant Physiol,*, 34, 26, 1975.
51. **Menhenett, R.**, Evidence that daminozide, but not two other growth retardants, modifies in the fate of applied gibberellin A_9 in *Chrysanthemum morifolium, Ramat, J. Exp. Bot.*, 31, 1631, 1980.
52. **Menhenett, R.**, Interaction of the growth retardants daminozide and piproctanyl bromide and gibberellins A_1, A_3, A_{4+7}, A_5 and A_{13}, in stem extension and inflorescence development in *Chrysanthemum morifolium, Ramat. Ann. Bot.*, 47, 359, 1981.
53. **Takeno, K., Legge, R. L., and Pharis, R. P.**, Effect of the growth retardant B-9 (SADH) on endogenous GA level, and transport and conversion of exogenously applied [^{3}H] GA_{20} in Alaska pea, *Plant Physiol.*, 67(Suppl.), 581, 1981.
54. **Lyons, J. M.**, Chilling injury in plants, *Ann. Rev. Plant Physiol.*, 24, 445, 1973.
55. **Christiansen, M. N. and St. John, J. B.**, The nature of chilling injury and its resistance in plants, in *Analysis and Improvement of Plant Cold Hardiness*, Olien, C. R. and Smith, M. N., Eds., CRC Press, Boca Raton, Fla., 1981, 1.
56. **Olien, C. R. and Smith, M. N.**, Protective systems that have evolved in plants, in *Analysis and Improvement of Plant Cold Hardiness*, Olien, C. R. and Smith, M. N., Eds., CRC Press, Boca Raton, Fla., 1981, 61.
57. **Siminovitch, D.**, Common and disperate elements in the processes of adaptation of herbaceous and woody plants to freezing — a perspective, *Cryobiology*, 18, 166, 1981.
58. **Larcher, W.**, Resistenzphysiologische Grundlagen der evolutiven Kälteakklimatisation von Sprosspflanzen, *Pl. Syst. Evol.*, 137, 145, 1981.
59. **Larcher, W.**, Effects of low temperature stress and frost injury on plant producitivity, in *Physiological Processes Limiting Plant Productivity*, Johnson, C. B., Ed., Butterworths, London, 1981, 253.
60. **Jung, J.**, Über den Einfluss von CCC auf die Überwinterung von Weizen und dessen Halmlänge, *Z. Acker- u. Pflanzenbaum*, 122, 5, 1965.
61. **Wünsche, U.**, Influence of 2-chloroethyl trimethylammonium chloride and gibberellin A_3 on frost hardiness of winter wheat, *Naturwissenschaften*, 53, 386, 1966.
62. **Toman, F. R. and Mitchell, H. L.**, Effects of cycocel and B-nine on cold hardiness of wheat plants, *J. Agric. Food Chem.*, 16, 771, 1968.
63. **Roberts, D. W. A.**, Effect of CCC (chlorocholin chloride) and gibberellins A_3 and A_7 on the cold hardiness of Kharkov 22 MC winter wheat, *Can. J. Bot.*, 49, 705, 1971.
64. **Vasil'eva, I. M., Rafikova, F. M., Khisamutdinova, V. I., Estrina, R. I., Smol'yaninov, S. N., and Galiev, N. A.**, Effect of chlorocholine chloride (CCC) on winter hardiness and yield of winter wheat, *Skh. Biol.*, 8, 532, 1973 (in Russian).
65. **Zadoncev, A. I., Pikus, G. R., and Grincenko, A. L.**, *CCC in der Pflanzenproduktion*, VEB Deutscher Landwirtschaftsverlag, Berlin, 1977 (German translation of the Russian original from 1973).
66. **Pikus, G. R. and Grincenko, A. L.**, in *CCC in der Pflanzenproduktion*, Zadoncev, A. I., Pikus, G. R., and Grincenko, A. L., Eds., VEB Deutscher Landwirtschaftsverlag, Berlin, 1977, 5 (German translation of the Russian original from 1973).
67. **De Silva, W. H., Bocion, P. F., Eggenberg, P., and de Mur, A.**, The plant growth regulating and herbicidal activity of 2-amino-6-methyl benzoic acid, *Z. Pflanzenkr. Pflanzenschutz*, 86, 546, 1979.
68. **Okii, M., Onitake, T., Kawai, M., Takematsu, T., and Konnai, M.**, Method for protecting crops from suffering damages, U.S. Patent, 4.231.789, 1980.
69. **Schlehuber, A. M. and Tucker, B. B.**, Culture of wheat, in *Wheat and Wheat Improvement*, Quisenbury, K. S. and Reitz, L. P., Eds., American Society of Agronomy, Madison, 1967, 117.
70. **Tekinel, O.**, Water stress and its implications (irrigation) in the future of agriculture, in *Plant Regulation and World Agriculture*, Scott, T. K., Ed., Plenum Press, New York, 1979, 457.
71. **Turner, N. C. and Kramer, P. J.**, *Adaptation of Plants to Water and High Temperature Stress*, John Wiley & Sons, New York, 1980.

72. **Gale, J. and Hagan, R. M.,** Plant antitranspirants, *Ann. Rev. Plant Physiol.*, 17, 269, 1966.
73. **Mansfield, T. A.,** Chemical control of stomatal movements, *Phil. Trans. R. Soc. Lond. B.*, 273, 541, 1976.
74. **Wright, S. T. C.,** *Phytohormones and stress phenomena, in Phytohormones and Related Compounds — A Comprehensive Treatise,* Vol. 2, Letham, D. S., Goodwin, P. B., and Higgins, T. J. V., Eds., Elsevier/North-Holland, Amsterdam, 1978, 495.
75. **Das, V. S. R. and Raghavendra, A. S.,** Antitranspirants for improvement of water use efficiency of crops, *Outlook Agr.*, 10, 92, 1979.
76. **Karamanos, A. J.,** Water stress: A challenge for the future of agriculture, in *Plant Regulation and World Agriculture,* Scott, T. K., Eds., Plenum Press, New York, 1979, 415.
77. **Jones, H. G.,** PGRs and plant water relations, in *Aspects and Prospects of Plant Growth Regulators,* Monogr. No. 6, Jeffcoat, B., Ed., British Plant Growth Regulator Group, Wessex Press, Wantage, 1981, 91.
78. **Jones, H. G.,** How plants respond to stress, *Nature (London),* 271, 610, 1978.
79. **Wright, S. T. C. and Hiron, R. W. P.,** (+)-Abscisic acid, the growth inhibitor induced in detached wheat leaves by a period of wilting, *Nature (London),* 224, 719, 1969.
80. **Raschke, K.,** Stomatal action, *Ann. Rev. Plant Physiol.*, 26, 309, 1975.
81. **Pierce, M. and Raschke, K.,** Correlation between loss of turgor and accumulation of abscisic acid in detached leaves, *Planta,* 148, 174, 1980.
82. **Pierce, M. and Raschke, K.,** Synthesis and metabolism of abscisic acid in detached leaves of *Phaseolus vulgaris* L. after loss and recovery of turgor, *Planta,* 153, 156, 1981.
83. **Mansfield, T. A. and Wilson, J. A.,** Regulation of gas exchange in water-stressed plants, in *Physiological Processes Limiting Plant Productivity,* Johnson, C. B., Ed., Butterworths, London, 1981, 237.
84. **Wellburn, A. R., Ogunkanmi, A. B., Fenton, R., and Mansfield, T. A.,** All-trans farnesol: a naturally occurring antitranspirant, *Planta,* 120, 255, 1974.
85. **Malloch, K. R. and Fenton, R.,** Reversible effects of farnesol on *Commelina communis* L., *New Phytol.*, 88, 249, 1981.
86. **Bearder, J. R.,** Plant hormones and other growth substances — their background, structures and occurrence, in *Hormonal Regulation of Development I, Molecular Aspects of Plant Hormones,* MacMillan, J., Ed., Springer-Verlag, New York, 1980, 9.
87. **Bittner, S., Gorodetsky, M., Har-Paz, I., Mizrahi, Y., and Richmond, A. E.,** Synthesis and biological effects of aromatic analogs of abscisic acid, *Phytochemistry,* 16, 1143, 1977.
88. **Carbonnier, J., Giraud, M., Hubac, C., Molho, D., and Valla, A.,** Activité antitranspirante d'analogues de l'acide abscissique, *Physiol. Plant.*, 51, 1, 1981.
89. **Miller, N. A.,** The effect of N-decenylsuccinic acid on the leaf water balance of *Zea mays* L., *Bot. Gaz.*, 142, 197, 1981.
90. **Hecht, S. M.,** Anticytokinins as probes for cytokinin utilization, in *Plant Growth Substances,* Mandava, N. B., Ed., American Chemical Society, Washington, D.C., 1979, 79.
91. **Livnè, A. and Vaadia, Y.,** Stimulation of transpiration rate in barley leaves by kinetin and gibberellic acid, *Physiol. Plant.*, 18, 658, 1965.
92. **Biddington, N. L. and Thomas, T. H.,** Influence of different cytokinins on the transpiration and senescence of excised oat leaves, *Physiol. Plant.*, 42, 369, 1978.
93. **Rademacher, W., Jung, J., Mockel, D., and Lang, P.-C.,** unpublished data.
94. **Quarrie, S. and Jones, H. G.,** Effects of abscisic acid and water stress on development and morphology of wheat, *J. Exp. Bot.*, 28, 192, 1977.
95. **Hall, K. H. and McWha, J. A.,** Effects of abscisic acid on growth of wheat (*Triticum aestivum* L.), *Ann. Bot.*, 47, 427, 1981.
96. **Baker, E. A. and Hunt, G. M.,** Cuticle development in stressed plants and response to exogenous growth substances, in *Responses of Plants to Environmental Stress and their Mediation by Plant Growth Substances,* Abstracts of papers presented at the Meet. Assoc. Appl. Biol. Br. Plant Growth Regul. Group, 1981, 5.
97. **Gaither, D. H., Lutz, D. H., and Farrence, L. E.,** Abscisic acid stimulates elongation of excised pea root tips, *Plant Physiol.*, 55, 948, 1975.
98. **Hartung, W., Ohl, B., and Kummer, V.,** Abscisic acid and the rooting of runner bean cuttings, *Z. Pflanzenphysiol.*, 98, 95, 1980.
99. **Glinka, Z. and Reinhold, L.,** Abscisic acid raises the permeability of plant cells to water, *Plant Physiol.*, 48, 103, 1971.
100. **Fiscus, E. L.,** Effects of abscisic acid on the hydraulic conductance of and the total ion transport through *Phaseolus* root systems, *Plant Physiol.*, 68, 169, 1981.
101. **Stewart, C. R. and Hanson, A. D.,** Proline accumulation as a metabolic response to water stress, in *Adaptation of Plants to Water and High Temperature Stress,* Turner, N. C. and Kramer, P. J., Eds., John Wiley & Sons, New York, 1980, 173.

102. **Aspinall, D., Singh, T. N., and Paleg, L. G.,** Stress metabolism. V. Abscisic acid and nitrogen metabolism in barley and *Lolium temulentum* L., *Aust. J. Biol. Sci.,* 26, 319, 1973.
103. **Stewart, C. R.,** The mechanism of abscisic acid-induced proline accumulation in barley leaves, *Plant Physiol.,* 66, 230, 1980.
104. **Gohlke, A. F. and Tolbert, N. E.,** Effect of 2-chloroethyl trimethylammonium chloride on phosphate absorption and translocation, *Plant Physiol.,* 37 (Suppl.), XII, 1962.
105. **El-Damaty, A. H., Kühn, H., and Linser, H.,** Water relations of wheat plants under the influence of 2-chloroethyl trimethylammonium chloride, *Physiol. Plant.,* 18, 650, 1965.
106. **Plaut, Z. and Halevy, A. H.,** Regeneration after wilting, growth and yield of wheat plants, as affected by two growth retarding compounds, *Physiol. Plant.,* 19, 1064, 1966.
107. **Aufhammer, W.,** Für die Ertragsbildung kritische Wachstumsstadien bei der Getreidepflanze, *DLG-Mitt.,* 14, 780, 1976.
108. **Aufhammer, W.,** Role of plant growth regulators in wheat yield, in *Aspects and Prospects of Plant Growth Regulators,* Monogr. No. 6, Jeffcoat, B., Ed., British Plant Growth Regulator Group, Wessex Press, Wantage, 1981, 131.
109. **Jenner, C. F.,** Physiological investigations on restrictions to transport of sucrose in ears of wheat, *Aust. J. Plant Physiol.,* 3, 337, 1976.
110. **Jenner, C. F. and Rathjen, A. J.,** Supply of sucrose and its metabolism in developing grains of wheat, *Aust. J. Plant Physiol.,* 4, 691, 1977.
111. **Stoy, V.,** Grain filling and the property of the sink, in *Physiological Aspects of Crop Productivity,* Proc. 15th Colloq. Int. Potash Inst., Wageningen/The Netherlands, 1980, 65.
112. **Michael, G. and Beringer, H.,** The role of hormones in yield formation, in *Physiological Aspects of Crop Productivity,* Proc. 15th Colloq. Int. Potash Inst., Wageningen/The Netherlands, 1980, 85.
113. **Patrick, J. and Wareing, P. F.,** Hormonal control of assimilate movement and distribution, in *Aspects and Prospects of Plant Growth Regulators,* Monogr. No. 6, Jeffcoat, B., Ed., British Plant Growth Regulator Group, Wessex Press, Wantage, 1981, 65.
114. **Bruinsma, J.,** The endogenous hormonal pattern and its interference by exogenous plant growth regulators, in *Physiological Aspects of Crop Productivity,* Proc. 15th Colloq. Int. Potash Inst., Wageningen/The Netherlands, 1980, 117.
115. **Humphries, E. C., Welbank, P. J., and Witts, K. J.,** Effect of CCC (chlorocholin chloride) on growth and yield of spring wheat in the field, *Ann. Appl. Biol.,* 56, 351, 1965.
116. **Kühn, H., Linser, H., and Schuster, H.,** CCC-Wirkung in Feldversuchen mit hoher Stickstoffdüngung bei einigen Sommer- und Winterweizensorten, *Z. Acker- u. Pflanzenbau,* 123, 356, 1966.
117. **Höfner, W. and Brückner, U.,** Einfluss von Wachstumsregulatoren auf Ertrag und Ertragskomponenten bei Getreide, *Kali-Briefe (Büntehof),* 15, 277, 1980.
118. **Primost, E.,** The Effect of CCC on the grain and straw yields of winter wheat during two years of contrasting wet and dry weather, *Z. Acker- u. Pflanzenbau,* 119, 211, 1964.
119. **Heyland, K.-U., Solansky, S., and Aufhammer, W.,** Einflüsse von CCC- und Gibberellinsäure-Behandlungen auf die Ertragsbildung der Sommergerste, *Z. Acker- u. Pflanzenbau,* 141, 109, 1975.
120. **Höfner, W., Feucht, D., and Brückner, U.,** Beeinflussung der Ähren- und Kornentwicklung von Sommerweizen durch Wachstumsregulatoren, *Z. Acker- u. Pflanzenbau,* 149, 177, 1980.
121. **Herzog, H. and Geisler, G.,** Der Einfluss von Cytokininapplikation auf die Assimilateinlagerung und die endogene Cytokininaktivität der Karyopsen bei zwei Sommerweizensorten, *Z. Acker- u. Pflanzenbau,* 144, 230, 1977.
122. **Ray, S. and Choudhuri, M. A.,** Effects of plant growth regulators on grain-filling and yield of rice, *Ann. Bot.* 47, 755, 1981.
123. **Tietz, A., Ludewig, M., Dingkuhn, M., and Dörffling, K.,** Effect of abscisic acid on the transport of assimilates in barley, *Planta,* 152, 557, 1981.
124. **Ohlrogge, A. J.,** The development of DNBP (dinoseb) as a biostimulant for corn, *Zea mays* L., in *Plant Growth Regulators,* Adv. Chem. Ser. 159, Stutte, C. A., Ed., American Chemical Society, Washington, D.C., 1977, 79.
125. **Ohlrogge, A. J., Oplinger, E. S., Abdel-Rahman, M., Roth, J. A., and Fulk-Bringman, S. S.,** DNBP (premerge 3) stimulatory effects on corn grain yield as affected by time of application, *Proc. 7th Annu. Meet. Plant Growth Regul. Working Group,* Dallas, 1980, 155.
126. **Ries, S. K., Wert, V., Sweeley, C. C., and Lewitt, R. A.,** Triacontanol: a new naturally occurring plant growth regulator, *Science,* 195, 1339, 1977.
127. **Eriksen, A. B.,Selldén, G., Skogen, D., and Nilsen, S.,** Comparative analyses of the effect of triacontanol on photosynthesis, photo respiration and growth of tomato (C_3-plant) and maize (C_4-plant), *Planta,* 152, 44, 1981.
128. **Knowles, N. R. and Ries, S. K.,** Rapid growth and apparent total nitrogen increases in rice and corn plants following applications of triacontanol, *Plant Physiol.,* 68, 1279, 1981.

129. **Ohlrogge, A. J. and Fulk-Bringman, S. S.,** Triacontanol effects on the grain yield of field corn, *Proc. 7th Annu. Meet. Plant Growth Regul. Working Group*, Dallas, 1980, 138.
130. **Bhalla, P. R.,** Triacontanol as a plant biostimulant, *Proc. 8th Annu. Meet. Plant Growth Regul. Working Group*, St. Petersburg, Fla., 1981, 184.

INDEX

D

E

F

G

H

I

K

L

M

N

O

P

Q

R

S

T

V

W

X

Y